T0302318

ADVANCED PROCESS CONTROL

ADVANCED PROCESS CONTROL
Beyond Single-Loop Control

Cecil L. Smith

AIChE®

WILEY

A JOHN WILEY & SONS, INC., PUBLICATION

Published by John Wiley & Sons, Inc., Hoboken, New Jersey
Published simultaneously in Canada

For general information on our other products and services or for technical support, please contact
our Customer Care Department within the United States at 877-762-2974, outside the United States
at 317-572-3993 or fax 317-572-4002.

Wiley also publishes its books in a variety of electronic formats. Some content that appears in print
may not be available in electronic formats. For more information about Wiley products, visit our
web site at www.wiley.com.

Library of Congress Cataloging-in-Publication Data:

Smith, Cecil L.
  Advanced process control : beyond single loop control / Cecil L. Smith.
      p. cm.
  Includes bibliographical references and index.
  ISBN 978-0-470-38197-7 (cloth)
 1. Chemical process control. I. Title.
  TP155.75.S583 2010
  660'.2815–dc22

                                                                    2009045870

Printed in Singapore

10 9 8 7 6 5 4 3 2 1

CONTENTS

PREFACE

Exactly what is advanced process control? My favorite definition is from an attendee to a continuing education course: Advanced control is what we should be applying in our plants but are not applying, for whatever reason. This definition lacks specificity, but it does reflect the reality that what seems advanced to some does not seem advanced to others.

To be categorized as advanced, a control configuration must have at least one of the following attributes:

- It relies on more than one measurement.
- It drives more than one final control element.
- It incorporates a process relationship of some form (which may be as simple as a characterization function).
- It incorporates functions such as constraint control that are intended to optimize process operations.
- It addresses interaction between process variables.
- It is beyond the capabilities of a technician (or at least all but the best of them).

One possible definition is anything other than simple feedback control, which is understood to be a configuration consisting of three elements:

- A final control element
- A PID controller that generates the output to the final control element
- A measurement device that provides the process variable input to the PID controller

If simple feedback control provides the required performance, it should definitely be used. Going beyond simple feedback control always incurs costs that must be justified by the returns from the improved performance. Advanced control should be pursued only when the improved performance translates into enhanced process performance.

Cascade is a good example of the difficulty of defining advanced process control. To most, a level-to-flow cascade is only slightly above simple feedback control on the scale of sophistication. Few would consider these to be advanced control. But consider a temperature-to-temperature cascade applied to a process consisting of interacting stages (as are most temperature processes). Most find these quite challenging and beyond the capabilities of all but the most experienced technicians. Given the importance of temperatures to process operations, arguments can be made to include such cascades in the advanced control category.

The term *advanced control* is sometimes used to refer to some form of model predictive control (MPC) technology. Model predictive control is definitely advanced control; however, other control technologies deserve to be included in the advanced control category.

The focus of this book is process control, not process safety. Process control must operate the process in the most effective manner, which often leads to considerable complexity. Process safety must avoid unsafe process operating conditions, usually by initiating a shutdown or trip. Although these two are largely separate issues, one requirement must be imposed on the process controls: The process controls must not take any action that would necessitate a reaction from the safety system. Such trips are unnecessary trips and must not happen.

In the process industries the P&I diagram is used almost universally to present the control configuration. This representation encompasses all normal control functions. But for smooth operations, the following requirements must be addressed:

Bumpless transfer. For control configurations that generate multiple outputs, an "all or none" option is not acceptable. The operators must be able to assume control of an individual output at any time. This must not in any way disrupt the other functions being provided by the control configuration. When the manually controlled output is returned to automatic control, there must be no abrupt change in the value of the output (or in any other output from the controls).

Windup protection. When the output of a PID controller ceases to affect its measured variable, the reset mode will drive the controller output to a limit. This is windup. Subsequently, the controller must "unwind," and this is where the consequences appear. A common cause of windup is when a limiting condition has been attained. Limits apply to all process control applications, the simplest manifestation being a fully open or fully closed valve. However, limits can arise within the process, a common example being heat transfer limiting conditions.

Addressing these issues is often as challenging as developing the configuration for the normal control functions. This book gives such topics appropriate attention.

What if these issues are ignored? Consequences that surface during periods of normal control activities are usually considered to be nuisances that the operators can easily handle (we say that the control configuration has some "warts"). Unfortunately, consequences are most likely to appear during process upsets when the operators are very busy. What would otherwise be a nuisance becomes a distraction that takes the operator's attention away from more pressing matters. Given the "right sizing" of operations staffs, such distractions become serious matters.

This is one aspect that commercial model predictive control packages generally address quite well. Most permit operators to assume control of any output without disrupting the remaining functions. Limiting conditions can be imposed on the outputs, on dependent variables, and so on. That such factors have received appropriate attention has certainly contributed to the success of these packages.

This book also reflects the "You have to understand the process" philosophy that dates from my early years in this business. Process control is appropriately a part of chemical engineering, and those with a process background have made important contributions to the advancement of process control. Even though model predictive control relies on certain principles of linear systems theory, those who pioneered the initial applications were firmly rooted in the process technology.

I am a firm proponent of the time domain. Absolutely no background in Laplace transforms is required to understand the presentations in this book. The word "Laplace" is not mentioned outside this preface, and the Laplace transform variable s is not used anywhere. I firmly believe that Laplace transforms should not be taught in a process control course that is part of the undergraduate chemical engineering curriculum.

CECIL L. SMITH

Taos, NM
September 24, 2009

1

INTRODUCTION

The vast majority of the control requirements in the process industries can be satisfied with a simple feedback control configuration that consists of three components:

- A measurement device for the controlled variable or process variable (PV)
- A proportional-integral-derivative (PID) controller
- A final control element, usually a control valve

The performance of any control configuration can be quantified by the variance in the control error, which is the difference between the set point (SP) and the PV. Control configurations more sophisticated than simple feedback offer the promise to reduce (or narrow) this variance. However, proceeding in this direction requires an incentive, the following two being the most common:

- The simple feedback configuration performs so poorly that it affects process operations negatively. Narrowing the variance in the control error translates directly into more consistent process operations.
- A significant economic incentive exists to operate the process more efficiently. Usually, this entails improving the control performance so that the process can be operated closer to a limiting condition. This is summarized as "narrow the variance, shift the target."

In this book we examine several control methodologies that can be applied to enhance the performance of the controls. The user has two options:

Advanced Process Control: Beyond Single-Loop Control   By Cecil L. Smith
Copyright © 2010 John Wiley & Sons, Inc.

- Replace the PID controller, usually with some version of model predictive control. Few regulatory control systems provide model predictive control as a standard feature, but the technology is readily available and easily purchased.
- Retain the PID controller, but incorporate additional logic to enhance the control performance. Most digital systems implement the PID controller as a function block. Numerous additional function blocks are supplied as part of the basic offering, making this approach relatively easy to pursue.

The choice is often dictated by economics. Significant benefits are required to justify model predictive control, so such controllers are often used in conjunction with optimization efforts. Otherwise, the capabilities of the controls must be enhanced by using other function blocks in conjunction with the PID controller.

1.1. IMPLEMENTING CONTROL LOGIC

As used in control systems, a *block* may encompass the following:

Input or measurement block. This block accepts a signal of some type from a field measurement device and converts the input to a numerical value of the measured variable in engineering units (°C, psi, lb/min, etc.).

Output or valve block. This block provides a signal of some type to a final control element. Most final control elements in process facilities are control valves, hence the term *valve block*.

Control block. Each block is described by an equation or algorithm that relates the output(s) of the block to its input(s). Some control systems provide a large number of very simple control blocks; others provide a smaller number of more complex control blocks, each with numerous options. Either approach is possible.

The processing of inputs and outputs can be implemented by other means, but for the control functions, the use of blocks is almost universal.

Input or Measurement Block. Although technically incorrect, the term *analog* is commonly used within digital systems. Prior generations of process controls were based on either electronic or pneumatic technology, and the term *analog* was appropriate. To ease the transition to digital controls, the initial versions of microprocessor-based process controls were designed specifically to closely emulate their analog predecessors. Hence, it should not be surprising that the term *analog* would be applied to corresponding signals within digital systems, and it is also used herein.

The correct term is *digital*. A digital signal is a finite arithmetic approximation to an analog signal. All digital values have a finite resolution: specifically, a change of 1 in the least significant number used in the representation. Here are two examples:

Decimal. A four-digit decimal representation with the format xxx.x has a resolution of 0.1. There are 10,000 possible values (0.0 through 999.9), so the resolution is often stated as 1 part in 10,000.

Binary. A 16-bit binary integer value (short integer) has a resolution of 1 bit. The number of possible values is 64,536 ($= 2^{16}$), either 0 through 64,535 for unsigned integers or $-32,768$ through 32,767 for signed integers. The resolution is 1 part in 64,536 or less, depending on the range of values that can occur.

In processing inputs from room-temperature devices (RTDs) and thermocouples, a common approach is for the input card to convert the input to engineering units in either °C or °F (this is specified via an option on the input card). The result is a short integer value (16 bits) but with the format understood to be xxxx.x. That is, 1074 is understood to be either 107.4°C or 107.4°F. Considering the accuracy of RTDs and thermocouples, a resolution of 0.1°C or 0.1°F is reasonable. But for narrow spans on displays and trends, the finite resolution will be evident. Some address this issue by smoothing or filtering the input value, but this adds undesirable lag to a control loop.

In all examples presented herein that involve temperature measurements, a resolution of either 0.1°C or 0.1°F is imposed. The objective is to illustrate the impact of finite resolution on the performance of various control configurations.

Output or Valve Block. Some control valves fail closed; others fail open. For an output of 0%, a fail-closed control valve is fully closed; for an output of 0%, a fail-open control valve is fully open. If the output to a fail-closed control valve is 60%, the control valve is 60% open. If the output to a fail-open control valve is 60%, the control valve is 60% closed or 40% open.

The failure behavior of the control valve is not really a control consideration. A control configuration that outputs to a fail-open control valve will perform just as effectively as a control configuration that outputs to a fail-closed control valve, and vice versa. The behavior of the control valve on failure is appropriately a decision for those doing the hazards analysis. Those that configure the controls need to know how the control valve is to behave on failure, but they have no reason to prefer a fail-closed valve to a fail-open valve, or vice versa.

In the past, the failure behavior of the control valve was reflected within the control configuration in various ways, depending on how the supplier implemented certain features. But with digital systems, the trend is to configure the controls to generate all outputs as percent open, that is, as if the controls always output to a fail-closed control valve. Herein it is assumed that the input to the valve block or its equivalent will always be percent open. The valve block will address the issues pertaining to fail-open or fail-closed. Consequently, the output of the controls will be referred to routinely as *valve opening*. In effect, the controls determine the output in terms of valve opening and then let the valve block do the rest.

For a fail-closed valve, the valve block merely transfers the value of its input to the final control element. But for a fail-open valve, the percent open value of the input must be converted to a percent closed value for the final control element. Where this is done depends on the physical interface to the control valve:

Current loop. A current flow of 4 mA or less must cause the control valve to be in the desired failure state. Therefore, the conversion from percent open to percent closed must be done before the current loop output is generated. It will be assumed herein that this is done by the valve block, but if not, one need only insert a $Y = 100 - X$ computation into the control configuration to convert input X as percent open to output Y as percent closed.

Fieldbus. When the output is transmitted to a smart valve via a network or communications interface, the output can always be transmitted as percent open. If the control valve is fail-open, the smart valve converts to percent closed. On loss of communications with the controls, the smart valve can be configured to drive the valve to its failure or "safe" state (equivalent to 4 mA or less from a current loop).

With time, fieldbus interfaces will replace current loops within industrial control systems.

Control Block. The configuration of a control block involves three categories of specifications:

Options. For example, the PID is either direct or reverse acting.

Parameters. For the PID, the parameters include the tuning coefficients, the controller output limits, and others.

Inputs. Each input to a control block is usually the output of another block. Some inputs are optional in the sense that designating a source for such an input is not mandatory.

Why configure by designating the source of each input to the control block? Why not configure by designating the destination of each output? For each input to a control block, there can be only one source. However, a given output from a block may be an input to more than one other block.

For configuration purposes, each output of a block must have a unique designation. This designation has two components:

Tag name. Each block is assigned a unique tag name, such as FT101 for a flow measurement and TC4011 for a temperature controller. The numerical designation is always site specific; however, the use of FT for flow transmitters, TC for temperature controllers, and so on, is widespread. For many of the examples in this book, the numerical designation is not needed to identify a function block uniquely; often, only "FT," "TC," and so on, suffices as the tag name.

Attribute. Each output of a control block has a unique designation that depends on the type of the control block. For the PID control block, the attribute "SP" designates the current value of the set point. Every PID control block provides an output for the current value of the set point, and this output is designated by "SP."

Herein these two components are combined into a single mnemonic <Tag Name>.<Attribute>, with the decimal point or period serving as the separator. That is, TC4011.SP is the current value of the set point of control block TC4011.

Some systems also use attributes to designate the inputs to a function block. Using the PID controller as the example, "PV" designates the process variable input, "RSP" designates the remote set point input, and so on. This approach is used herein.

Process and Instrumentation (P&I) Diagram. Figure 1.1 presents the P&I diagram for a level-to-flow cascade configuration for controlling the level in a vessel. The output of the level controller is the set point for the discharge flow controller. This is conveyed explicitly in the P&I diagram, with the output of the vessel level controller connected to the set point [actually, the remote set point (RSP) input] of the discharge flow controller.

P&I diagrams such as in Figure 1.1 convey the requirements for normal operation of the controls. For the level-to-flow cascade in Figure 1.1, these requirements are as follows:

- The vessel level transmitter provides the PV input to the vessel level controller.
- The discharge flow transmitter provides the PV input to the discharge flow controller.

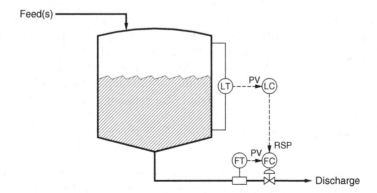

Figure 1.1 P&I diagram of a level-to-flow cascade.

- The output of the vessel level controller is the set point for the discharge flow controller.
- The output of the discharge flow controller is the opening of the control valve on the vessel discharge.

But for smooth operation, other requirements must be incorporated into the control configuration.

Bumpless Transfer and Windup Protection. When implementing the control configuration for an application, the requirements for normal operation of the controls take precedence. However, capabilities are also required to address the following:

- *The transition from manual to automatic must be smooth or "bumpless".* The PID block provides for bumpless transfer from manual to automatic. But what about switching the discharge flow controller in Figure 1.1 from automatic to remote? To achieve a smooth transition, functions in addition to those illustrated in Figure 1.1 are required. Similar requirements apply to all control configurations and usually increase in complexity with the complexity of the control configuration.
- *The PID controller must not be allowed to wind up.* Windup is a phenomenon associated with the reset mode and is often referred to as *reset windup*. The PID block invokes windup protection when the controller output is driven to either of the controller output limits. However, there are external factors that can result in windup. The condition for windup to occur is stated as follows:

  Reset windup occurs in a controller when changes in the controller output have no effect on the process variable.

This statement will be used repeatedly in subsequent chapters. Using the cascade control configuration in Figure 1.1 as an example, suppose that the measurement range of the discharge flow controller is 0 to 100 gpm, but when the control valve is fully open, the discharge flow is 70 gpm. Does increasing the set point above 70 gpm have any effect on the flow? Once the control valve is fully open, additional increases have no effect on the variable being controlled. The condition for windup exists in the vessel level controller. There are three capabilities for avoiding such windup:
- Integral tracking
- External reset
- Inhibit increase/inhibit decrease

Implementations of the PID block must provide at least one of these, but configuring such features is not normally represented on P&I diagrams such as Figure 1.1.

The logic required to address these issues can easily exceed the logic for the normal control functions. Ignoring the requirements for bumpless transfer and windup protection will have consequences. Rarely do consequences arise during normal production operations, but commonly arise when situations such as the following occur:

• During startup and shutdown.
• The process is driven to a limiting condition, such as maximum heat transfer in an exchanger or operating a fired heater at the minimum firing rate.
• Temporary disruptions to production operations, such as operating a column on total reflux (feed is stopped, but boil-up and reflux continue).
• Switching between modes of operation, such as regenerating the catalyst in a fluidized bed.

The importance of addressing the consequences depends on how frequently such events occur. If they arise only during startup and shutdown, the consequences can be addressed by incorporating appropriate actions into the operating procedures for startup and shutdown. But if they occur routinely during process operations, the controls must cope with any consequences without depending on intervention by the operators.

One approach is to switch the controls to manual should conditions arise where windup would occur. The operator must subsequently return the controls to automatic when such conditions no longer exist. This approach is certainly preferable to permitting windup to occur. To use this approach, bumpless transfer from manual to automatic is essential. However, the burden imposed on the process operators would be acceptable only when such conditions arise infrequently. Instead of switching the PID controller to manual, emphasis herein is placed on approaches that initiate appropriate windup protection via the inputs to the PID block.

Softwiring. In single-loop controllers, hardware terminals are provided for each input and output. For a PID controller, the signal from the measurement device is connected to the terminals for the PV input. The controller output is available via the terminals for the controller output. The control configuration is determined by the physical wiring for these terminals. Softwiring involves using an analogous approach in software, specifically, software emulation of hardwiring. Instead of physical connections, the source of each input is specified in the software configuration for each block. Graphical development facilities permit these connections for softwiring to be specified on the graphical representation of the control logic.

Figure 1.2 presents the configuration for a level-to-flow cascade. Two liberties have been taken:

• The customary P&I diagram representations are used for the controllers. Subsequently, a rectangular representation for the PID block is presented with all inputs on the left and all outputs on the right. Older configuration

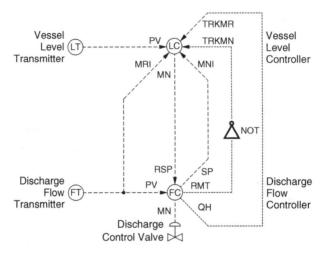

Figure 1.2 Control logic diagram of a level-to-flow cascade.

tools generally used a fixed representation for each type of block, but newer tools are far more flexible.

- Only those inputs for which wiring is actually provided are shown in Figure 1.2. For example, an RSP input is available for the vessel level controller, but since no source is specified for this input, it is not included in Figure 1.2. When used in a configuration tool, the block representation must include every possible input and output.

When using a graphical configuration tool, lines are constructed for signals to connect the appropriate output of one block to the appropriate input to another block. The possibilities are as follows:

- The value for the input is an output of another block. In Figure 1.2, input MRI to the level controller is the PV input to the flow controller.
- The value for the input must be computed from one or more outputs of other function blocks. In Figure 1.2, input TRKMN to the level controller is the inverse (logical NOT) of output RMT of the flow controller.

The graphical configuration tool must provide for both possibilities.

Control Logic Diagram. A descriptive term for diagrams such as Figure 1.2 is *control logic diagram*. In addition to the connections required for the normal control functions that are indicated on a P&I diagram, Figure 1.2 includes connections for the following:

Bumpless transfer. Output tracking is configured for the vessel level controller so that the transition from automatic to remote will be smooth. Two connections are involved:

- Output RMT from the discharge flow controller is inverted (logical NOT) and softwired to input TRKMN to the vessel level controller. Output tracking is to be active if the discharge flow controller is not on remote.
- Output SP from the discharge flow controller is softwired to input MNI to the vessel level controller. When output tracking is active, the vessel level controller must set its output to the value of input MNI.

Windup protection. Integral tracking is configured to prevent windup in the vessel level controller should the discharge flow controller drive the discharge control valve fully open. Two connections are involved:

- Output QH from the discharge flow controller is softwired to input TRKMR to the vessel level controller. Integral tracking is to be active if the discharge flow controller has fully opened the discharge control valve.
- The PV for the discharge flow controller is softwired to input MRI to the vessel level controller. When integral tracking is active, the vessel level controller sets its controller output bias to the value of input MRI.

These connections are explained in detail in Chapter 2. Similar requirements arise in other control configurations and are discussed is the chapters devoted to those control configurations.

Track or Initialization Request. The softwiring approach requires very explicit specifications for the tracking ("what you see is what you get"). The *track request* is an alternative approach that is largely hidden, with the actions performed mostly "under the hood." In general, any time that a function block does not use the value from one of its inputs, tracking or initialization is required. At some time in the future, the function block will again use the value of this input. The objective of tracking is to achieve a smooth transition from not using the input to using the input.

In a cascade configuration such as in Figure 1.1, track requests are generated under certain conditions. The PID controller uses the remote set point input only when the mode is remote. Consequently, the inner loop controller of a cascade must issue a track request whenever it is not on remote. In Figure 1.1 the discharge flow controller (the inner loop) issues the track request to the source of its RSP input, which is the vessel level controller. The track request must include a value that is the target of the tracking. For cascade control configurations, this value is the current value of the inner loop set point.

To generate a track request, four questions must be answered. To achieve bumpless transfer for the level-to-flow cascade, the questions and their answers are as follows:

- *What type of tracking is required?* For bumpless transfer from local to remote, output tracking is required. Other types of tracking are required in certain situations. For example, integral tracking is required when the inner loop has driven its output to either controller output limit.

- *To which block is the track request directed?* The inner loop obtains this information from the configuration for the remote set point. Specifically, the RSP input to the discharge flow controller is the MN output of the vessel level controller. Consequently, the discharge flow controller directs the track request to the vessel level controller.
- *When should the track request be generated?* The answer is: when the discharge flow controller is not on remote.
- *What value should accompany the track request?* The answer is: the current value of the set point for the discharge flow.

In a sense, the connection from output MN of the vessel level controller to input RSP to the discharge flow controller is bidirectional. The actions depend on the mode of the discharge flow controller:

Remote. The value of the MN output of the vessel level controller is copied to the set point location of the discharge flow controller.

Manual or automatic. The discharge flow controller issues a track request to the vessel level controller to set its MN output to the SP value that accompanies the track request.

The track request mechanism is different from the softwiring mechanism, but the actions are equivalent. Generation of the track request is equivalent to setting the TRKMN input to "true"; the track value that accompanies the track request is the same value that is provided to the MNI input. The PID block, the integrator/totalizer block, and a few others can process a track request by setting an internal coefficient (the controller output bias M_R for the PID block). But suppose that a multiplier block receives a track request. Normally, it must back-calculate the value of one of its inputs and then propagate the track request to the source of that input. How does it know which input? There are two possibilities:

1. Always propagate the track request to input X1.
2. The user specifies the input as part of the block configuration.

Sometimes the impression is given that the track request approach frees the user from all issues associated with initialization and tracking. Unfortunately, this is not quite the case. Perhaps one day, control systems will be able to do this, but at this point, those configuring the controls must be cognizant of how track requests will be issued and propagated.

Logic Statements. Within the process industries, P&I diagrams (such as in Figure 1.1) are used almost universally to represent the control configuration. However, the logic associated with bumpless transfer and windup protection is not normally included in a P&I diagram. The issue is how to express this logic.

When implementing a control configuration, one must use whatever facilities are provided by the control system supplier. But in developing this book, the issue is how to present the logic in a form that is most easily comprehended by someone learning about controls. The possibilities include the following:

Softwiring. Graphical configuration tools that rely on the softwiring approach are very popular with control system suppliers, the result being control logic diagrams such as in Figure 1.2. The objective of the graphical approach is to permit control logic of any complexity to be implemented without programming of any type. But as control configurations become more complex, the logic diagrams also become more complex.

Track request. Performing much of the logic "under the hood" has advantages in implementing a control configuration, but someone new to process control is likely to be perplexed by this approach.

Logic statements. The approach is summarized as follows:
- P&I diagrams for the normal control logic
- Statements for the logic for bumpless transfer and windup protection

Using this approach, the P&I diagram in Figure 1.1 is supplemented by the following logic statements:

```
LC.MNI = FC.SP
LC.TRKMN = !FC.RMT
LC.MRI = FT.PV
LC.TRKMR = FC.QH
```

The combination is equivalent to the control logic diagram in Figure 1.2.

The logic statements approach is used herein. The inputs and outputs for the control blocks as used in this book will be explained shortly. This enables the logic statements for the level controller (LC) to be read as follows:

- The value for output tracking (input MNI) is the set point of the flow controller.
- Output tracking is active when the flow controller is not on remote.
- The value for integral tracking (input MRI) is the current value of the flow.
- Integral tracking is active when the output of the flow controller is at its upper output limit (the flow control valve is fully open).

Composing the statements for the inputs depends on the nature of the input:

Analog (or actually digital). Analog values are required for inputs LC.MNI and LC.MRI. In constructing the statements, the usual arithmetic operators will be used, but functions such as MAX and MIN will also be allowed.

Discrete. Discrete values are required for inputs LC.TRKMN and LC.TRKMR. For discrete expressions, the discrete operators from C++ will be used:

Operator	Explanation
!	Logical NOT
|	Logical OR
&	Logical AND
~	Exclusive OR

1.2. CONTROL BLOCKS FOR PROCESS CONTROL

The control blocks provided by control systems generally include the following:

PID Controller. All control systems provide a block for the PID controller. Most provide a variety of options. The PID block will be described in detail shortly.

Arithmetic Computations. In the following descriptions, the notation is as follows:

$$X_i = \text{input } i \text{ to the block}$$
$$Y = \text{output of block}$$
$$k_i = \text{coefficient } i$$

Control blocks of this type include the following:

Summer. The usual equation is

$$Y = k_0 + k_1 X_1 + k_2 X_2$$

Since any coefficient can be negative, the summer also provides subtraction.
Multiplier. The usual equation is

$$Y = k_0 + k_1 X_1 X_2$$

Often, a power is provided on one of the inputs.
Divider. The usual equation is

$$Y = k_0 + k_1 \frac{X_1}{X_2}$$

Characterization function. Sometimes referred to as a *function generator block*, this block is described by the following relationship:

$$Y = f(X)$$

Usually, the function is defined by specifying individual points and using linear interpolation between the points.

There is considerable variability from one control system to another. Many implementations provide more than two inputs to summers and multipliers. Some provide a general arithmetic expression that provides multiplication, division, and power. For true object-oriented implementations, a block can be provided that accepts an arithmetic expression similar to one that can be programmed in C, Pascal, Fortran, and so on.

Logic Gates. In the following descriptions, the notation is as follows:

$$X_i = \text{analog input } i \text{ to the block}$$
$$Q_i = \text{discrete input } i \text{ to the block}$$
$$Z = \text{output of block}$$

Blocks of this type include the following:

NOT. The usual equation is

$$Z = !Q_1$$

The output Z is the logical complement of the input.
OR. The usual equation is

$$Z = Q_1 + Q_2$$

The output Z is the logical OR of the two inputs.
AND. The usual equation is

$$Z = Q_1 \ \& \ Q_2$$

The output Z is the logical AND of the two inputs.
XOR. The usual equation is

$$Z = Q_1 \sim Q_2$$

The output Z is the exclusive OR of the two inputs.

Comparator. The usual equation is

$$Z = X_1.\text{op}.X_2$$

The relational operator ".op." may be greater than, less than, equal to, greater than or equal to, less than or equal to, or not equal to. But for real-time analog values, comparisons for equality must be approached with caution.

Unit delay. The usual equation is

$$Z = Q_0$$

The output Z is the value Q_0 of the input Q on the preceding scan or sampling instant.

One-shot. The output of this function is true for one scan or sampling instant following a transition of the input. Most implementations provide the option of detecting 0-to-1 transitions only, 1-to-0 transitions only, or all transitions. If the one-shot is not supported but the unit delay is supported, the one-shot can be implemented using the following equations:

All transitions : $Z = Q \sim Q_0$

Only 0-to-1 transitions : $Z = (Q \sim Q_0) \& Q$

Only 1-to-0 transitions : $Z = (Q \sim Q_0) \& Q_0$

where Q is the current value of the input and Q_0 is the value of the input on the preceding scan or sampling instant.

Symbols often referred to as *gates* are in common use for logic operations such as AND, OR, XOR, and NOT. To obtain the state of a discrete input to a block, simple logic must be applied to the outputs from one or more other blocks. In control logic diagrams, the logic gates are one approach to expressing such logic. A logical NOT is used in the control logic diagram in Figure 1.2.

Figure 1.3 presents both the traditional symbols and the IEC (International Electrotechnical Commission) symbols for various gates (AND, OR, etc.). A small circle on either the input or the output designates logic inversion or NOT. Figure 1.3 illustrates adding such a circle to the AND, OR, and XOR gates to obtain the NAND (NOT AND), NOR (NOT OR), and NXOR (NOT XOR) gates. A small circle on the input to a gate means that the input is inverted before the logic operation is performed.

Dynamic Functions. These blocks are described by a differential equation or difference equation. Blocks of this type include the following:

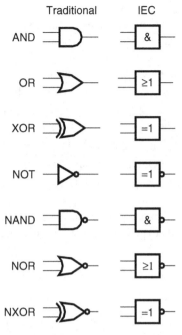

Figure 1.3 Logic gates.

Integrator (totalizer). Although it is occasionally encountered elsewhere, the most common application of this block is to totalize a flow. This block is described in detail in a later section of the chapter.

Lead-lag. This block is described in a subsequent section of the chapter. A special case is a pure lag.

Dead time. This block is described in a subsequent section of this chapter.

Moving average. The usual equation is

$$Y_k = \frac{1}{N} \sum_{j=0}^{N-1} X_{k-j}$$

where
X_k = value of input X at sampling instant k
Y_k = value of output Y at sampling instant k
N = number of input samples for computing the average
T_S = sampling time (time between input values)
$T_A = NT_S$ = time span of the arithmetic average

The value of the average is updated each sampling instant, the output being the arithmetic average of the previous N values of the input (hence the term *moving*).

This function block can be used to provide filtering or smoothing. However, for $T_A = \tau_F/2$, the results of the moving average are approximately the same as that of an exponential filter whose time constant is τ_F.

Sample-and-hold. This block has two modes:
- *Sample.* The output of the block is equal to the input.
- *Hold.* The output of the block retains its last value (the value of the input has no effect on the output of the block).

The input TRK determines the mode, the relationship being as follows:
- *Input TRK is true.* Output Y equals input X.
- *Input TRK is false.* Output Y retains the last value.

In a control system that does not provide a sample-and-hold function block, the equivalent functionality can be implemented using the integrator/totalizer block, as explained in a subsequent section.

Special Functions. These include the following, all of which will be described in more detail shortly:

Selector (auctioneer). This block (or its equivalent) is required to implement the override control.

Cutoff. Two- and three-stage cutoff blocks are commonly provided, although true object-oriented implementations can provide an unlimited number of stages.

Hand station. The purpose of a hand station is to enable the process operator to specify the value of an output from the control system. In older control systems, the hand station was a hardware item. In modern controls, it is usually implemented in software, but its purpose is the same.

1.3. PID CONTROLLER

All function block implementations provide a block for a PID controller. Options such as direct or reverse action are provided by all implementations. Most provide options that pertain to the PID control equation (such as proportional on E vs. proportional on PV), but the exact options that are provided differ from one implementation to the next. Table 1.1 lists the attributes used for each input and each output. A description of each is provided, including whether the input or output is analog (actually, digital) or discrete. All outputs are available, but inputs are designated as required (meaning must be configured) or optional.

Figure 1.4 presents a rectangular representation for the PID block. All inputs are on the left; all outputs are on the right. Graphical configuration tools require some type of representation for each type of block. Figure 1.4 is provided only as an example; such representations are not used in this book. In texts on process

Table 1.1   Inputs and outputs for the PID block

Mnemonic	Signal	Value	Purpose
PV	Input (required)	Analog	Process variable.
RSP	Input (optional)	Analog	Remote set point.
TRKMN	Input (optional)	Discrete	On true, set output M to the current value of input MNI and initialize for bumpless transfer. On false (or not configured), input MNI is not used.
MNI	Input (optional)	Analog	Value for output M when input TRKMN is true. If input MNI is not configured, the value is assumed to be zero.
TRKMR	Input (optional)	Discrete	On true, set controller output bias M_R to the current value of input MRI and then compute the output M using the proportional-plus-bias equation. On false (or not configured), input MRI is not used.
MRI	Input (optional)	Analog	Value for controller output bias M_R when input TRKMR is true. If input MRI is not configured, the value is assumed to be zero.
XRS	Input (optional)	Analog	External reset input to the reset mode in the reset feedback implementation of the PID. If not configured, the controller output M is used.
NOINC	Input (optional)	Discrete	If true, increases in the controller output M are not allowed.
NODEC	Input (optional)	Discrete	If true, decreases in the controller output M are not allowed.
FMANL	Input (optional)	Discrete	If true, force mode to manual.
FAUTO	Input (optional)	Discrete	If true, force mode to automatic.
FRMT	Input (optional)	Discrete	If true, force mode to remote.
MN	Output	Analog	Controller output.
SP	Output	Analog	Current value of controller set point.
AUTO	Output	Discrete	If true, controller is on automatic (either local automatic or remote automatic).
RMT	Output	Discrete	If true, controller is on remote.
SPH	Output	Discrete	True if the set point is at or above the upper set point limit.
SPL	Output	Discrete	True if the set point is at or below the lower set point limit.
QH	Output	Discrete	True if the controller output has been driven to the upper output limit.
QL	Output	Discrete	True if the controller output has been driven to the lower output limit.

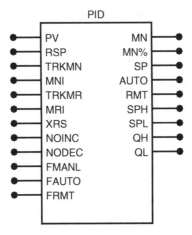

Figure 1.4   PID block with the inputs and outputs required for various forms of tracking.

control, the output of the PID controller is known as the *manipulated variable* and is usually designated as either M or $M(t)$ to designate that it is a continuous function of time. In the discrete equations that will be presented subsequently, the output is designated as M_n, which is the value of output M at sampling instant n. To be consistent with this notation, attribute MN will be used to designate the output.

Tuning coefficients. The following are used herein:

K_C = controller gain (%/%); the proportional band PB (in %) is $100/K_C$

T_I = reset time (min); the reset rate R_I is $1/T_I$ and the reset gain
$\quad K_I$ is K_C/T_I

T_D = derivative time (min); the derivative gain K_D is $K_C T_D$.

So that the units for the controller gain K_C can be %/%, the PID block configuration includes two ranges:

Input range. Input PV and input SP are in engineering units. Before performing the PID calculations, these two inputs are converted to percent of span based on the input range specified. Normally, this input range is the same as the measurement range of the transmitter that provides the PV input, but this is not mandatory.

Output range. The result of the PID calculations is the controller output as a percent of span. Herein, attribute MN% designates this value. The output in engineering units is computed from the output as a percent of span and the output range. Herein, attribute MN designates this value. For the percent opening of a control valve, the output range would be 0 to 100%, giving output MN and output MN% the same value.

An alternative approach is to compute the controller gain $K_{C,EU}$ in engineering units as follows:

$$K_{C,EU} = K_C \frac{\text{span of output}}{\text{span of input}}$$

The following are equivalent:

- Gain as K_C and control error in % of span
- Gain as $K_{C,EU}$ and control error in engineering units

Herein, the control equations are expressed in terms of K_C, the exception being that $K_{C,EU}$ is used in the equations for numerical examples of control calculations.

PID Control Equations. When all modes are based on the control error, the PID controller is formulated as follows:

1. *Control error E.* The controller action determines the sign of the control error:

$$E = \begin{cases} PV - SP & \text{if direct acting} \\ SP - PV & \text{if reverse acting} \end{cases}$$

2. *Derivative mode equation.* The projected error $\hat{E}$ is computed by projecting the current rate of change of the control error E for one derivative time into the future:

$$\hat{E} = E + T_D \frac{dE}{dt}$$

Most implementations include a small amount of smoothing in the derivative mode equation, with the time constant for the smoothing being αT_D. The coefficient α is the derivative mode smoothing factor; the derivative gain limit is the reciprocal of α. A typical value for α is 0.1.

3. *Proportional mode equation.* The equation is actually proportional plus bias:

$$M = K_C \hat{E} + M_R$$

where

$$M = \text{controller output}$$
$$M_R = \text{controller output bias}$$

4. *Integral mode equation.* The possibilities for the integral mode equation include the following:

Parallel	Series

$$M_R = \int \frac{K_C}{T_I} E \, dt \qquad M_R = \int \frac{K_C}{T_I} \hat{E} \, dt$$

$$\frac{dM_R}{dt} = \frac{K_C}{T_I} E \qquad \frac{dM_R}{dt} = \frac{K_C}{T_I} \hat{E}$$

$$T_I \frac{dM_R}{dt} + M_R = M$$

The terms *noninteracting* and *ideal* are often used in lieu of *parallel*; *interacting* and *nonideal* are often used in lieu of *series*. The third equation is for the reset feedback implementation of the control equation and is only applicable to the series form of the control equation.

Output Tracking. A common use of this feature is to achieve bumpless transfer in various control applications (such as switching the inner loop of a cascade from automatic to remote). There are two inputs that pertain to output tracking:

Input TRKMN. This discrete input determines whether or not output tracking is active. Output tracking is active when the state of this input is true.

Input MNI. If output tracking is active, the controller output M is set equal to the value of input MNI. In addition, the controller output bias M_R is initialized in the same manner as on a manual-to-auto mode transition. If output tracking is not active, input MNI is not used.

Overrange. To assure that the valve can be driven fully open and fully closed, some overrange is provided on the output to the control valve. The options are:

- In the examples herein, the overrange is provided by specifying -2% and 102% for the controller output limits.
- For PID blocks that do not provide such an overrange, the span adjustments at the valve are such that the valve is fully closed at a value slightly above 0% and fully open at a value slightly below 100%.

The overrange is more essential for "valve closed." A valve that is not quite fully open may go unnoticed, but a valve that is not quite fully closed will attract attention.

In the ideal world, the following statements are equivalent:

- *Control valve fully closed (open).* This is when windup begins.

- *Controller output driven to its lower (upper) output limit.* This is when windup protection is invoked.

In the practical world they differ because of the overrange provided. Unless the overrange is excessive, the effect on performance is insignificant.

Windup Protection. Within the PID block, some form of windup protection is always provided, two common options being:

Bias limits. Apply the controller output limits to the controller output bias M_R as well as to the controller output.

Bias freeze. When the output of the controller attains either output limit, disable the reset mode, which "freezes" the value of M_R.

For most loops, these suffice. But for some, limiting conditions external to the PID controller are occasionally encountered. For these loops, there are three possibilities for preventing windup: integral tracking, external reset, and inhibit increase/inhibit decrease. Although all three can be implemented within the same PID block, few commercial systems do so. All three are unnecessary; any one will suffice.

Integral Tracking. The following two inputs from Table 1.1 pertain to integral tracking:

Input TRKMR. This discrete input determines whether or not integral tracking is active. Integral tracking is active when the state of this input is true.

Input MRI. If integral tracking is active, the controller output bias M_R is set equal to the value of input MRI. If integral tracking is not active, input MRI is not used.

When integral tracking is active in a controller (input **TRKMR** is true), the following computations are performed.

1. The controller output bias M_R is set equal to the value of input MRI. The normal reset mode calculations are not performed; that is, the reset action is effectively disabled.
2. The output of the controller is computed by the proportional-plus-bias equation:

$$M = K_C \hat{E} + M_R$$

External Reset (Input XRS). This feature was provided in conventional pneumatic and electronic controls to provide windup protection, so it actually predates the other windup protection mechanisms. As illustrated in Figure 1.5, external reset applies specifically to the reset feedback implementation of the PID. Input

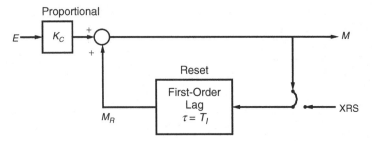

Figure 1.5 Feedback structure for PI controller with external reset.

XRS is the input to the lag circuit used to implement the reset action. If nothing is configured for input XRS, the default is to use the output of the controller, which could be thought of as "internal reset."

Unlike integral tracking, which can be active at times and not active at others (depending on the state of input TRKMR), the value of input XRS is always used in the control calculations. However, the nature of input XRS determines whether or not the reset mode is functioning in the usual manner:

- If input XRS is the controller output or some variable determined by the controller output, the reset mode functions in the usual manner and attempts to drive the process variable to the set point.
- If input XRS is not affected by changes in the controller output, the reset mode does not function in the usual manner, and is essentially disabled.

If input XRS is the controller output M, the reset feedback control calculations are as follows:

$$\text{Proportional:} \quad M = K_C E + M_R$$

$$\text{Reset:} \quad T_I \frac{dM_R}{dt} + M_R = \text{input XRS} = M$$

The two equations can be expressed as follows:

$$\text{Proportional:} \quad M - M_R = K_C E$$

$$\text{Reset:} \quad \frac{dM_R}{dt} = \frac{M - M_R}{T_I} = \frac{K_C E}{T_I}$$

Integrating the reset equation reveals that the reset feedback form is equivalent to integrating the control error:

$$M_R = \int \frac{K_C}{T_I} E \, dt$$

Windup can occur in a controller only when the reset mode is effectively integrating the control error. When input XRS to the controller is the output M from the controller, the reset mode is integrating the control error. Otherwise, the reset mode in the reset feedback form of the PID is not integrating the control error, and windup cannot occur. When input XRS is other than the output of that controller, the controller output is effectively tracking input XRS. The controller output bias is input XRS lagged by the reset time. In essence, the controller output bias is tracking input XRS. As the controller output is computed by the proportional-plus-bias equation, the controller output is also tracking input XRS.

Inhibit Increase/Inhibit Decrease. (Inputs NOINC and NODEC). This provides windup protection via two discrete inputs:

Input NOINC. If true, the control equation is not permitted to increase its output. Decreases in the controller output are allowed regardless of the state of input NOINC.

Input NODEC. If true, the control equation is not permitted to decrease its output. Increases in the controller output are allowed regardless of the state of input NODEC.

1.4. INTEGRATOR OR TOTALIZER

A common application of the integrator block is to totalize a flow. There are other applications of the integrator block, but certain features of the integrator block reflect the requirements of totalizer applications. The common ones are the following:

Nearly zero flows. When a flow is stopped, ideally the output of the flow measurement device should be exactly zero. However, this is not assured. Especially in current loop applications, a flow of zero will not translate into a current signal of exactly 4 mA. Consequently, a flow of exactly zero provides a measured value that is either slightly positive or slightly negative. To address this issue, many totalizers provide a deadband on the integrand. If the value of the integrand is less than the deadband, the integrand is considered to be zero.

Difference in flows. One way to implement *leak detection* is to compute the difference between flow in and flow out for an item of process equipment, a section of a pipeline, and so on. Unfortunately, instantaneous values of flow measurements tend to be noisy, and taking the difference between two flows that are nearly equal greatly amplifies the noise. Either instead of or in addition to monitoring the instantaneous value of the difference, the difference is integrated or totalized over some period of time. If there is no loss of fluid, the totalized value should be zero, but due to measurement errors and other factors, the totalized value will be nearly zero. An integrator block that accepts two inputs and integrates their difference facilitates such applications.

Preset and pre-preset. In automating batch processes, a common requirement is to add a specified amount of material to a vessel. One approach is to measure the fluid flow rate and totalize this flow to determine the amount transferred. The flow is to be stopped when the totalized value attains the target, which is often referred to as the *preset*. Some integrator block implementations provide an input for the preset value and a discrete output to indicate when the totalized value has attained the preset. So that the specified amount can be charged more accurately, some charge systems provide for a *dribble flow*. A fast feed rate is used to transfer most of the material, but near the end, the flow is reduced. The switch to the dribble flow occurs when the totalized value attains the pre-preset. Some integrator block implementations provide an input for the pre-preset and a discrete output to indicate when the totalized value has attained the pre-preset.

Reset.   Most totalizers are reset at specified times or on specified events. The following considerations occasionally arise:

- In continuous applications, the totalizers may be reset to zero at a specific time of day, such as midnight. It seems simple to read all totalizers and then reset the totalizers. However, this requires that these two actions be tightly synchronized: The time lapse between when the totalizers are read and when they are reset must be very short (any flow during this period is not taken into account). One approach is to provide two outputs, one being the current totalized value and the other being the totalized value at the time the totalizer was last reset. In this way, the totalizers can be reset at precisely the appropriate time, and the totalized value can subsequently be read.

- In batch applications, the totalizer is usually reset to zero prior to the start of a material transfer. But in some applications, the same material is fed on two separate occasions. The specification is often the total amount of material to transfer. For the second feed, the totalizer should be reset to the amount of material transferred on the first feed.

The usual equation for the integrator or totalizer is as follows:

$$Y = \int k_1(X_1 - X_2)dt$$

where

$X_1 =$ input 1 (required)
$X_2 =$ input 2 (optional)
$Y =$ output of the block
$k_1 =$ coefficient of integration
DB = deadband; if $|X_1 - X_2| <$ DB, $X_1 - X_2$ is considered to be zero

A common application of the coefficient k_1 is to convert engineering units. In control applications, the time base for the integrator is usually minutes, but is occasionally seconds. If the time base is minutes, the flow must be in units such as lb/min or gpm. But what if the flow is lb/hr? The value of coefficient k_1 must be 1/60, which effectively converts the flow from lb/hr to lb/min.

Table 1.2 defines each input and output. To be consistent with the notation for other function blocks, input TRKMN is effectively the input used to reset the totalizer.

In control systems that do not provide a sample-and-hold function block, the equivalent functionality can be provided by configuring an integrator bock as follows:

Inputs X_1 and X_2. Not configured. If the configuration tool requires that an input be configured for X_1, specify any input for X_1 and specify zero for the coefficient of integration k_1.

Input TRKMN. Counterpart to input TRK for the sample-and-hold.

Input MNI. Counterpart to the input to the sample-and-hold.

If input TRKMN is true, the output is set equal to the value of input MNI. If input TRKMN is false, output Y does not change because the integrand is zero (either input X_1 is not configured or the coefficient of integration k_1 is zero).

Table 1.2 Inputs and outputs for the integrator/totalizer block

Mnemonic	Signal	Value	Purpose
X1	Input (required)	Analog	Input 1 to integrator/totalizer.
X2	Input (optional)	Analog	Input 2 to integrator/totalizer.
PSET	Input (optional)	Analog	Value for preset.
PPSET	Input (optional)	Analog	Value for pre-preset.
TRKMN	Input (optional)	Discrete	On true, set output Y to the current value of input MNI.
MNI	Input (optional)	Analog	Value for output Y when input TRKMN is true. If input MNI is not configured, value is assumed to be zero.
HOLD	Input (optional)	Discrete	If true, freeze the value of Y to its current value.
Y	Output	Analog	Current value.
Y0	Output	Analog	Value of Y at the time integrator/totalizer was last reset, that is, when input TRKMN made a false-to-true transition.
Q1	Output	Discrete	True if $Y <$ PSET.
Q2	Output	Discrete	True if $Y <$ PPSET.

1.5. LEAD-LAG ELEMENT

In Chapter 6 we use a lead-lag element to provide dynamic compensation. In this section, the concept of lags and leads is presented, which is a key to understanding how a lead-lag element functions. In Chapter 6 we look at how to incorporate this block into ratio and feedforward control configurations.

Lag. The differential equation for a first-order lag is as follows:

$$\tau_{LG}\frac{dY}{dt} + Y = X$$

where

$X =$ input to the first-order lag
$Y =$ output of the first-order lag
$t =$ time (min)
$\tau_{LG} =$ time constant for the lag (min)

The behavior of a first-order lag is most frequently represented by its response to a step change in input X. As illustrated in Figure 1.6, the response attains 63.2% of the total change in one time constant. However, the notion of a lag is more apparent from the response of the first-order lag to a ramp change in the input X, as illustrated in Figure 1.7. There is a brief transient period (between three and five time constants), but thereafter the response of the first-order lag is also a ramp. However, this ramp is delayed with respect to the input by time equal to the time constant. The response basically lags the input by τ_{LG}.

The first-order lag arises frequently in process control:

- The first-order lag is used in representing the dynamics of most processes. Very few are adequately described by a single first-order lag, but two lags

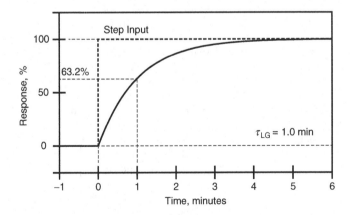

Figure 1.6   Step response of a first-order lag.

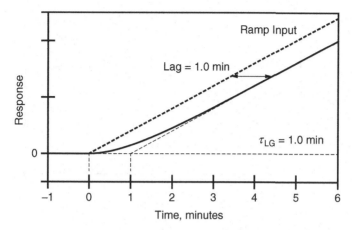

Figure 1.7 Ramp response of a first-order lag.

in combination with dead time provides an adequate approximation of the dynamic behavior of most processes.

- Filtering in the form of exponential smoothing is implemented using a first-order lag.
- The reset feedback structure presented in Figure 1.5 for the PID controller utilizes a first-order lag to implement the reset action.
- A pure lag suffices for dynamic compensation in some ratio and feedforward control configurations.

Lead. Dynamically, lead is the inverse of lag. The differential equation for a pure lead is as follows:

$$Y = \tau_{LD}\frac{dX}{dt} + X$$

where

$X =$ input to the lead
$Y =$ output of the lead
$t =$ time (min)
$\tau_{LD} =$ time constant for the lead (min)

Figure 1.8 presents the response to a ramp change in the input X. The effect of the lead is exactly the opposite of a lag; the response Y leads the input X by τ_{LD}. The response in Figure 1.8 is the theoretical response of a pure lead to a ramp change in its input. In practice, a pure lead cannot be implemented. A lead must always be accompanied by a lag, the result being the lead-lag element.

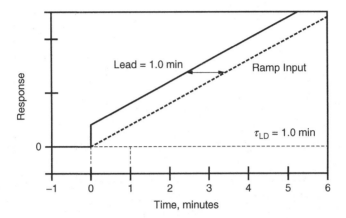

Figure 1.8 Ramp response of a lead.

Lead-Lag Element. The differential equation for a lead-lag is a combination of the equations presented previously for a pure lag and a pure lead:

$$\tau_{LG}\frac{dY}{dt} + Y = \tau_{LD}\frac{dX}{dt} + X$$

where

X = input to lead-lag
Y = output of lead-lag
t = time (min)
τ_{LG} = lead time (min)
τ_{LG} = lag time (min)

The lag time τ_{LG} must be a positive value (it can be neither zero nor negative), but the lead time τ_{LD} can be positive, negative, or zero.

Figure 1.9 presents several responses of a lead-lag element to a step change in the input X. The time axis is normalized by the lag time τ_{LG}. Responses are presented for various values of the lead time τ_{LD}, expressed as the ratio τ_{LD}/τ_{LG} of the lead time to the lag time.

In a sense, the lead-lag element either retards or advances its input. The behavior depends on the relative values of the lead time τ_{LD} and lag time τ_{LG}:

- $\tau_{LD} = \tau_{LG}$. The input is a step change; the output is a step change. The lead-lag element has no effect.
- $\tau_{LD} < \tau_{LG}$. Except for $\tau_{LD} = 0$ (a pure lag), there is some immediate change in the output, but the change is less than the size of the step change in the input. The remaining change is then implemented in the exponential decaying fashion typical of a first-order lag. In a sense, the lead-lag element is retarding its input.

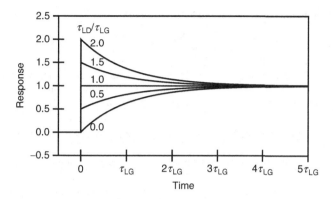

Figure 1.9 Step response of a lead-lag element.

- $\tau_{LD} > \tau_{LG}$. The immediate change in the output exceeds the size of the step change in the input. The excess change is then backed out in the exponential decaying fashion typical of a first-order lag. In a sense, the lead-lag element is advancing its input.

Suppose that the input to the lead-lag is the required steady-state corrective action. Consider the output of the lead-lag to be relative to this corrective action. For $\tau_{LD} < \tau_{LG}$, the compensator undercorrects during the transient period. For $\tau_{LD} > \tau_{LG}$, the compensator overcorrects during the transient period. The duration of the transient period is determined solely by the lag time τ_{LG}.

Inputs and Outputs. Table 1.3 defines each input and output. The lead-lag equation is a first-order differential equation, so the complete formulation must include an initial condition. Input TRKMN activates initialization or tracking. In addition to setting the current value of the output Y to the value of input MNI, the internal storage locations used to integrate the differential equation are also initialized appropriately.

1.6. DEAD TIME

Dead time or transportation lag occurs to some extent in almost all processes. Simulation of a dead time is required in the following control applications:

Ratio or feedforward control. Occasionally, the dynamic compensation requires a dead time in addition to a lead-lag.

Dead-time compensation. In loops whose dynamic behavior is dominated by dead time, a technique known as *dead-time compensation* should be applied.

The following equation describes dead time:

$$Y(t) = X(t - \theta)$$

Table 1.3   Inputs and outputs for the lead-lag block

Mnemonic	Signal	Value	Purpose
X	Input (required)	Analog	Input to lead-lag.
TLEAD	Input (optional)	Analog	Value for lead time. All function block implementations permit the lead time to be zero, and most also permit the lead time to be negative.
TLAG	Input (optional)	Analog	Value for lag time. The lag time must be a positive value.
TRKMN	Input (optional)	Discrete	On true, set output Y to the current value of input MNI and also appropriately initialize the internal storage locations.
MNI	Input (optional)	Analog	Value for output Y when input TRKMN is true. If input MNI is not configured, the current value of input X is used for MNI.
Y	Output	Analog	Output of lead-lag.

where

$$X(t) = \text{value of input } X \text{ at time } t$$
$$Y(t) = \text{value of output } Y \text{ at time } t$$
$$t = \text{time (min)}$$
$$\theta = \text{dead time or transportation lag (min)}$$

The current value of the output Y is the value of the input X at θ units of time in the past.

Simulation of a dead time requires a storage array. If the dead time θ is constant, the simplest approach is to allocate one element for each sampling instant within the dead time and then use the following logic on each sampling instant:

$$S_k = S_{k+1} \qquad k = 2, 3, \ldots, N-1$$
$$S_N = X$$
$$Y = S_1$$

where

$$S_k = \text{storage array of } N \text{ elements, designated } S_1, S_2, \ldots, S_N$$
$$N = \theta/\Delta t = \text{number of elements in the storage array}$$
$$\Delta t = \text{sampling interval}$$

Instead of shifting the elements, it is more efficient to consider S_k a circular array and manage the pointers appropriately.

What if the dead time θ varies? One's initial proposal might be as follows:

- Specify a maximum allowable value θ_{max} for the dead time.
- Allocate the storage array for θ_{max}; that is, let $N = \theta_{max}/\Delta t$.
- Use the following logic on each sampling instant:

$$S_k = S_{k+1} \qquad\qquad k = 1, 2, \ldots, N - 1$$

$$S_N = X$$

$$Y = S_{N+1-j} \qquad \text{where } j = \theta/\Delta t$$

For small sampling intervals, the nearest integer value can be used for j, but if desired, linear interpolation between points can be incorporated into the logic.

Unfortunately, process dead times or transportation lags do not behave in this manner. The belt conveyor illustrated in Figure 1.10 is typical of process material transport systems; fluid flowing in pipes behaves in a similar manner. At any instant of time, the dead time or transport time is the length of the conveyor L divided by the belt velocity V. Changing the speed changes the belt velocity V (for fluid flowing pipes, this is equivalent to changing the flow rate). The foregoing logic for a variable dead time does not describe the consequences of changing the speed of the conveyor in Figure 1.10 or the flow rate of a fluid flowing in a pipe.

The belt conveyor in Figure 1.10 will be used as the basis for simulating a variable dead time or transportation lag. A storage array S is allocated on the following basis:

- Consider the length of the conveyor to be 1 unit.
- Reserve $N + 1$ storage elements for storage array S. As illustrated in Figure 1.11, S_k is the current weight of the material at location k on the conveyor. Distance is measured from the discharge end of the conveyor,

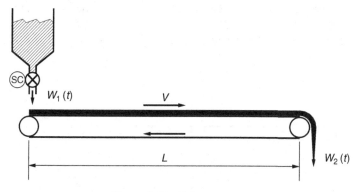

Figure 1.10 Belt conveyor.

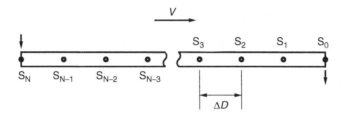

Figure 1.11 Location of storage elements along the conveyor.

that is, in the direction opposite to its travel. Element S_0 is where the material falls off the conveyor. Element S_N is where the material falls onto the conveyor.

- In the units used for the length of the conveyor, the distance ΔD between the storage elements is

$$\Delta D = \frac{1}{N}$$

The current value of the dead time reflects the belt velocity V. Since the length is 1 unit, the current velocity V is as follows:

$$V = \frac{1}{\theta}$$

An infinitely fast conveyor ($\theta = 0$) and conveyor stopped ($V = 0$ or $\theta = \infty$) must be treated as special cases, as we explain shortly.

Unless the number of storage elements reserved is extremely large, the distance traveled by the conveyor over one sampling instant Δt will be much less than ΔD. In other words, it takes several sampling instants for a point on the conveyor to move from location $(k + 1)\Delta D$ to location $k\ \Delta D$. Consequently, the storage elements will not be shifted on every sampling instant.

As the belt moves, the points move with the belt, as illustrated in Figure 1.12. Let d be the distance that the conveyor has traveled since the points in the storage array were last shifted. When d is equal to or exceeds ΔD, the points will be shifted and the value of d adjusted accordingly. Consequently, $0 \le d < \Delta D$.

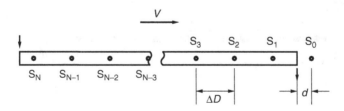

Figure 1.12 Simulation of variable dead time by shifting storage locations.

The following variables are used in the dead-time simulation:

X_0 = value of input X on preceding sampling instant

d_0 = distance traveled after preceding sampling instant

$\Delta d = V\Delta t = \Delta t/\theta$ = distance traveled over preceding sampling instant

X = current value of input to dead time

$d = d_0 + \Delta d$ = distance traveled after current sampling instant

Using this notation, the following calculations are normally performed on each sampling instant:

1. Compute the value for Δd and add to d_0 to obtain d.
2. If $d \geq \Delta D$, do the following:
   a. Shift the storage array:

   $$S_k = S_{k+1} \qquad k = 0, 1, 2, \ldots, N - 1$$

   Using logic for a circular buffer avoids the overhead of shifting values in memory.
   b. Use linear interpolation to determine the value of S_N:

   $$S_N = \frac{(X - X_0)(d - d_0)}{\Delta d}$$

   c. Subtract the distance ΔD between storage points from the current value of d.
   d. Repeat this step until $d < \Delta D$.
3. Use linear interpolation to compute the value of Y from S_0 and S_1:

   $$Y = \frac{(S_1 - S_0)d}{\Delta D}$$

Two cases require special handling:

1. *Conveyor moving at high speed.* If $\theta \leq \Delta t$, the dead time is negligible, so do the following:

   $$S_k = X \qquad k = 0, 1, 2, \ldots, N$$

   $$Y = X$$

   $$d = 0$$

2. *Conveyor stopped.* If $\theta \geq 1/\Delta t$, perform the normal calculations, except use a value of zero for Δd.

Table 1.4 Inputs and outputs for dead time block

Mnemonic	Signal	Value	Purpose
X	Input (required)	Analog	Input to dead time.
DTIME	Input (optional)	Analog	Value for dead time. Negative values for the dead time are considered to be zero.
TRKMN	Input (optional)	Discrete	On true, set output Y to the current value of input MNI, store the current value of input MNI in all storage locations, and set d to zero.
MNI	Input (optional)	Analog	Value for output Y when input TRKMN is true. If input MNI is not configured, the current value of input X is used for MNI.
Y	Output	Analog	Output of dead time.

Inputs and Outputs. Table 1.4 defines each input and output. Input TRKMN activates initialization or tracking. In addition to setting the current value of the output Y to the value of input MNI, the internal storage array is initialized to the value of input MNI and d is set to zero.

1.7. SELECTOR BLOCK

Although the block has other applications, the primary purpose of the selector block or *auctioneer* is to implement override controls. This application requires a selector block with the following capabilities:

- The block has two inputs.
- The block can be configured as follows:
  - *Low select.* Output is the smaller of the two inputs.
  - *High select.* Output is the larger of the two inputs.
- For each input, a discrete output is provided to indicate that the respective input is currently selected.
- For each input, a discrete input is provided to force the selector block to select the respective input.

Table 1.5 explains the purpose of each input and output.

Although not reflected in Table 1.5, most implementations provide even more capabilities:

- More than two inputs
- Can be configured as a median select (requires three inputs)

Table 1.5 Inputs and outputs for the selector block

Mnemonic	Signal	Value	Purpose
X1	Input (required)	Analog	Input 1.
X2	Input (required)	Analog	Input 2.
F1	Input (optional)	Discrete	Force input 1 to be selected.
F2	Input (optional)	Discrete	Force input 2 to be selected.
Y	Output	Analog	Value of selected input.
Q1	Output	Discrete	True if input 1 is currently selected.
Q2	Output	Discrete	True if input 2 is currently selected.

The relationship for the selector block is as follows:

$$Y = \begin{cases} \max(X_1, X_2) & \text{if a high select} \\ \min(X_1, X_2) & \text{if a low select} \end{cases}$$

where

Y = output of the selector block
X_1 = input 1 to the selector block
X_2 = input 2 to the selector block

The selector block provides two discrete outputs:

- Q_1: "True" if input 1 is selected ($Y = X_1$).
- Q_2: "True" if input 2 is selected ($Y = X_2$).

The selector block can also serve as a comparator.
Discrete inputs are provided to force the selection of a specific input:

- F_1: If "true," the output of the function block is input 1.
- F_2: If "true," the output of the function block is input 2.

In some implementations, the operator can also force the selection of a specified input (but inputs F_1 and F_2 usually take priority over the operator's specification).

1.8. CUTOFF BLOCK

The relationship for a two-stage cutoff block is as follows:

$$Y = \begin{cases} Y_0 & \text{if } X < X_C \\ Y_1 & \text{if } X \geq X_C \end{cases}$$

Table 1.6 Inputs and outputs for the two-stage cutoff block

Mnemonic	Signal	Value	Purpose
X	Input (required)	Analog	Input to cutoff logic.
Y0	Input (optional)	Analog	Value for output Y for stage 0 (X< XC).
Y1	Input (optional)	Analog	Value for output Y for stage 1 (X ≥ XC).
XC	Input (optional)	Analog	Value for cutoff.
Y	Output	Analog	Output of cutoff logic.
Q0	Output	Discrete	Current stage is stage 0 (X< XC).
Q1	Output	Discrete	Current stage is stage 1 (X ≥ XC).

where

Y = output of the cutoff block
X = input to cutoff block
Y_0 = value of output for cutoff stage 0 $(X < X_C)$
Y_1 = value of output for cutoff stage 1 $(X \geq X_C)$
X_C = cutoff value

Table 1.6 provides details on the inputs and outputs for this block. In configuring the cutoff block, there are two options for both Y_0 and Y_1:

1. Constant value
2. Output of another function block, which is why Y_0 and Y_1 are indicated as inputs in Table 1.6

The two-stage cutoff block also provides two discrete outputs:

- Q_0: "True" if cutoff block is in stage 0; that is, $X < X_C$ and $Y = Y_0$.
- Q_1: "True" if cutoff block is in stage 1; that is, $X \geq X_C$ and $Y = Y_1$.

The cutoff block can also serve as a comparator.

Most control systems also provide a three-stage cutoff block. True object-oriented implementations can provide a cutoff block with any number of stages.

1.9. HAND STATION

In the process industries, the customary practice is to provide the capability for the process operators to manually specify the value of the output to any final control element. This capability is used routinely during plant startup. Thereafter, its use should be the exception but is necessary on the occurrence of events such as a measurement device failure. The hand station addresses this requirement. Table 1.7 provides details on the inputs and outputs for this block.

Table 1.7   Inputs and outputs for the hand station block

Mnemonic	Signal	Value	Purpose
RMN	Input (optional)	Analog	Remote input. If not configured, the remote mode is not allowed.
MN	Output	Analog	Output of hand station
RMT	Output	Discrete	True if mode is remote.
QH	Output	Discrete	True if the hand station output is at the upper output limit.
QL	Output	Discrete	True if the hand station output is at the lower output limit.

In older control systems, the hand station was a hardware item that permitted the operator to assume manual control. In modern controls, the hand station is usually a software function block that provides two modes of operation:

Local. The process operator specifies the value of the output to the final control element. The value of input RMN (the remote input to the hand station) is not used.

Remote. The output is the value of input RMN.

Output RMT from the hand station indicates its remote/local status, with a true value indicating remote.

Most hand station blocks provide a lower output limit and an upper output limit. These limits are imposed on the value from input RMN and also on the values specified by the operators. The function block for the hand station also provides outputs QL and QH, which indicate that the output has been driven to the respective limit.

Not all control systems provide a hand station, but most provide the equivalent functionality in manners such as the following:

Valve block. When a valve block provides the output to the final control element, that block could provide the remote/local option. In remote, the value of the input to the valve block determines the opening of the control valve. In local, the process operator specifies the control valve opening.

Manual mode for function blocks. The PID controller block provides a manual mode of operation whereby the process operator can directly specify the controller output. This capability can be extended to all control blocks. In auto, the block output is computed using the appropriate relationships. In manual, the operator specifies a value for the block output.

2

CASCADE CONTROL

Cascade control appeared in the 1950s, being implemented using pneumatic controls. The earliest article on cascade control was by Wills [1] at the Brown Instrument Company (now part of Honeywell). With the advent of digital systems, the use of cascade control configurations increased significantly. As compared to simple feedback configurations, only one additional measurement is required. The potential benefits of cascade include:

- Isolating the controller for a key process variable from a problem element, such as a control valve with stiction or hysteresis
- Responding faster to certain disturbances
- Providing more consistent performance over a range of process conditions

A good starting point is to examine a common and very successful application of cascade control, specifically to control the temperature in a jacketed vessel with a recirculation system on the jacket.

2.1. JACKETED REACTOR

The reactor illustrated in Figure 2.1 is equipped with a recirculation system on its jacket (sometimes called a *tempered water system*). Cooling water is admitted at the suction of the recirculation pump; excess water is returned to the cooling tower. The reaction is exothermic, so only cooling is provided. However, cascade

Advanced Process Control: Beyond Single-Loop Control   By Cecil L. Smith
Copyright © 2010 John Wiley & Sons, Inc.

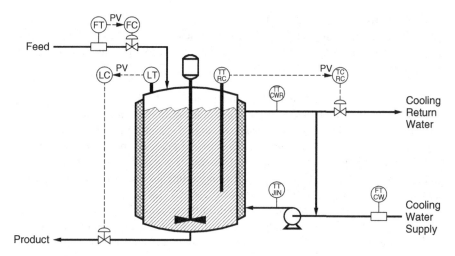

Figure 2.1   Jacketed reactor, simple feedback control.

is equally effective for endothermic reactions that require heating by media such as hot oil. The reactor in Figure 2.1 is a continuous reactor, but recirculating jackets are also applied to batch reactors.

Compared to once-through jackets, recirculating jackets provide a more uniform temperature difference between the reactor contents and the cooling media. However, recirculating jackets also require additional capital and operating costs for the pump. Therefore, recirculating jackets are usually installed only where the benefits to the reacting system justify these extra costs. Although the issues are different than for recirculating jackets, cascade control can be applied to vessels with once-through jackets, but generally with less success.

Instrumentation. The focus herein is exclusively on control of reactor temperature. In Figure 2.1, controllers are also provided for the reactor feed flow and the reactor level, but these loops receive no attention herein.

The control configurations presented subsequently will use some combination of the following measurements, illustrated in Figure 2.1:

Reactor temperature. Control of reactor temperature is of utmost importance. Sometimes more than one reactor temperature transmitter is installed.
Jacket outlet temperature/cooling water return temperature. The temperature transmitter is installed on the outlet of the jacket, but within the recirculation loop. The cooling water return temperature is the same as the jacket outlet temperature.
Jacket inlet temperature. The transmitter is installed at the inlet to the jacket. Being within the recirculation loop, this temperature is not the same as the cooling water supply temperature.

Cooling water flow. Especially in batch reactors, this measurement can be challenging. Many batch reactors experience a large change in the heat removal rate between the early and late stages of the batch. A turndown ratio of 50 or more is often necessary.

All temperature measurements have a resolution of 0.1°F.

Both jacket inlet and jacket outlet temperatures are often measured for reasons other than control. In recirculating jackets, the temperature rise from jacket inlet to jacket outlet should be in the range 2 to 5°F. Larger temperature rises suggest that the recirculation flow rate is too low. Such a small temperature rise also permits the assumption that the jacket temperature is uniform and that "jacket temperature" applies to whichever temperature is measured.

For this example no limits are imposed on the acceptable values for the jacket outlet temperature. In practice, this is sometimes not the case. When the cooling medium is tower cooling water, the conditioning of the cooling water imposes an upper limit on the cooling water return temperature. Exceeding this temperature leads to scaling of the heat transfer surfaces.

With the control valve on the cooling water return as in Figure 2.1, the jacket pressure is the cooling water supply pressure. If a lower jacket pressure is desired, the control valve can be installed on the cooling water supply and a backpressure regulator installed on the cooling water return.

Process Equations. This example is based on an industrial polymerization reaction that involves an activator. The following statements apply:

- The reaction rate is determined by the activator feed rate.
- The reactor temperature has no effect on the reaction rate.

Such reactions require multiple feeds that are maintained at precise ratios. Figure 2.1 illustrates only one feed, which is the total feed rate for all feeds.

At steady-state, the following must be equal:

- Heat liberated by the reaction
- Heat transferred to the jacket
- Heat removed from the jacket by the cooling water

The heat generated by the reaction is

$$Q = F \Delta H_F$$

where

Q = heat released by reaction (Btu/hr)
 = heat transfer rate to jacket (Btu/hr)
 = heat removed by the cooling water (Btu/hr)

F = reactor feed rate (lb/hr)
ΔH_F = heat released by reaction per unit of feed (Btu/lb)

The heat transfer rate from reactor to jacket is

$$Q = UA(T - T_J)$$

where

U = heat transfer coefficient (Btu/hr-ft^2-$^\circ$F)
A = heat transfer area (ft^2)
T = reactor temperature ($^\circ$F)
T_J = jacket temperature ($^\circ$F)

The heat removed from the jacket by the cooling water is

$$Q = Wc_P(T_J - T_{\text{CWS}})$$

where

W = cooling water flow (lb/hr)
c_P = cooling water heat capacity (Btu/lb-$^\circ$F)
T_{CWS} = cooling water supply temperature ($^\circ$F)
T_J = jacket temperature ($^\circ$F)
     = cooling water return temperature ($^\circ$F)

Control Objective. The reactor temperature is to be maintained at or near its target by adjusting the opening of the control valve on the cooling water supply. The following performance issues are relevant:

- Response to changes in the reactor temperature set point
- Response to disturbances originating with the reacting medium, including
  - Reactor feed flow rate
  - Reactor feed temperature
- Response to disturbances originating with the cooling medium, specifically,
  - Cooling water supply pressure
  - Cooling water supply temperature
- Impact of problem elements within the control loop: specifically, a control valve that exhibits
  - Hysteresis
  - Stiction
- Performance of the controls at high heat transfer rates and at low heat transfer rates

The performance will be examined for two cascade configurations in addition to the simple feedback configuration.

Simple Feedback. Figure 2.1 presents the simple feedback control configuration. The measured value of the reactor temperature is the PV input to the reactor temperature controller; the output of the reactor temperature controller is the cooling water valve opening. The simple feedback control configuration requires only one measurement device, one controller, and one control valve. The advantages of the simple feedback configuration are:

• The simplest possible control configuration
• The fewest process measurements (one)

However, its performance is inferior to the cascade configurations in several respects.

Temperature-to-Flow Cascade. Figure 2.2 presents the temperature-to-flow cascade control configuration. This configuration requires two measurement devices and two controllers:

• Reactor temperature
• Cooling water flow

These controllers are connected such that the set point for the cooling water flow controller is provided by the reactor temperature controller.

If the reactor temperature is increasing, the reactor temperature controller increases the set point for the cooling water flow, which causes the cooling water flow controller to open the cooling water valve to increase the cooling water flow.

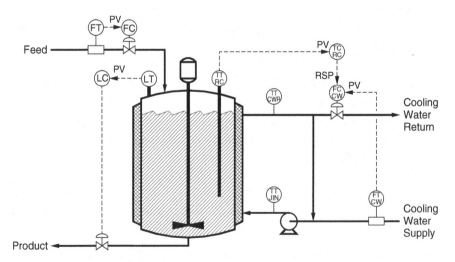

Figure 2.2   Jacketed reactor, temperature-to-flow cascade.

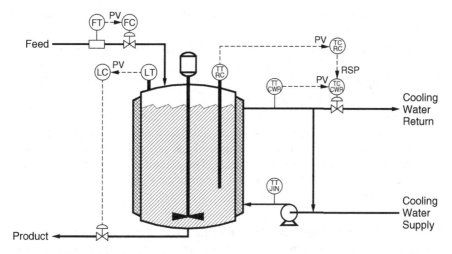

Figure 2.3   Jacketed reactor, temperature-to-temperature cascade.

Temperature-to-Temperature Cascade. Figure 2.3 presents the temperature-to-temperature cascade control configuration. This configuration also requires two measurement devices and two controllers:

- Reactor temperature
- Jacket temperature

These controllers are connected such that the set point for the jacket temperature controller is provided by the reactor temperature controller.

If the reactor temperature is increasing, the reactor temperature controller decreases the set point of the jacket temperature controller, causing the jacket temperature controller to open the cooling water valve to reduce the jacket temperature.

2.2. BLOCK DIAGRAMS

The simplified block diagram in Figure 2.4 corresponds to the P&I diagram in Figure 2.3 for the temperature-to-temperature cascade. As explained subsequently, the block diagram in Figure 2.4 is a bit too simplified for the temperature-to-temperature cascade. But for most cascades, it is quite appropriate.

Terminology. The following terminology is applied to cascade control configurations:

Outer/inner. These terms are best understood from the block diagram in Figure 2.4. There are two loops, one contained within the other, hence the terms *inner loop* and *outer loop*. The reactor temperature loop is the outer loop; the jacket temperature loop is the inner loop. The outer loop always provides the set point to the inner loop.

Master/slave. By providing the set point, the reactor temperature controller is in a sense telling the jacket temperature controller what to do. The reactor temperature controller is the *master controller*; the jacket temperature controller is the *slave controller*.

Primary/secondary. These terms pertain to the relative importance of the controlled variables. From a process perspective, the reactor temperature is of primary importance. The importance of the jacket temperature is through its effect on the reactor temperature. The reactor temperature controller is the primary controller; the jacket temperature controller is the secondary controller.

Disturbances with the Cooling Media. One of the inputs to the block diagram in Figure 2.4 represents the disturbances associated with the cooling medium: specifically, cooling water supply temperature or cooling water supply pressure. These disturbances are to the inner loop, the sequence of events being as follows:

1. The disturbance affects the jacket temperature.
2. The jacket temperature controller responds by changing the cooling water valve opening so as to maintain the jacket temperature at the set point provided by the reactor temperature controller.

If the inner loop is fast relative to the outer loop, the excursions of the jacket temperature from its set point will be small, and the effect of a cooling medium

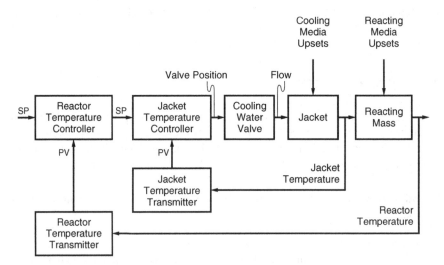

Figure 2.4 Simplified block diagram of the temperature-to-temperature cascade.

disturbance on the reactor temperature will be nominal. Cascade control is very effective for a disturbance that enters within a fast inner loop.

Disturbances with the Reacting Media. Another input to the block diagram in Figure 2.4 represents disturbances associated with the reacting medium, including reactor feed temperature, reactor feed flow rate, and the like. These disturbances are to the outer loop. The reactor temperature controller must initiate the response to these disturbances:

> *Temperature-to-temperature cascade.* The reactor temperature controller responds by changing the set point of the jacket temperature controller.
>
> *Simple feedback.* The reactor temperature controller responds by changing the cooling water valve opening.

Will the cascade configuration perform any better than the simple feedback configuration? Two aspects must be considered:

> *Dynamics.* The reactor temperature controller in the cascade configuration has no significant advantages over the reactor temperature controller in the simple feedback configuration.
>
> *Steady-state.* The operating lines for the two configurations can be different. In some cases (including the jacketed reactor) this difference gives the cascade configuration an advantage. But in other cases, cascade has no advantage.

More on this shortly.

2.3. PROBLEM ELEMENT

The most frequently encountered problem element is the control valve. Two characteristics degrade the performance of the control loop:

> *Hysteresis.* Valves are mechanical in nature. With age, mechanical parts exhibit wear. Originally tight connections loosen with wear, resulting in a characteristic known as *hysteresis*.
>
> *Stiction.* The packing around the valve stem provides resistance to any movement of the valve. For the actuator to produce a sufficient force to overcome the frictional effects of the packing, the difference between the current valve position and the target provided by the controller must exceed some threshold. But once this threshold is exceeded, most valves move to a position very close to the target specified by the controller.

Equipping the valve actuator with a positioner minimizes such effects, but never eliminates them totally.

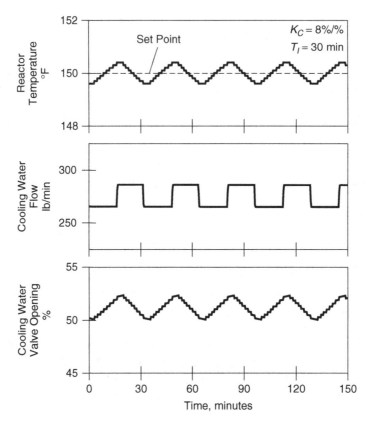

Figure 2.5 Effect of stiction in control valve on simple feedback control.

Simple Feedback. With hysteresis and/or stiction in the control valve, the reactor temperature will not line-out at the set point but, instead, will cycle about the set point. Such cycles are called *limit cycles*. Figure 2.5 presents the performance for a 2% stiction in the control valve. Limit cycles are evident in:

Reactor temperature. The resolution of 0.1°F is apparent but has little effect on performance.

Controller output. The controller output must cycle sufficiently to overcome the stiction in the control valve.

Cooling water flow. The cooling water flow basically switches between two values.

The limit cycles in Figure 2.5 are reasonably symmetrical, but this is not assured. Nor are the limit cycles repeatable; they may be symmetrical at times but nonsymmetrical at other times. Tuning adjustments will affect the limit cycle, but the limit cycle cannot be "tuned out."

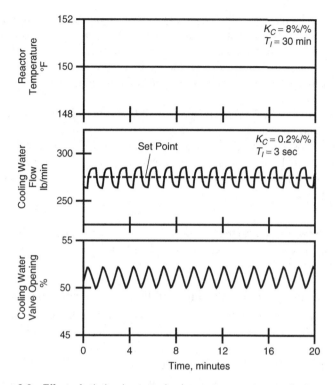

Figure 2.6 Effect of stiction in control valve on temperature-to-flow cascade.

Temperature-to-Flow Cascade. Stiction in the control valve has an adverse affect on the performance of the cooling water flow controller. As illustrated in Figure 2.6, the cooling water flow does not line-out at the set point but, instead, cycles about the set point. For 2% stiction in the control valve, the limit cycle is evident in both the cooling water flow and in the output of the cooling water flow controller.

A flow loop is much faster than a temperature loop. This is reflected in the period of the limit cycles:

- *Simple feedback:* Approximately 45 min (Figure 2.5).
- *Temperature-to-flow cascade:* Approximately 1.3 min (Figure 2.6).

The reactor temperature is not affected significantly by high-frequency cycles in cooling water flow. In Figure 2.6 for the temperature-to-flow cascade, the reactor temperature exhibits no perceptible deviation from its set point.

Long-Term Issues. The temperature-to-flow cascade provides far better performance in regard to controlling the reactor temperature. However, the higher-frequency cycle results in a higher rate of wear on the control valve, which translates into increased maintenance. The ideal solution is to address any stiction (or hysteresis) problem at its source, that is, at the valve itself. But sometimes problems must be addressed as quickly as possible, such as during plant startup. "Quick fixes" are justifiable to get the plant to produce satisfactory product. This also permits the quick fixes to be replaced with permanent fixes on a more reasonable schedule.

So that a quick fix will last as long as possible, the tuning for the flow controller can be adjusted so that the period of the limit cycle in the cooling water flow is as long as possible. As the period of the limit cycle becomes longer, a limit cycle will eventually appear in the reactor temperature. The limiting factor on the period of the limit cycle is the acceptable amplitude of the limit cycle in the reactor temperature.

Temperature-to-Temperature Cascade. With a stiction of 2% in the control valve, highly irregular limit cycles are evident in Figure 2.7 for the following variables:

Reactor temperature. The amplitude of the cycle is approximately 0.2°F.

Jacket temperature set point. The 0.1°F resolution in the reactor temperature contributes to the somewhat erratic behavior exhibited in this cycle.

Jacket temperature. The stiction in the control valve causes the jacket temperature to cycle about its set point.

Cooling water flow. The cooling water flow changes almost abruptly between two values. The time at the higher flow is much longer than that at the lower flow.

Jacket temperature controller output. (cooling water valve opening) The noisy appearance in this cycle is due to the 0.1°F resolution in the temperature measurements.

The period of all these cycles is a little over 30 min (compared to 45 min for simple feedback and 1.3 min for the temperature-to-flow cascade). Although some consequences are evident, the overall performance is not affected significantly by the 0.1°F resolution in the temperature measurements. The limit cycles in Figure 2.7 are highly irregular, but this is not always the case.

For the responses in Figure 2.7, the jacket temperature controller is tuned conservatively. More aggressive tuning would reduce the period of all the cycles and would reduce the amplitude of the reactor temperature limit cycle. However, this comes at the expense of increased wear on the control valve. In most applications, a 0.2°F amplitude cycle in the reactor temperature would be tolerable, but if necessary, tighter tuning in the jacket temperature controller would reduce this amplitude.

For problems such as stiction in the final control element, the temperature-to-flow cascade (Figure 2.2) is definitely capable of performance superior to that of the temperature-to-temperature cascade (Figure 2.3). But this comes at the

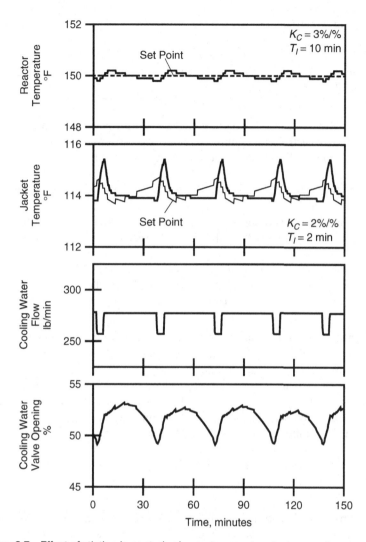

Figure 2.7 Effect of stiction in control valve on temperature-to-temperature cascade.

expense of increased wear on the control valve. To minimize this wear, the tuning in the flow controller must be relaxed until a small-amplitude cycle appears in the reactor temperature. Usually, the end result is essentially the same as for the temperature-to-temperature cascade.

2.4. COOLING MEDIA DISTURBANCES

In many batch plants, the demand on the cooling water system varies substantially, the result being disturbances in both:

- Cooling water supply pressure
- Cooling water supply temperature

In this section we examine how the control configurations in Figures 2.1 to 2.3 respond to these disturbances.

Simple Feedback. In the simple feedback control configuration illustrated in Figure 2.1, the reactor temperature controller has to respond to all disturbances. The cooling media disturbance quickly affects the jacket temperature:

- A change in the cooling water supply temperature directly affects the jacket temperature.
- A change in the cooling water supply pressure affects the flow through the valve, which in turn affects the jacket temperature.

However, the reactor temperature controller does not respond directly to a change in the jacket temperature. No corrective action will be taken until the disturbance affects the reactor temperature. This eventually occurs, but much later.

Temperature-to-Flow Cascade. The temperature-to-flow cascade control configuration in Figure 2.2 responds very effectively to changes in cooling water supply pressure. The cooling water flow controller responds rapidly to any change that affects the cooling water flow. Consequently, even large disturbances to the cooling water supply pressure would have little effect on either the jacket temperature or the reactor temperature.

For changes in the cooling water supply temperature, the temperature-to-flow cascade offers no advantages over the simple feedback control configuration. Neither will take corrective action until the change in cooling water supply temperature affects the reactor temperature.

Temperature-to-Temperature Cascade. Both cooling media disturbances first affect the jacket temperature. In the temperature-to-temperature cascade configuration in Figure 2.3, the jacket temperature controller responds to any change in the jacket temperature. Whereas the simple feedback configuration does not respond until the disturbance affects the reactor temperature, the temperature-to-temperature cascade takes corrective action as soon as a change appears in the jacket temperature.

The responses in Figure 2.8 are to an increase of 10°F in the cooling water supply temperature. With simple feedback control, the peak in reactor temperature is almost 2°F. But with a temperature-to-temperature cascade, the peak is less than 0.5°F. The faster the jacket temperature loop relative to the reactor temperature loop, the greater the difference in performance.

For changes in cooling water supply pressure, the response of the temperature-to-temperature cascade will be comparable to its response to cooling water supply

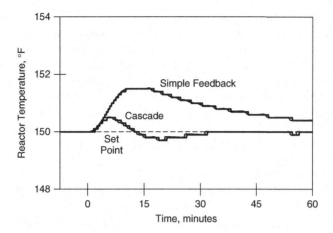

Figure 2.8 Response to cooling water temperature increase.

temperature changes. The temperature-to-flow cascade would respond faster, but in most applications the performance of the temperature-to-temperature cascade is adequate.

If such an improvement in performance is required, a temperature-to-temperature-to-flow cascade can be installed. This cascade configuration requires three controllers:

Reactor temperature controller.

Jacket temperature controller. The set point is provided by the reactor temperature controller.

Cooling water flow controller. The set point is provided by the jacket temperature controller.

There is no limit on the number of loops in a cascade configuration. However, few plants have a cascade with three loops.

2.5. EFFECT OF VARYING HEAT TRANSFER RATE

For the jacketed reactor, the relationship between the heat transfer rate and the cooling water flow is extremely nonlinear. At low cooling water flow rates, a change in the cooling water flow has a significant effect on the heat transfer rate (the sensitivity is high). But at high cooling water flow rates, a change in the cooling water flow has a much smaller effect on the heat transfer rate (the sensitivity is low). Such nonlinearities often cause tuning problems. The major potential for tuning problems is in reactors that must operate at varying heat transfer rates. Batch reactors are notorious for exhibiting large changes. The heat transfer rate at the end of the batch is often less that one-tenth of the heat transfer rate at the beginning of the batch.

Control Valve Characteristics. The nonlinearity between the cooling water flow and the heat transfer rate is of the decreasing sensitivity type—the sensitivity decreases as the cooling water flow increases. Some improvement can be realized by coupling a decreasing sensitivity component with an increasing sensitivity component. For flow systems, installing an equal-percentage valve does just that. Normally, the increasing sensitivity nonlinearity of the equal-percentage valve at least partially offsets the decreasing sensitivity nonlinearity of the cooling water flow to heat transfer rate relationship. However, this is unlikely to be sufficient for batch reactors; the range of heat transfer rates experienced in most batch reactors is too great.

   In this section the focus is on the nonlinearity of the heat transfer rate to the cooling water flow, which is the basic cause of the tuning difficulties. To remove the characteristics of the control valve from the relationships, the reactor temperature and the jacket temperature will be related to the cooling water flow rather than the cooling water valve opening.

Simple Feedback. Figure 2.9 presents the operating line that relates the reactor temperature to the cooling water flow. This operating line exhibits strong nonlinearities of a decreasing sensitivity nature. For the simple feedback configuration, the process gain (sensitivity) is the slope of the operating line in Figure 2.9. Between cooling water flows of 100 and 500 lb/min, the process gain decreases by a factor of 10, more or less. For reactors that operate over a range of heat transfer rates, tuning difficulties will arise, usually resulting in very conservative tuning for the reactor temperature controller.

Temperature-to-Temperature Cascade. The cascade configuration consists of two loops. For each loop, the operating line is a plot of the controlled variable as a function of the manipulated variable:

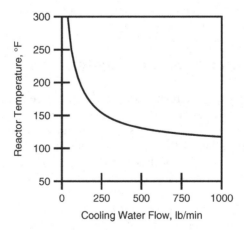

Figure 2.9 Operating line, simple feedback control.

Outer loop (reactor temperature controller). The controlled variable is the reactor temperature. Technically, the manipulated variable is the set point for the jacket temperature controller. But as the jacket temperature must equal its set point at steady-state, the jacket temperature is used to construct the operating line in Figure 2.10(a).

Inner loop (jacket temperature controller). Technically, the manipulated variable is the cooling water valve opening. But to focus on the process nonlinearities, cooling water flow is used to construct the operating line in Figure 2.10(b).

Process Operating Line for the Outer Loop. The operating line for the outer loop presented in Figure 2.10(a) is linear. As noted in the process description,

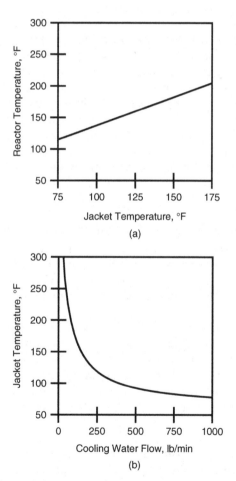

Figure 2.10 Operating lines for temperature-to-temperature cascade: (a) reactor temperature loop; (b) jacket temperature loop.

the reaction rate is determined by the feed rate of an activator. Consequently, reactor temperature has no effect on reaction rate, so the operating line is linear. For most other reaction mechanisms, the reactor temperature affects the reaction rate, which imparts some nonlinearity to the operating line.

As the reactor temperature is linearly related to the jacket temperature, the process sensitivity for the reactor temperature controller is constant. This yields two major benefits:

- The controller will perform consistently over a wide operating range.
- The controller need not be tuned conservatively.

In effect, the linear operating line permits the reactor temperature controller to be tuned more effectively, which yields improved performance both for disturbances to the outer loop and for set point changes.

Process Operating Line for the Inner Loop. The operating line for the inner loop presented in Figure 2.10(b) is highly nonlinear. For the reactor, the major nonlinearity is the relationship between the cooling water flow and the heat transfer rate. For the temperature-to-temperature cascade, this nonlinearity appears in relationships for the jacket temperature loop.

The *process gain* (sensitivity) is the slope of the operating line. Between cooling water flows of 100 and 500 lb/min, the process gain decreases by a factor of 10, more or less. Tuning difficulties can certainly arise in the jacket temperature controller, the usual result being a conservatively tuned controller. Although the same was said for the simple feedback configuration, there is a big difference:

Simple feedback. The conservative tuning is in the reactor temperature controller, which means poorer control of the reactor temperature.

Temperature-to-temperature cascade. The conservative tuning is in the jacket temperature controller (the secondary loop), which means poorer control of the jacket temperature. The question is to what extent this degrades the performance of the reactor temperature controller (the primary loop).

For a cascade configuration to function, the inner loop must be dynamically faster than the outer loop. When the inner loop is faster by a factor of 5 or more, conservative tuning (if not excessive) in the inner loop will have little effect on the performance of the outer loop. The issue is not the performance of the inner loop as judged from its own time frame, but the performance of the inner loop as judged from the time frame of the outer loop. In a subsequent discussion of tuning, we revisit this issue.

Use of the Temperature-to-Temperature Cascade. The temperature-to-temperature cascade is routinely applied to vessels with a recirculating jacket. The easiest advantage to understand is the improved performance for disturbances entering with the cooling medium. But in practice, this is not the primary benefit.

In the temperature-to-temperature cascade, the reactor temperature controller will deliver more consistent performance over a range of conditions. This advantage is especially important in batch reactors.

Such an improvement is not realized in all cascade configurations. Recognizing situations in which cascade control can provide such an advantage requires a thorough understanding of the process. As process engineers have the best understanding of the process, they need to understand cascade control sufficiently that they can recognize such opportunities.

2.6. CASCADE CONTROL MODES

The depiction in Figure 2.11 of the PID controller contains two mode switches:

Auto/manual. In cascade control, the auto/manual selection has exactly the same significance as that for simple feedback control:
- *Manual.* The process operators specify the controller output. The PID calculations are not performed.
- *Auto.* The PID calculations determine the controller output.

Remote/local. The remote/local selection specifies the source of the set point for the PID calculations:
- *Local.* The value for the set point is specified by the process operator.
- *Remote.* The value for the set point is the current value of the input configured for the remote set point.

Valid Operational Modes. The two switches indicated in Figure 2.11 suggest four possible operational modes:

- Local−manual
- Local−automatic

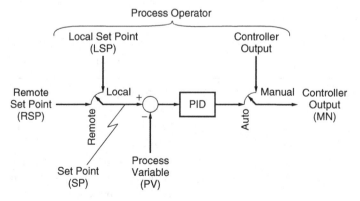

Figure 2.11 Mode switches for PID controller.

- Remote–manual
- Remote–automatic

When the controller is in manual, the PID calculations are not performed. The value of the set point is not used, so the modes "local–manual" and "remote–manual" are equivalent. Consequently, there are only three operational modes:

Operational Mode	Remote/Local Switch	Auto/Manual Switch	PID Calculations	Operator Specifies
Manual	Irrelevant	Manual	Not performed	Controller output
Automatic	Local	Auto	Performed	Local set point
Remote	Remote	Auto	Performed	Nothing

The *remote mode* is often referred to as the *cascade mode*. While this makes sense in the context of cascade control, the remote set point feature has applications other than cascade control.

Controller Configuration. For a controller whose set point is to be supplied by another component of the control system, a source must be configured for the remote set point (RSP) input. For the temperature-to-temperature cascade configuration, the remote set point for the jacket temperature controller is the output of the reactor temperature controller. When the remote set point input is not configured, the remote or cascade mode is not allowed. For the temperature-to-temperature cascade, the remote or cascade mode is available for the jacket temperature controller. As no remote set point is configured, the remote or cascade mode is not available for the reactor temperature controller.

Operational Mode Combinations. For the temperature-to-temperature cascade, there are four possibilities for the operational modes of the two loops:

1. *Inner loop on manual*

Loop	Controller	Mode	Explanation
Inner	Jacket temperature	Manual	• Operator specifies the cooling water valve opening. • PID control calculations are not performed. • PID initialization calculations for bumpless transfer from manual to automatic are performed.

Outer	Reactor temperature	Immaterial	• Inner loop is not using the remote set point input. • PID control calculations are not performed. • Output of the outer loop is tracking the local set point for the inner loop (so that switching the mode of the inner loop from manual to cascade will be bumpless).

2. *Inner loop on automatic*

Loop	Controller	Mode	Explanation
Inner	Jacket temperature	Automatic	• Operator specifies the set point for the inner loop. • PID control calculations are performed using the local set point. • The remote set point input is not being used.
Outer	Reactor temperature	Immaterial	• Inner loop is not using the remote set point input. • PID control calculations are not performed. • Output of the outer loop is tracking the local set point for the inner loop (so that switching the mode of the inner loop from automatic to cascade will be bumpless).

3. *Inner loop on remote or cascade; outer loop on manual*

Loop	Controller	Mode	Explanation
Inner	Jacket temperature	Remote	• PID control calculations are performed using the remote set point. • Operator cannot change set point or controller output.

Outer	Reactor temperature	Manual	• Inner loop is using the remote set point input. • PID control calculations are not performed. • Operator specifies the value for the output of the outer loop, which becomes the remote set point for the inner loop.

4. *Inner loop on remote or cascade; outer loop on automatic*

Loop	Controller	Mode	Explanation
Inner	Jacket temperature	Remote	• PID control calculations are performed using the remote set point. • Operator cannot change set point or controller output.
Outer	Reactor temperature	Automatic	• Inner loop is using the remote set point input. • PID control calculations are performed. • Operator specifies the value for the set point of the outer loop.

The following mode combinations are very similar:

- *Inner loop on automatic.* Operator changes jacket temperature set point via the jacket temperature controller.
- *Inner loop on remote; outer loop on manual.* Operator changes jacket temperature set point by changing the output of the reactor temperature controller.

Either gets the job done, but most operators prefer the former.

2.7. REMOTE SET POINT

The depiction of the PID controller in Figure 2.11 provides for two sources of the set point:

Local set point. The value specified by the process operator for the PID controller set point is called the *local set point*.

Remote set point. This set point is usable only when the configuration of the PID controller includes an entry for the remote set point input.

Their use depends on the remote/local status:

Local. The set point of the PID controller is the value of the local set point. The value of the remote set point input is not retrieved. So that the switch from local to remote will be bumpless, tracking calculations must be initiated that cause the value of the remote set point input to be the same as the current value of the local set point.

Remote. The set point of the PID controller is the value of the remote set point. So that the switch from remote to local will be bumpless, the value of the remote set point is also written to the local set point.

Controller Output in Percent. For the temperature-to-temperature cascade, suppose that the measurement and output ranges for the two controllers are as follows:

Loop	Process Variable	Measurement Range	Controller Output Range
Inner	Jacket temperature	0–200°F	0–100%
Outer	Reactor temperature	0–300°F	0–100%

When the controller output is the opening of a control valve (as is the case for the inner loop), the appropriate range for the controller output is 0 to 100%. But for the outer loop, the controller output in percent must be converted to a jacket temperature set point in °F.

To make this conversion, the reactor temperature controller output is considered to be the percent of the span of the measurement device for the jacket temperature. Suppose that the reactor temperature controller output is 62%. This is considered to be 62% of the 0 to 200°F measurement range for the jacket temperature. The jacket temperature set point would be 124°F. The output of all conventional pneumatic and electronic controllers was basically in percent of span. Some digital systems do the same, but the trend is for the controller output to be in engineering units.

Controller Output in Engineering Units. Instead of 0 to 100%, let the output range for the reactor temperature controller be 0 to 200°F:

Loop	Process Variable	Measurement Range	Controller Output Range
Inner	Jacket temperature	0–200°F	0–100%
Outer	Reactor temperature	0–300°F	0–200°F

With the output of the reactor temperature controller in °F, the value of the controller output can be used directly as the set point for the jacket temperature. This is the trend in digital control systems. The controller output is in engineering units in all examples in this book. In this example, the measurement range for the jacket temperature (the PV for the inner loop) and the output range for the reactor temperature controller (the outer loop) are both 0 to 200°F. This is the normal practice, but most digital systems do not impose such a restriction.

2.8. OUTPUT TRACKING

Output tracking is a feature that causes the output of the controller to be set to a specified value under certain conditions. In cascade configurations, output tracking is used to achieve bumpless transfer when changing the mode of the inner loop from automatic (specifically, local–automatic) to remote (specifically, remote–automatic).

Bumpless Transfer in Cascade Control Configurations. Suppose that the jacket temperature controller is currently on automatic or, more specifically, local–automatic. The controller is using the local set point provided by the process operator. When the mode is switched to remote, the controller will switch immediately to using the remote set point. Bumpless transfer means that the value of the jacket temperature controller set point just after the switch to remote is the same as the value just prior to the switch to remote. For this to be the case, the value of the remote set point must "track" the value of the local set point when the jacket temperature controller is not in remote. The remote set point is the output of the reactor temperature controller. When the jacket temperature controller is not in remote, the output of the reactor temperature controller must continuously be updated to the current value of the local set point of the jacket temperature controller.

The logic required is stated as follows:

If the inner loop is not on remote, the output of the outer loop must be set equal to the current value of the local set point for the inner loop.

This logic is implemented by configuring output tracking in the outer controller.

Configuring Cascade Control. To implement the temperature-to-temperature cascade, two PID blocks must be configured. A PID block is required for each "TC" element of the P&I diagram in Figure 2.3. In configuring these blocks, the designations in Table 1.1 for the PID block inputs and outputs will be used. Most P&I diagrams include only those signals required for normal control functions. For the temperature-to-temperature cascade these are as follows:

Loop	Signal	Configuration
Outer	Input PV	Measured variable for the reactor temperature
	Input RSP	Not configured
	Output MN	Provides the remote set point for the inner loop
Inner	Input PV	Measured variable for the jacket temperature
	Input RSP	Output MN of the reactor temperature controller
	Output MN	Provides the cooling water valve opening

Configuring Output Tracking. To implement output tracking to achieve bumpless transfer, inputs TRKMN and MNI must be configured as follows in the reactor temperature controller of the temperature-to-temperature cascade:

Input TRKMN. Output RMT from the jacket temperature controller is inverted (logical NOT) to provide input TRKMN to the reactor temperature controller. This logic commands the reactor temperature controller to track when the jacket temperature controller is not on remote.

Input MNI. Output SP from the jacket temperature controller provides input MNI to the reactor temperature controller. When the jacket temperature controller is not on remote, output SP is the local set point. Thus, when the reactor temperature controller is tracking, the controller output is set to the current value of the local set point of the jacket temperature controller.

The following statements implement output tracking for the reactor temperature controller:

```
TCRC.TRKMN = !TCCWR.RMT
TCRC.MNI = TCCWR.SP
```

The tag name TCRC designates the reactor temperature controller; the tag name TCCWR designates the jacket temperature controller (jacket temperature is the same as the cooling water return temperature).

When output tracking is active (input TRKMN is true), the controller is also initialized as if the controller were on manual. Specifically, the actions are as follows:

1. If PV tracking is enabled, the current value of the PV is written to the local set point.
2. The controller output bias M_R is initialized to provide a smooth transition from manual to automatic.

When the TRKMN input is true, a few systems force the mode of the PID to manual. However, most permit the mode to remain in automatic, but the tracking calculations are performed instead of the customary PID calculations (in effect, the controller behaves as if it were on manual).

Tracking Indication. When the output of the reactor temperature controller is tracking the set point of the jacket temperature controller, it makes no sense for the process operator to change the output of the reactor temperature controller. Even if such a change were accepted, the tracking logic would immediately over-write the operator-entered value with the current set point of the jacket tempera-ture loop. When a controller is tracking, some indication to the process operator is essential. All digital control systems provide such a capability; however, the mechanism for presenting this information differs from one system to another.

2.9. CONTROL MODES

The utilization of modes in the two controllers in a cascade configuration is most frequently as follows:

Inner loop. The inner loop is normally proportional-integral. Using derivative in the inner loop is not recommended.

Outer loop. The outer loop is at least proportional-integral, and if a tempera-ture loop, is likely to be proportional-integral-derivative.

As there are occasional exceptions, the role of each mode is examined next in more detail.

Proportional Mode in the Inner Loop. When the inner loop is a flow loop, it is tuned in the usual manner: low controller gain and short reset time, making the reset mode do most of the work. But for other loops, the proportional mode should be the primary mode of control in the inner loop, the trade-offs being:

- Using as much proportional action as possible in the inner loop leads to the fastest response speed for the inner loop, giving the greatest separation of dynamics between the inner loop and the outer loop.
- Increasing the controller gain decreases the margin of stability, which could lead to problems where the inner loop process is nonlinear and thus suscep-tible to changes in the process sensitivity.

Although rarely the case for temperature loops, significant noise is sometimes present on the measured variable for the inner loop. Smaller values of the con-troller gain must be used in such loops.

Integral Mode in the Inner Loop. The integral mode eliminates offset. But is offset in the inner loop really a problem? For the temperature-to-temperature cascade for the jacketed reactor, the presence of offset in the inner loop means that the jacket temperature does not line-out at the set point provided by the reactor temperature controller. What if the jacket temperature set point is 88°F and the jacket temperature lines-out at 85°F? Provided the reactor temperature is at its set point, this is perfectly acceptable.

Does offset in the inner loop degrade the performance of the outer loop? Although one might suspect the opposite, the presence of offset in the inner loop does not degrade the performance of the outer loop. Provided that the reset mode is tuned properly (specifically, the reset time is not exceedingly short), the effect of integral on response speed is minor.

If a cascade configuration is tuned using proportional-integral in the inner loop and then the reset action is removed, some degradation in the performance of the outer loop should be expected. With reset in the inner loop, the closed-loop gain for the inner loop is exactly 1.0. When the reset is removed, the value of the closed-loop gain will be less than 1.0. To compensate for this change, the controller gain in the outer loop must be increased. Making this adjustment will restore the performance of the outer loop to its original value.

Does offset in the inner loop raise issues for the process operators? Unfortunately, the answer is often yes, and not necessarily just to the operators. To anyone with limited understanding of automatic control, it seems that every loop should line-out with the measured variable equal to the set point. That the inner loop of a cascade is an exception requires some explanation. Usually, it is easier to tune the reset mode in the inner loop than to explain (often, again and again) that offset in the inner loop is not really a problem.

Reset in the inner loop causes no problems provided that it is properly tuned. However, a common misconception is that reset action must be used to attain fast response. But when the inner loop is tuned with a heavy reliance on reset, the actual result is a slower inner loop, which then complicates the tuning of the outer loop.

Derivative Mode in the Inner Loop. Derivative is not recommended in the inner loop of a cascade. The usual argument for omitting derivative is to keep the tuning task as simple as possible.

Proportional Mode in the Outer Loop. When fast response is desired in the outer loop, proportional action is certainly appropriate, and indeed, the controller should be tuned with emphasis on the proportional action. In the temperature-to-temperature cascade for the jacketed reactor, close control of reactor temperature is usually essential, and the proportional mode must be properly tuned in such loops.

Integral Mode in the Outer Loop. The presence of offset in the outer loop means that the primary controlled variable does not line-out at its set point. In the temperature-to-temperature cascade for the jacketed reactor, this means that the reactor temperature does not line-out at its set point. This is unacceptable, so the integral mode should definitely be used. Where the outer loop is controlling the level in an integrating process, the integral mode is occasionally omitted from the outer loop. When the process is integrating, too much reset in the level controller leads to stability problems. If reset is used in such loops, the reset time must be set to a relatively long value.

Derivative Mode in the Outer Loop. The case for adding derivative to the outer loop is basically the same as the case for adding derivative to a simple feedback loop. For those loops where the derivative mode is able to enhance the stability margin, a higher value of the controller gain can be used, which in turn leads to a faster response. This is most likely to be successful for temperature loops, which are usually also free of noise. Thus, derivative is most frequently used in those temperature loops (such as reactor temperatures) whose performance has a direct impact on plant profitability.

2.10. INTERACTING STAGES

Cascade is often applied to processes that can be considered to consist of stages. The jacketed reactor can be treated in this manner, the stages being the jacket and the reacting mass. The cooling water flow first affects the jacket, which in turn affects the reacting mass. Thus, it is reasonable to consider the jacket to be the first stage and the reacting mass to be the second.

Events in the first stage clearly affect the second stage. Anything that affects the jacket temperature will also affect the reactor temperature. But processes differ in how events in the second stage affect the first stage:

Noninteracting. Events in the second stage have no effect on the first stage.

Interacting. Events in the second state also affect the first stage.

The consequences impact the tuning of the inner loop of a cascade. However, cascade control is equally effective for both arrangements.

Noninteracting Stages. Two tanks in series can be interacting or noninteracting, depending on the piping between the two tanks. Heat transfer stages such as in the jacketed reactor are always interacting. Figure 2.12 illustrates two tanks in the noninteracting stages configuration. The feed stream enters the first tank; a recycle stream enters the second tank. Gravity is the driving force for all flows.

For a given valve opening, the discharge flow from the first tank depends on only the level in the first tank. The discharge flow from the first tank is not affected by the level in the second tank. The recycle stream flow affects the level in the second tank but has no effect on the level in the first tank. The two tanks in Figure 2.12 meet the criteria for noninteracting stages. Events in the first stage affect the second stage, but events in the second stage have no effect on the first stage.

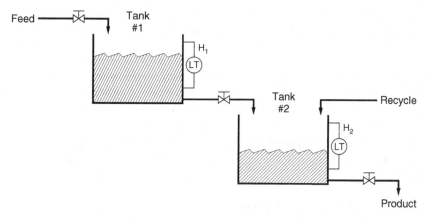

Figure 2.12 Noninteracting stages.

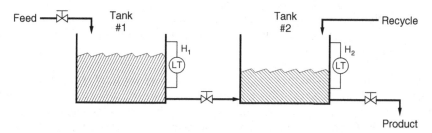

Figure 2.13 Interacting stages.

Interacting Stages. Figure 2.13 illustrates the same two tanks, but in the interacting stages configuration. The only difference is that the flow between the two tanks depends on the difference in the levels in the two tanks. Note a similarity to the stages in the jacketed reactor—the heat flow depends on the difference in the temperatures. In the configuration in Figure 2.13, a change in the recycle flow affects the level in the second tank, which affects the flow between the two tanks and, consequently, the level in the first tank. A momentary reversal of the flow between the two tanks is even possible. The two tanks in Figure 2.13 meet the criteria for interacting stages. Events in the first stage affect the second stage, but events in the second stage also affect the first stage.

Temperature-to-Flow Cascade (Figure 2.2). The first stage is the cooling water flow system; the second stage is the reacting mass (including the jacket). These stages meet the criteria for noninteracting stages:

1. *Events in the first stage affect the second stage.* The cooling water valve opening affects the cooling water flow, which in turn affects the reactor temperature.

2. *Events in the second stage have no effect on the first stage.* With the cascade configuration on manual, changes in the reactor temperature have no effect on the cooling water flow.

Suppose that there is an increase in the feed rate. As the reaction is exothermic, this increases the reactor temperature. But with the cascade configuration on manual, there would be no change in cooling water flow.

Temperature-to-Temperature Cascade (Figure 2.3). The first stage is the jacket; the second stage is the reacting mass. These stages meet the criteria for interacting stages:

1. *Events in the first stage affect the second stage.* The cooling water valve opening affects the jacket temperature, which in turn affects the reactor temperature.
2. *Events in the second stage affect the first stage.* Even with the cascade configuration on manual, changes in the reactor temperature will affect the jacket temperature (the heat transfer rate depends on the temperature difference).

Suppose that there is an increase in the feed rate. As the reaction is exothermic, this increases the reactor temperature. In turn, this increases the heat transfer rate from the reacting mass to the jacket, which in turn increases the jacket temperature. Clearly, events in the second stage (the reacting mass) affect the first stage (the jacket).

Why is this significant for cascade systems? The behavior of interacting stages complicates the interpretation of the response of the inner loop controlled variable (the jacket temperature). The trend in Figure 2.14 presents the responses of the jacket temperature and the reactor temperature to a decrease in the cooling water valve opening from 51.2% to 46.2%. Both responses require about the same time to line-out. Superficially, the jacket and reactor appear to have similar dynamics. However, their dynamics are actually very different.

The jacket temperature response exhibits a very rapid initial rise of approximately 4°F that is followed by a more gradual increase in jacket temperature. The initial rise is due to the jacket dynamics; the subsequent gradual increase reflects the reactor dynamics. As the reactor temperature increases, the heat transfer to the jacket increases, which in turn increases the jacket temperature. In interpreting the jacket temperature response, the critical issue is to separate the jacket dynamics from the reactor dynamics. When the stages are interacting, interpretation of the inner loop response must be approached very carefully, or one will arrive at the wrong conclusions.

Block Diagram for Interacting Stages. The simplified block diagram in Figure 2.4 was used in a previous explanation of the advantages of the temperature-to-temperature cascade for responding to cooling media

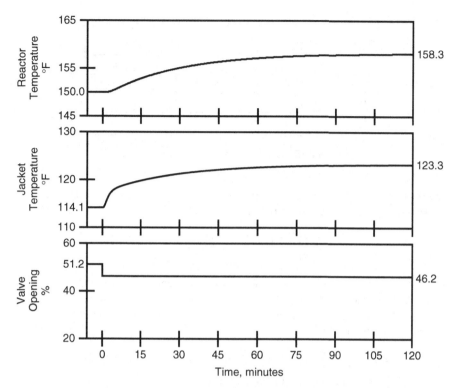

Figure 2.14 Response of reactor temperature and jacket temperature to a decrease in the cooling water valve position.

disturbances. It was also noted that the block diagram was not quite accurate. The block diagram in Figure 2.4 accurately depicts the relationships for noninteracting stages, but for interacting stages, there is a missing component.

The change in cooling water flow W affects the jacket temperature T_J, which in turn affects the reactor temperature T. But because the stages are interacting, the reactor temperature affects the jacket temperature. As depicted in Figure 2.15, this adds an additional path to the block diagram. It is this path that makes it difficult to interpret jacket temperature responses.

The jacket temperature response in Figure 2.14 is the result of two effects:

1. *Effect of cooling water flow on the jacket temperature.* The tuning of the jacket temperature controller depends only on this effect.
2. *Effect of the reactor temperature on the jacket temperature.* This effect is merely a disturbance to the jacket temperature loop and does not affect the tuning.

The challenge is to separate these effects when interpreting the jacket temperature responses.

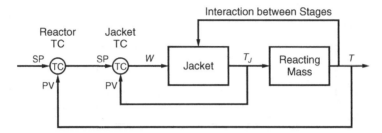

Figure 2.15   Block diagram for interacting stages.

Gain or Sensitivity. From the step response in Figure 2.14, the process gain or sensitivity K_J is normally computed as follows:

$$K_J = \frac{\Delta T_J}{\Delta M} = \frac{123.3°\text{F} - 114.1°\text{F}}{46.2\% - 51.2\%} = 1.84°\text{F}/\%$$

But for interacting stages, this is not the process gain to which the jacket temperature controller is tuned.

Since the stages are interacting, there are two contributors to the change in the jacket temperature:

- ΔM: Change in the cooling water valve opening.
- ΔT: Change in the reactor temperature.

Using a linear approximation gives the following expression for the change in jacket temperature:

$$\Delta T_J = \left.\frac{\partial T_J}{\partial M}\right|_T \Delta M + \left.\frac{\partial T_J}{\partial T}\right|_M \Delta T$$
$$= K_W \Delta M + K_T \Delta T$$

Figure 2.16 presents the block diagram for the jacket temperature loop only. With regard to tuning the jacket temperature controller, the following observations apply:

Cooling water flow. The element with gain K_W is within the jacket temperature loop. The tuning of the jacket temperature controller is affected by K_W.

Reactor temperature. The reactor temperature is a disturbance to the jacket temperature loop. The gain K_T applies only to this disturbance and has no effect on the tuning of the jacket temperature controller.

For noninteracting stages, the value of the gain K_W would be the same as the gain K_J computed from the step response data in Figure 2.16. But for

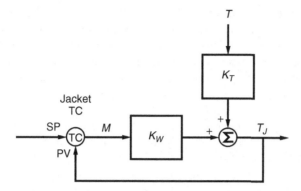

Figure 2.16 Block diagram for inner loop of cascade.

interacting stages, these gains are not equal. Unfortunately, the gain K_W cannot be determined form the step response in Figure 2.16.

2.11. TUNING CASCADES

Tuning a cascade control configuration is conceptually a straightforward endeavor: Tune the inner loop and then tune the outer loop. For the temperature-to-temperature cascade for the jacketed reactor in Figure 2.3, first tune the jacket temperature loop and then tune the reactor temperature loop.

In most cases, tuning a cascade configuration is no more difficult than tuning two simple feedback configurations: In each case, two controllers must be tuned. But two issues can potentially lead to complications:

Inadequate dynamic separation. This causes the outer controller to be untunable.

Interacting stages. This complicates the tuning of the inner loop but otherwise does not impair the performance of the cascade configuration.

Dynamic Separation. The inner loop of a cascade must be faster than the outer loop. The usual desire is that the inner loop be five times faster than the outer loop. Smaller differences complicate the tuning of the outer loop. From this statement, some conclude that it is necessary to tune the inner loop to respond as rapidly as possible. Usually, this is not necessary. Once a five-to-one separation in dynamics is achieved, tuning the inner loop to respond even faster has little impact on the performance of the outer loop. This is certainly the case for flow loops. Flow loops are inherently very fast, and rarely is there a need to make them even faster.

When the process is nonlinear, a tightly tuned inner loop could become unstable should the sensitivity of the process increase. A more conservatively tuned inner loop would be more tolerant of process sensitivity changes.

Tuning Cascade Controls. The temperature-to-temperature cascade for the jacketed reactor will be used to illustrate the procedure for tuning cascade controls. Both controllers will be proportional-integral. Tuning will be by trial-and-error procedures. The controllers must be tuned one at a time, first the inner loop and then the outer loop. In other words, start at the valve and work up or out. For the temperature-to-temperature cascade, the approach is as follows:

1. Switch the jacket temperature controller to automatic. Since the remote set point input is not being used, the reactor temperature controller is tracking, so its mode is irrelevant.
2. Tune the jacket temperature controller.
3. Switch the jacket temperature controller to remote and the reactor temperature controller to automatic.
4. Tune the reactor temperature controller.

Proportional Mode, Inner Loop. The adjustment of the proportional mode in the inner loop must reflect the following two observations:

1. The desire is for the inner loop to be five times faster than the outer loop.
2. The proportional mode determines the speed of response of a loop.

Unless some prior experience with tuning the jacket temperature controller is available, start with a low value for the controller gain and then increase the gain until the desired performance is achieved. Starting with controller gain of 1.0%/% and no reset or derivative action, the jacket temperature controller will be tuned by making $10°F$ changes in its set point, alternating between 114 and $124°F$.

The trend in Figure 2.17 presents the responses for controller gains of 1, 2, 4, and 8%/% in the inner loop. All of these responses exhibit unusual behavior—none of the responses line-out at a constant value of the jacket temperature. This is a consequence of interacting stages.

The following observations apply to the response for a controller gain of 2%/% to the decrease of $10°F$ in the set point:

- The initial rapid drop in the jacket temperature reflects the dynamics of the jacket.
- For noninteracting stages, the inner loop would line-out. But because the stages are interacting, the jacket temperature continues to decrease, but much more slowly.
- The decrease in jacket temperature causes the reactor temperature to decrease slowly, which reflects the dynamics of the outer loop.
- The decrease in reactor temperature also affects the jacket temperature, a consequence of interacting stages.

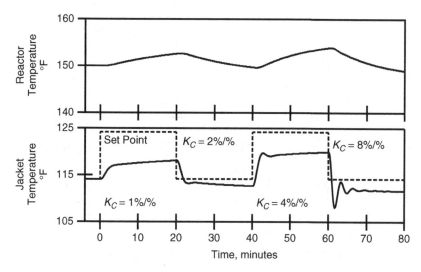

Figure 2.17 Tuning the proportional mode in the jacket temperature loop.

With sufficient time between set point changes, the reactor temperature would line-out, and the jacket temperature would also line-out. However, waiting for the reactor temperature to attain equilibrium simply takes too much time.

As there is no compelling need to speed up the inner loop for the jacketed reactor, a conservative value of 2%/% will be used for the controller gain, the considerations being:

• The larger the gain, the more likely it is that problems would arise, due to the nonlinearities in the jacket temperature loop.

• Overly conservative tuning in the inner loop would degrade the performance of the outer loop.

As part of the tuning effort for the outer loop, the performance of the inner loop must be assessed. By including both jacket temperature and jacket temperature set point in all trends, such an assessment can be made both during and after the tuning effort.

Integral Mode, Inner Loop. The cycle in the response in Figure 2.17 for a gain of 8%/% reflects the dynamics of the jacket. The period of this cycle is approximately 3 min, which is a reasonable starting value of the reset time. But instead, a conservatively long reset time of 8.0 min will be used as the starting point.

The trend in Figure 2.18 provides responses to a 10°F set point change for the following reset times:

• *8 min.* Reset time is much too long.

• *4 min.* Reset time is still too long.

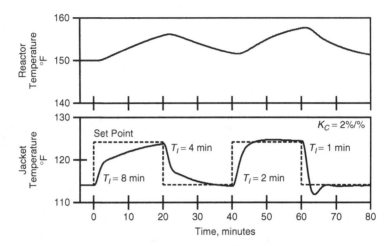

Figure 2.18 Tuning the integral mode in the jacket temperature loop.

- *2 min.* Technically, this response exhibits a small overshoot. But this overshoot is probably not due to the controller tuning. The reactor temperature is also increasing, and since the stages are interacting, this causes the jacket temperature to increase. No doubt this is contributing to most, and probably all, of the overshoot.
- *1 min.* Response exhibits more overshoot than is present for proportional-only control. The reset time is too short.

The appropriate tuning for the jacket temperature controller is a controller gain of 2%/% and a reset time of 2 min.

Some effort could be expended to refine the reset setting; however, the benefits are questionable for the jacketed reactor. The jacket dynamics are much faster than the reactor dynamics. In such cases, fine-tuning the inner loop will not significantly affect the performance of the outer loop.

Proportional Mode, Outer Loop. Although close control of the reactor temperature is desired, process engineers often prefer a cautious approach. Consequently, the objective will be a response with a modest overshoot (on the order of 10%) with no oscillations. But in meeting this performance objective, the reactor temperature loop should respond as rapidly as possible. Since the proportional mode determines the speed of response of a loop, the adjustment of the controller gain for the outer loop of a cascade is critical.

The reactor temperature controller will be tuned by making 5°F changes in its set point, alternating between 150 and 155°F. Figure 2.19 presents the responses for the following values of the controller gain:

- *1%/%.* The response exhibits no overshoot.
- *2%/%.* The overshoot is barely discernible.

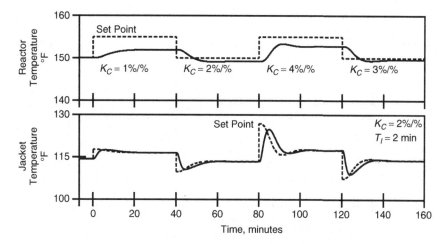

Figure 2.19 Tuning the proportional mode in the reactor temperature loop.

- *4%/%.* The overshoot exceeds the stated objective of 10%.
- *3%/%.* The overshoot is approximately 10%.

A controller gain of 3%/% will be used.

Integral Mode, Outer Loop. The integral mode should be tuned so as to remove the offset expeditiously but without introducing significantly more over-shoot than is present when only the proportional mode is in use. The transient period of the response in Figure 2.19 for $K_C = 3\%/\%$ is approximately 20 min. This is usually a conservative starting point for the reset time. Figure 2.20 presents the responses for the following reset times:

- *20 min.* The reset time is clearly too long.
- *10 min.* The reset time looks about right.
- *8 min.* More overshoot is present than in the response in Figure 2.19 for $K_C = 3\%/\%$.
- *12 min.* The reset time is a little too long.

A controller gain of 3%/% and a reset time of 10 min will be used for the outer loop.

Evaluation of Inner Loop Tuning. Figure 2.20 includes a trend for the jacket temperature and the jacket temperature set point, the purpose being as follows:

- When tuning the outer loop of a cascade, one must also assess the perfor-mance of the inner loop.

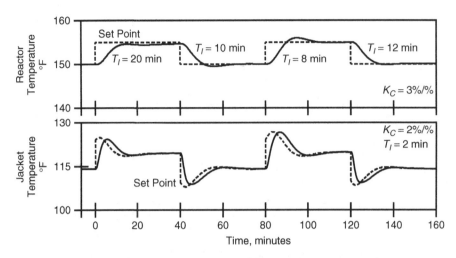

Figure 2.20   Tuning the integral mode in the reactor temperature loop.

- If, subsequently, a problem arises with the cascade configuration, one must first determine if the problem originates within the inner loop or within the outer loop.

From a trend of the jacket temperature and its set point, how does one assess the performance of the inner loop? Qualitatively, the question is simple: Is the inner loop controlled variable adequately tracking the set point changes from the outer loop? In this context, *adequate* is from the perspective of the outer loop. The controlled variable will always be lagging the set point, but is it so far behind that the performance of the outer loop is degraded?

In the trends in Figure 2.20, the inner loop performance is adequate, but probably only barely. Speeding up the inner loop by increasing the controller gain improves the performance of the outer loop by only a nominal degree. For Figure 2.21 the controller gain for the inner loop is 4%/% instead of 2%/%. The jacket temperature is tracking its set point better than in Figure 2.20. However, the improvement in performance in the outer loop is insignificant.

The inner loop must be fast enough to "keep up" with changes to its set point, but once this is achieved, there is no benefit in further increasing the speed of response in the inner loop. Tuning the inner loop to respond as rapidly as possible is not necessary, and can be counterproductive.

2.12. WINDUP IN CASCADE CONTROLS

In Chapter 1 the condition for windup was stated as follows:

Reset windup occurs in a controller when changes in the controller output have no effect on the process variable.

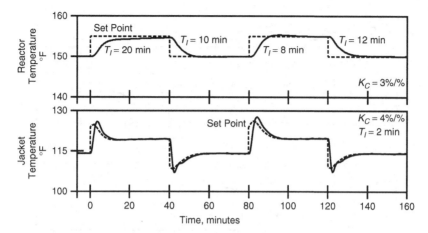

Figure 2.21   Impact of faster inner loop on outer loop performance.

In this section we explain how windup can arise within a cascade control configuration; in subsequent sections we present the three ways to provide windup protection.

Set Point for the Inner Loop.   For the temperature-to-temperature cascade for the jacketed reactor, the measurement and output ranges are as follows:

Loop	Process Variable	Measurement Range	Controller Output Range
Inner	Jacket temperature	0–200°F	0–100%
Outer	Reactor temperature	0–300°F	0–200°F

With these ranges, the reactor temperature controller can reduce the set point for the jacket temperature to 0°F. The cooling medium is cooling water, so clearly this cannot be achieved. The minimum possible jacket temperature is the cooling water supply temperature. But even this cannot be achieved. With heat being transferred to the jacket, the jacket temperature will be higher than the cooling water supply temperature. This provides the potential for windup. Windup occurs if the reset action in the reactor temperature controller is permitted to drive the jacket temperature set point below the minimum jacket temperature that can be attained.

Going into Windup.   At the start of the trend in Figure 2.22, the conditions are as follows:

- The reactor feed rate is 85 lb/min.
- The reactor temperature is lined-out at its set point of 150°F.

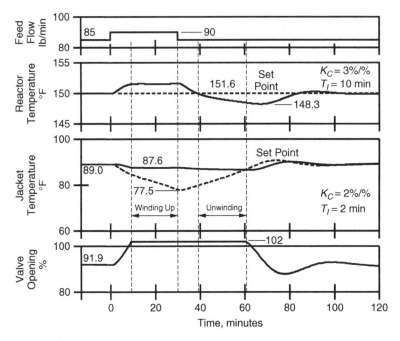

Figure 2.22   Windup in a cascade control configuration.

- The jacket temperature is lined-out at its set point of 89.0°F.
- The cooling water valve is 91.9% open.

The following events then occur:

1. The reactor feed rate is increased to 90 lb/min.
2. The reactor temperature rises above its set point.
3. The reactor temperature controller responds by lowering the set point for the jacket temperature.
4. The jacket temperature controller drives the cooling water valve fully open. As a small overrange is provided, the controller increases its output to the upper output limit of 102%.
5. Even with maximum cooling, the reactor temperature remains above its set point, lining-out at 151.6°F.
6. The jacket temperature does not track its set point, but lines-out at 87.6°F.
7. Because of the reset action, the reactor temperature controller continues to lower the jacket temperature set point. The controller is winding up throughout the time marked "Winding Up" in Figure 2.22.

For a feed rate of 90 lb/min, a reactor temperature of 150°F cannot be attained even with full cooling. For this feed rate and full cooling, the reactor temperature lines-out at 151.6°F. That the reactor temperature is not maintained at its set point is due to a process limitation, not a deficiency of the controls.

Recovering from Windup. After 30 min at a feed rate of 90 lb/min, the feed flow is reduced to 85 lb/min. The events in Figure 2.22 are as follows:

1. The reduced feed rate causes the reactor temperature to decrease, eventually dropping below the set point of 150°F.
2. The control error in the reactor temperature controller is increasing (actually, is becoming less negative). The proportional action in the reactor temperature controller increases the set point to the jacket temperature controller.
3. Even after the reactor temperature drops below 150°F, the jacket temperature set point is considerably below the jacket temperature (this is due to windup). As long as the jacket temperature is below its set point, the cooling water valve remains fully open.
4. The duration of the time designated "Unwinding" in Figure 2.22 is approximately 20 min. Throughout this time, a situation exists that makes no sense:
   - The reactor temperature is below its set point, essentially lining-out at 148.3°F.
   - The cooling water valve is fully open.
   This is a consequence of windup.
5. At the end of the time marked "Unwinding," the jacket temperature set point crosses the jacket temperature and the cooling water valve begins to close.

In Figure 2.22 the windup was allowed to continue until the jacket temperature set point decreased to 77.5°F. Had the feed flow rate remained at 90 lb/min, the set point would have eventually attained the lower output limit of −4°F. This would have significantly lengthened the time to recover from the windup condition.

Why Windup Occurs. In the example of windup presented in Figure 2.22, increasing the feed flow to 90 lb/min essentially caused the reactor temperature to line-out at 151.6°F with the cooling water valve fully open. The reactor temperature controller is reverse acting, so the control error is SP − PV, or −1.6°F.

To understand the problem, consider the following representation of the proportional-integral control equation:

$$\text{Proportional:} \quad M = K_{C,\text{EU}} E + M_R$$

$$\text{Reset:} \quad M_R = \int \frac{K_{C,\text{EU}}}{T_I} E \, dt$$

A constant control error of $-1.6°F$ causes the controller output bias M_R to continue to decrease. The windup protection provided within the PID block prevents this from continuing indefinitely. The bias limits approach for windup protection within the PID would allow the controller output bias to decrease to $-4°F$.

Windup Prevention. To prevent windup, the integrator in the reactor temperature controller must stop when the jacket temperature controller drives its output to its upper output limit. But this must be done in such a way that the cooling water valve remains fully open. During the period designated as "Winding Up" in Figure 2.22, the conditions are as follows:

- The reactor temperature is above its set point.
- The cooling water valve has been driven fully open.

The maximum cooling rate has been attained, so lowering the jacket temperature set point further will have no effect. However, as long as the reactor temperature is above its set point, the cooling water valve should remain fully open. To assure this, the reactor temperature controller must specify a jacket temperature set point that is below the current jacket temperature.

2.13. INTEGRAL TRACKING

When the jacket temperature controller output has been driven to its upper output limit, consider doing the following in the reactor temperature controller:

1. Set the controller output bias to the current value of the jacket temperature (the measured variable for the inner loop):

$$M_R = T_J$$

   This stops the reset integrator.
2. Compute the controller output (the jacket temperature set point) using the usual proportional-plus-bias equation:

$$T_{J,\text{SP}} = M = K_{C,\text{EU}}\ E + M_R = K_{C,\text{EU}}\ E + T_J$$

   Provided that the reactor temperature remains above its set point, the control error will be negative and the value computed for the jacket temperature set point will be less than the jacket temperature.

This approach to windup prevention is known as *integral tracking*.

Example. Figure 2.23 illustrates the performance of integral tracking for a feed flow increase to 90 lb/min for 30 min. As before, the jacket temperature controller

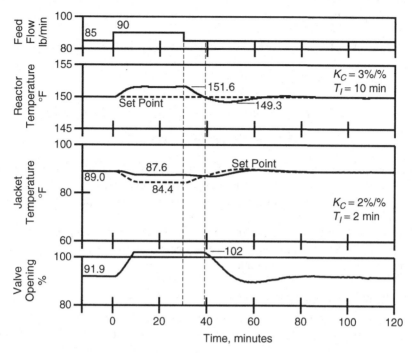

Figure 2.23 Performance of temperature-to-temperature cascade with windup protection via integral tracking.

quickly drives the cooling water valve fully open. The following conditions are the same as during the time marked "Winding Up" in Figure 2.22:

Variable	Value
Jacket temperature controller output	102%
Jacket temperature	87.6°F
Reactor temperature set point	150.0°F
Reactor temperature	151.6°F
Reactor temperature controller K_C	3%/%
Reactor temperature controller $K_{C,\mathrm{EU}}$	2°F/°F

But with integral tracking configured for the reactor temperature controller, the controller does not continue to lower the jacket temperature set point. The following calculations are being performed in the reactor temperature controller:

1. Set the controller output bias to the current value of the jacket temperature:

$$M_R = 87.6°F \ (43.8\% \text{ of } 200°F \text{ span})$$

2. Compute the control error for the reactor temperature:

$$E = \text{SP} - \text{PV} = 150.0^\circ\text{F} - 151.6^\circ\text{F} = -1.6^\circ\text{F} \ (-0.533\% \text{ of } 300^\circ\text{F span})$$

3. Compute the controller output:

$$M = K_{C,\text{EU}} \ E + M_R = 2^\circ\text{F}/^\circ\text{F} \times (-1.6^\circ\text{F}) + 87.6^\circ\text{F}$$
$$= 84.4^\circ\text{F} \ (42.2\% \text{ of } 200^\circ\text{F span})$$

With no windup protection (Figure 2.22), the jacket temperature set point continues to decrease. But with integral tracking, the jacket temperature set point only decreases to 84.4°F. As this is less that the current jacket temperature of 87.6°F, the cooling water valve remains fully open.

When the feed flow is reduced to 85 lb/min, the reactor temperature begins to decrease. Note that the following events occur simultaneously (approximately 10 min after the feed flow is reduced to 85 lb/min):

- The reactor temperature crosses its set point.
- The jacket temperature set point crosses the jacket temperature.
- The jacket temperature controller output comes off the upper output limit.

The reactor temperature decreases to 149.3°F (compared to 148.3°F with no windup protection). The recovery period is much shorter.

The behavior of integral tracking depends on the sign of the control error in the reactor temperature controller:

- $E < 0$ ($T_{\text{SP}} < T$). The jacket temperature set point is below the current value of the jacket temperature by an amount that depends on the magnitude of the control error and the value of the controller gain in the reactor temperature controller. The cooling water valve remains fully open. The value specified for the jacket temperature set point cannot be attained. However, the proportional action, not the integral action, determines how much the jacket temperature set point is below the current value of the jacket temperature. This is not windup; windup is caused only by reset.
- $E = 0$ ($T_{\text{SP}} = T$). The jacket temperature set point is equal to the current value of the jacket temperature.
- $E > 0$ ($T_{\text{SP}} > T$). The jacket temperature set point is above the current value of the jacket temperature. The jacket temperature controller begins to close the cooling water valve, which is appropriate since the reactor temperature is below its set point.

Configuration. The following specifications are required to activate integral tracking within the reactor temperature controller when the output from the jacket temperature controller is at the upper output limit (attributes are defined in Table 1.1):

> *Input TRKMR.* Output QH from the jacket temperature controller provides the value for input TRKMR to the reactor temperature controller. Integral tracking is active within the reactor temperature controller when the output of the jacket temperature controller is at its upper output limit.
>
> *Input MRI.* The measured value of the jacket temperature provides input MRI to the reactor temperature controller.

For the reactor temperature controller, the following specifications provide both output tracking (for bumpless transfer) and integral tracking (for windup protection):

```
TCRC.TRKMN = !TCCWR.RMT
TCRC.MNI = TCCWR.SP
TCRC.TRKMR = TCCWR.QH
TCRC.MRI = TTCWR.PV
```

For applications that require windup protection when the cooling water valve is fully closed, input TRKMR is set true when output QL of the jacket temperature controller is true.

2.14. EXTERNAL RESET

External reset appeared during the era of conventional pneumatic and electronic controls. As discussed in Chapter 1, external reset applies to the reset feedback form of the PID control equation presented in Figure 1.5. Some digital implementations provide the reset feedback form of the PID along with external reset, but many do not.

Use in Cascade Configurations. The external reset input to the controller for the outer loop is the PV input to the controller for the inner loop. The following specifications provide both output tracking (for bumpless transfer) and external reset (for windup protection) for the reactor temperature controller:

```
TCRC.TRKMN = !TCCWR.RMT
TCRC.MNI = TCCWR.SP
TCRC.XRS = TTCWR.PV
```

Unlike integral tracking, where tracking is initiated on certain conditions, the external reset input is used in the PID calculations at all times.

If the external reset input is not configured (sometimes referred to as *internal reset*), the input to the reset mode is the controller output M, which is the jacket temperature set point. When the external reset input is configured, the input to the reset mode is the jacket temperature. If the cooling water valve is not fully open, the jacket temperature will be close to its set point, so there is little difference between internal reset (reset based on the jacket temperature set point) and external reset (reset based on the jacket temperature). But when the cooling water valve is fully open, using external reset prevents windup.

Example. Figure 2.24 illustrates the performance of external reset for a feed flow increase to 90 lb/min for 30 min. As before, the jacket temperature controller quickly drives the cooling water valve fully open. The following conditions are the same as during the time marked "Winding Up" in Figure 2.22:

Variable	Value
Jacket temperature controller output	102%
Jacket temperature	87.6°F
Reactor temperature set point	150.0°F
Reactor temperature	151.6°F
Reactor temperature controller K_C	3%/%
Reactor temperature controller $K_{C,\text{EU}}$	2°F/°F

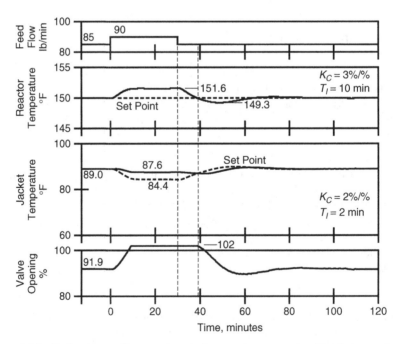

Figure 2.24 Performance of temperature-to-temperature cascade with windup protection via external reset.

During this period, the behavior of external reset is as follows:

1. The input to the exponential lag used to implement the reset mode (refer to Figure 1.5) is a constant value ($87.6°F$). Consequently, the controller output bias M_R approaches this value. After about three reset times have elapsed, the following can be assumed:

$$M_R = 87.6°F \ (43.8\% \text{ of } 200°F \text{ span})$$

2. Compute the control error for the reactor temperature:

$$E = \text{SP} - \text{PV} = 150.0°F - 151.6°F = -1.6°F \ (-0.533\% \text{ of } 300°F \text{ span})$$

3. Compute the controller output:

$$M = K_{C,\text{EU}} \ E + M_R = 2°F/°F \times (-1.6°F) + 87.6°F$$
$$= 84.4°F \ (42.2\% \text{ of } 200°F \text{ span})$$

With no windup protection (Figure 2.22) the jacket temperature set point continues to decrease. But with external reset, the jacket temperature only decreases to $84.4°F$. As this is less than the jacket temperature set point, the cooling water valve remains fully open.

When the feed flow is reduced to 85 lb/min, the reactor temperature begins to decrease. The following events occur simultaneously (at approximately 10 min after the feed flow is reduced to 85 lb/min):

- The reactor temperature crosses its set point.
- The jacket temperature set point crosses the jacket temperature.
- The jacket temperature controller output comes off the upper output limit.

The reactor temperature decreases to $149.3°F$ (compared to $148.3°F$ with no windup protection). The recovery period is much shorter.

The performance of external reset and that of integral tracking are very similar, but not identical. The trends in Figure 2.24 (external reset) and Figure 2.23 (integral tracking) are not identical, but with the resolution of these trends, the differences are impossible to detect.

2.15. INHIBIT INCREASE/INHIBIT DECREASE

During the time marked "Winding Up" in Figure 2.22, the reactor temperature controller continues to decrease the jacket temperature set point, even though the cooling water valve is fully open. One way to prevent this is as follows:

- Detect when the jacket temperature controller (the inner loop) has driven its output to the upper output limit.

- As long as the jacket temperature controller output is at its upper output limit, prohibit the reactor temperature controller from decreasing its output.

This is the essence of the inhibit increase/inhibit decrease windup prevention mechanism.

Use in Cascade Configurations. The following specifications provide both output tracking (for bumpless transfer) and inhibit increase/inhibit decrease (for windup protection) for the reactor temperature controller:

```
TCRC.TRKMN = !TCCWR.RMT
TCRC.MNI = TCCWR.SP
TCRC.NODEC = TCCWR.QH
```

What if the cooling water valve could be driven fully closed? To prevent windup should this occur, output QL from the jacket temperature controller (inner loop) must provide input NOINC to the reactor temperature controller. The specifications are as follows:

```
TCRC.TRKMN = !TCCWR.RMT
TCRC.MNI = TCCWR.SP
TCRC.NOINC = TCCWR.QL
TCRC.NODEC = TCCWR.QH
```

The jacket temperature controller is direct acting (on an increase in jacket temperature, the controller increases the opening of the cooling water valve). If the inner loop is a reverse-acting controller, output QL provides the NODEC input to the outer loop and output QH provides the NOINC input to the outer loop.

Example. Figure 2.25 illustrates the performance of inhibit increase/inhibit decrease for a feed flow increase to 90 lb/min for 30 min. As before, the jacket temperature controller quickly drives the cooling water valve fully open. The following conditions are the same as during the time marked "Winding Up" in Figure 2.22:

Variable	Value
Jacket temperature controller output	102%
Jacket temperature	87.6°F
Reactor temperature set point	150.0°F
Reactor temperature	151.6°F

When the jacket temperature controller output attains the upper output limit, the jacket temperature set point is 84.7°F. The reactor temperature controller is

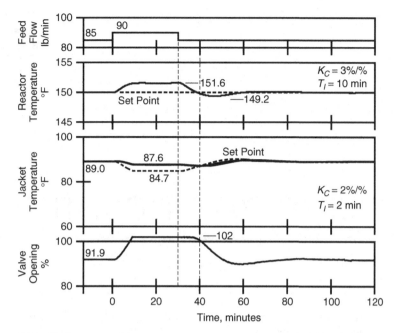

Figure 2.25 Performance of temperature-to-temperature cascade with windup protection via inhibit increase/inhibit decrease.

then inhibited from further decreasing its output, which is the jacket temperature set point. The reactor temperature controller can increase its output, but it cannot decrease it. For this reason, this windup prevention mechanism is sometimes referred to as *one-direction control*.

Unlike the integral tracking and external reset windup prevention mechanisms, it is not possible to compute the value that inhibit increase/inhibit decrease will impose on the jacket temperature set point. However, it will be less than the jacket temperature, so the cooling water valve remains fully open.

When the feed flow is reduced to 85 lb/min, the reactor temperature begins to decrease. This makes the control error less negative. In response, the reactor temperature controller increases the jacket temperature set point. Only decreases are inhibited, so these increases are allowed. Just prior to the time that the jacket temperature crosses its set point, the jacket temperature controller begins to decrease its output. The reactor temperature decreases to 149.2°F (as compared to 149.3°F with integral tracking or external reset). The recovery period is also about the same.

LITERATURE CITED

1. Wills, D. M., Cascade Control Applications and Hardware, Technical Bulletin TX119-1, Minneapolis-Honeywell Regulator Company, Philadelphia, 1960.

3

SPLIT-RANGE CONTROL

Some control applications involve a dual mode of operation. The two that are most frequently encountered are:

Heat/cool. Sometimes the temperature must be maintained by adding heat to the vessel, but at other times the temperature must be maintained by removing heat from the vessel. Many reactors impose such requirements.

Vent/bleed. Sometimes the pressure must be maintained by venting gases from the vessel, but at other times the pressure must be maintained by adding gases to the vessel. Such a capability is required for some storage tanks.

In most such applications, a separate final control element is provided for each mode. In some applications, the modes are exclusive in that only one must be in operation at a given time. However, this is not always the case.

There are two approaches to providing control in such applications:

Separate controllers for each operating mode. This normally requires that the set points for the individual controllers be separated sufficiently so that only one controller is active at a given time, the other having driven its final control element to a limit.

Split range. A single controller is provided, but its output range is "split" such that one mode of operation is active from 0 to 50% and the other is active from 50 to 100%.

Advanced Process Control: Beyond Single-Loop Control   By Cecil L. Smith
Copyright © 2010 John Wiley & Sons, Inc.

The use of separate controllers is common in pressure control applications, but most temperature control applications require split-range control.

3.1. STORAGE TANK PRESSURE CONTROL

Figure 3.1 illustrates a storage tank for which the pressure is to be maintained within acceptable limits, although not necessarily at a fixed target. The storage tank pressure is affected by the following:

Vent. Gas from the tank is released into the plant vent system, which is maintained below atmospheric pressure by a blower. Any vapors are removed from the vent gas before release to the environment.

Bleed. An inert gas is added to the tank. The inert gas may be anything that does not react with the contents of the tank. Some applications require nitrogen, but when possible, less expensive gases such as carbon dioxide or methane are used.

Material pumped to the process. Removing material from the storage tank causes the pressure to drop, which in turn requires that inert gas be added to the tank.

Material delivery. The material in the tank is replenished from either tank trucks or railcars. Pumping material into the tank causes the pressure to rise, and gas must be vented from the tank to avoid excess pressures within the tank.

Leaks. Few storage tanks are perfectly sealed. The leak rate should be small, but increases with the difference between tank pressure and atmospheric pressure.

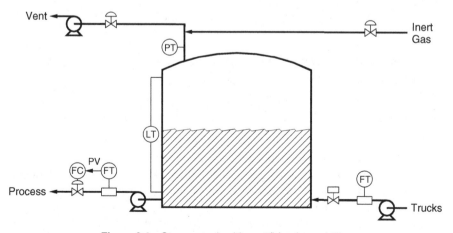

Figure 3.1   Storage tank with vent/bleed capability.

The configuration in Figure 3.1 permits either of the following objectives to be achieved:

- Maintain the pressure slightly below atmospheric so that no vapors from the tank leak into the environment.
- Maintain the pressure slightly above atmospheric pressure so that no oxygen enters the vapor space within the storage tank.

The control issues are basically the same, so herein only the latter are discussed.

With the piping arrangement illustrated in Figure 3.1, the vent line and the inert gas line are connected via a manifold to the same vessel nozzle. The pressure is sensed in this manifold. If the vent valve and the inert gas valve are open at the same time, the inert gas is added only to be released into the vent system, with no benefit. An occasional vent/bleed application requires an operating mode known as *purge*. The inert gas is swept through the vessel to the vent, the typical objective being to lower the oxygen concentration in the atmosphere within the vessel. But to be effective, the nozzles for the inert gas and the vent must be separated and arranged so that the gas sweeps through the vessel.

The following limiting conditions apply to storage tanks for which the pressure must be above atmospheric:

Low. To avoid admitting oxygen to the tank, the pressure within the storage tank must never drop below atmospheric pressure.

High. An excessive pressure will result in a mechanical failure of the tank.

Pressure switches (not shown in Figure 3.1) detect these conditions. Equipment protection is provided by the safety system, not the process controls. The objective of the process controls is to maintain the pressure within the limits; the objective of the safety system is to initiate an appropriate response if a limit is violated for any reason. Herein only issues pertaining to the process controls are examined.

Two Pressure Controllers. Figure 3.2 presents a control configuration with two pressure controllers, one that manipulates the inert gas valve and one that manipulates the vent valve. The set points for the two controllers are determined as follows:

Inert gas pressure controller. This set point must be above atmospheric pressure by a sufficient amount that negative pressures do not occur during normal operations (pumping material to the process and delivering material to the tank). But as the loss of gas through leaks increases with pressure, the set point should be as low as possible. This controller must be reverse acting—if the pressure is increasing, the controller decreases the inert gas valve opening.

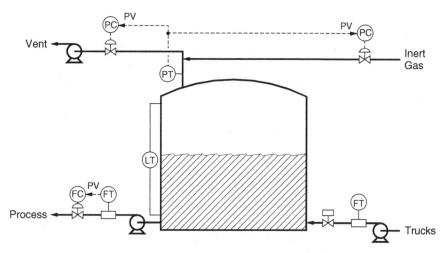

Figure 3.2 Controlling tank pressure using two pressure controllers.

Vent pressure controller. This set point must be below the maximum pressure that the storage tank is designed to withstand. Increasing the pressure set point reduces the gas vented during deliveries but increases the loss of gas through leaks. This controller must be direct acting: If the pressure is increasing, the controller increases the vent valve opening.

One other factor usually affects the addition and release of gas: namely, the day–night variation in temperature:

Day. Most storage tanks warm up, causing the gas to expand and the pressure to increase.

Night. Most storage tanks cool down, causing the gas to contract and the pressure to decrease.

It is not productive to add inert gas during the evening only to release the gas during the next day. In most cases, the set points for the two controllers in Figure 3.2 can be separated adequately to avoid this.

Mode Selection and Tuning. When the controls are implemented in digital systems, both pressure controllers are likely to be proportional-integral. Questions can be raised as to the need for integral and will be examined after the issues pertaining to using proportional-integral are discussed.

The following observations apply to the tuning coefficients for storage tank pressure control:

Controller gain. The valve sizes, pressure drops, and so on, are not the same. As the addition rate during a delivery is substantially larger than the flow rate to the process, the capacity of the vent valve must be larger than the capacity of the inert gas valve. The sensitivity of storage tank pressure to the vent valve is likely to be higher than the sensitivity to the inert gas valve. Consequently, the controller gain for the vent pressure controller should be lower than the controller gain for the inert gas pressure controller.

Reset time. The process dynamics for adding inert gas and for venting are dominated by the storage tank volume and should be approximately the same for venting and adding inert gas. The two controllers should have the same reset time.

There is one complication to consider during tuning. The rate of response of the process is inversely proportional to the total gas volume. The most rapid response will occur when the tank is nearly full of liquid; the slowest response will occur when the tank is nearly empty of liquid. The effect on the control configuration in Figure 3.2 is illustrated in the following trends:

- *Figure 3.3.* Tank is 20% full of liquid.
- *Figure 3.4.* Tank is 80% full of liquid.

For each trend, the material flows are as follows:

- $t = 0$ *min.* A process flow of 75 lb/min occurs for 15 min.
- $t = 30$ *min.* A delivery of 800 lb/min occurs for 15 min.

The set point and the tuning of the two controllers are as follows:

Controller	Set Point	Controller Gain	Reset Time
Adding inert gas	2 mmHg	10.0%/%	3.0 min
Venting	6 mmHg	5.0%/%	3.0 min

The tank pressure measurement range is 0 to 20 mmHg.

Following the material delivery, the tank pressure quickly drops below the set point for venting. The subsequent slow decrease to the set point for adding inert gas is due to the leaks. However, the rate of the decrease is noticeably faster when the tank is nearly full (Figure 3.4) than when it is nearly empty (Figure 3.3).

In general, the controller tuning is relatively conservative. Maintaining the tank pressure exactly at the respective set point is not essential. As long as the tank pressure is positive but not excessively so (that is, the tank is not overpressurized), no adverse consequences accrue. Expending time and effort to improve the performance of these controllers cannot be justified.

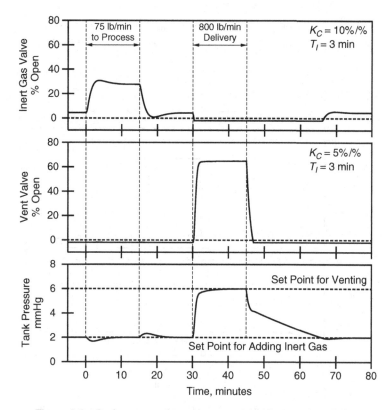

Figure 3.3 Performance of two pressure controllers, tank 20% full.

Issues with the Valves. With the piping arrangement in Figure 3.2, simultaneously opening the inert gas valve and the vent valve is merely releasing inert gas into the vent system. Often, preventing the two valves from being open at the same time is taken as an absolute must. This could be enforced by configuring output tracking as follows: (PCVENT is the tag name of the vent pressure controller; PCINERT is the tag name of the inert gas pressure controller):

```
PCVENT.TRKMN = !PCINERT.QL
PCVENT.MNI = -2.0
PCINERT.TRKMN = !PCVENT.QL
PCINERT.MNI = -2.0
```

For each controller, the logic is as follows:

- Output tracking is active if the output of the other controller is not at its lower output limit.

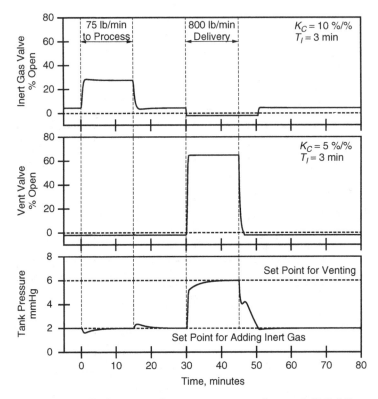

Figure 3.4 Performance of two pressure controllers, tank 80% full.

- If output tracking is active in a controller, the controller output is forced to its lower output limit (-2%).

Is it absolutely necessary to prevent the two valves from being open at the same time? The result is a waste of inert gas, but no safety issues arise (otherwise, appropriate logic would have to be added to the safety system). Should both valves be slightly open for a few seconds or perhaps a minute, the consequences are tolerable, but this must not continue for hours.

As will be explained shortly, the set points of the two pressure controllers can be adequately separated so that the two valves cannot be open at the same time. However, increasing the separation between the two set points increases the pressure at which venting occurs, which increases the rate of leaks from the tank. It is certainly possible that the loss of gas through the higher leak rate exceeds the loss of gas should both valves be slightly open for a short period of time. Consequently, excessive separation of the set points is not advisable.

Separation of the Set Points. The proportional mode adjustment on most conventional controllers was the proportional band, usually as a percent of the measured variable span but occasionally in engineering units. The proportional band (PB) is related to the controller gain K_C as follows (S_{PV} is the measured variable span, which is 20 mmHg for the tank pressure measurement):

$$PB = \frac{1}{K_C} \qquad \text{as a fraction of measured variable span}$$

$$= \frac{100}{K_C} \qquad \text{as a percent of measured variable span}$$

$$= \frac{S_{PV}}{K_C} \qquad \text{in engineering units}$$

The values of the controller gain K_C and the proportional band PB for each controller are as follows:

Controller	Controller Gain	Proportional Band
Inert gas	10.0%/%	10% or 2.0 mmHg
Vent	5.0%/%	20% or 4.0 mmHg

A change in the tank pressure equal to the proportional band is sufficient to change the controller output from 100% to 0%. For the pressure controller for adding inert gas, the inert gas valve will be fully closed if the tank pressure is above the set point by 2.0 mmHg or more. For the pressure controller for venting, the vent valve will be fully closed if the tank pressure is below the set point by 4 mmHg or more. To guarantee that both valves will never be open at the same time, the set points must be separated by the sum of the proportional bands.

In most applications, summing the proportional bands gives a larger separation for the set points than is necessary. In applications where there is an incentive to reduce the separation between the set points, a separation equal to the larger of the two proportional bands is usually adequate. But with this separation, both valves could be slightly open for a short period of time. For Figures 3.3 and 3.4, the set points are separated by 4 mmHg, which is the larger of the two proportional bands.

Proportional Plus Bias. In central control systems, the controllers are almost always proportional-integral or proportional-integral-derivative. The use of reset action in a controller is rarely questioned. However, there are cases where proportional-only should be considered. Windup protection for the pressure controllers is essential, but not all conventional electronic and pneumatic controllers provided windup protection. The consequence was problems during the transition from adding inert gas to venting, and vice versa. When the control configuration in Figure 3.2 is implemented using two pressure regulators, windup cannot occur.

There are applications where the integral mode is not really required, and storage tank pressure control is one of them. When venting, it is not necessary to maintain the tank pressure at the set point for venting. Provided that the storage tank is not overpressurized, any tank pressure is acceptable. Conversely, when adding inert gas, it is not necessary to maintain the tank pressure at the set point for adding inert gas. Any positive tank pressure is acceptable.

The proportional-only control equation is actually a proportional-plus-bias equation that can be expressed as follows:

$$M = K_C E + M_R$$

where

M = controller output (%)
E = control error (%)
K_C = controller gain (%/%)
M_R = controller output bias (%)

The controller output bias M_R is the value of the controller output when the process variable equals the set point, that is, when the control error E is zero. Consider imposing the following requirements:

- The inert gas valve is fully open at a tank pressure of 1 mmHg and closes as the pressure rises above 1 mmHg.
- The vent valve is fully open at a tank pressure of 7 mmHg and closes as the pressure drops below 7 mmHg.

This can be achieved by configuring the controllers as follows:

Controller	Set Point	Bias M_R	Proportional Band
Inert gas	1 mmHg	100%	2.0 mmHg ($K_C = 10.0\%/\%$)
Vent	7 mmHg	100%	4.0 mmHg ($K_C = 5.0\%/\%$)

Figure 3.5 presents the output of each controller as a function of the tank pressure. When the tank pressure is 3 mmHg, both valves are closed. On increasing pressure, the inert gas valve closes before the vent valve opens. On decreasing pressure, the vent valve closes before the inert valve opens.

Figure 3.6 presents the performance of the controls. As the controllers are proportional-only, neither controller lines out at its set point. However, the performance is quite acceptable. The pressure never drops below the set point of 1 mmHg for the inert gas pressure controller and never exceeds the set point of 7 mmHg for the vent pressure controller. Furthermore, the following statements can be made:

- The tank pressure should always be below 7 mmHg, which is the set point for venting. If the tank pressure ever exceeds 7 mmHg, material is being added to the tank faster than gas can be vented from the tank. The vent valve is fully open, so the controls are doing all that is possible.

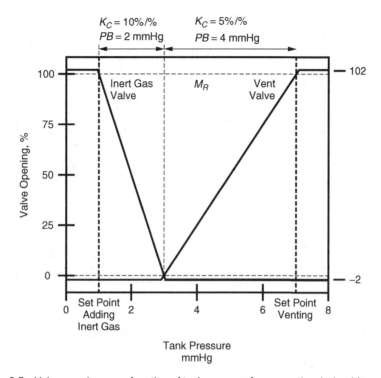

Figure 3.5 Valve openings as a function of tank pressure for proportional-plus-bias control.

- The tank pressure should always be above 1 mmHg, which is the set point for adding inert gas. If the tank pressure falls below 1 mmHg, material is being removed from the tank faster than gas can be replenished. The inert gas valve is fully open, so the controls are doing all that is possible.

Should either set point be exceeded, the culprit is a process problem, not a control problem.

The use of proportional-only control within digital systems is so unusual that some manufacturers do not really support it. It is not sufficient to be able to disable the reset action and the derivative action. To specify the proportional-plus-bias equation completely, one needs to be able to specify a value for the controller output bias M_R.

3.2. SPLIT RANGE

If the requirement is to operate at the same pressure when adding inert gas and when venting, a single pressure controller must be used. The output from this controller must drive two final control elements using logic generally referred to as *split range*.

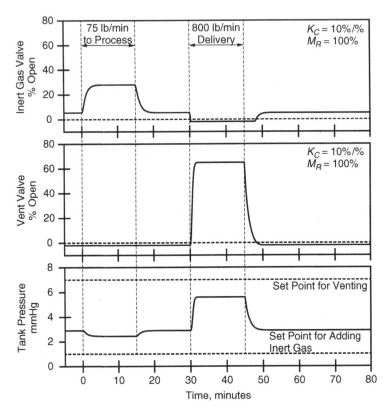

Figure 3.6 Performance of two pressure controllers tuned with proportional and bias.

Ideal Split Range. Figure 3.7 presents the logic for driving two valves using only one controller output. For the storage tank, the logic is as follows:

- At a controller output of 50% (midrange), both valves are closed.
- As the controller output increases above 50%, the vent valve opens but the inert gas valve remains fully closed.
- As the controller output drops below 50%, the inert gas valve opens but the vent valve remains fully closed.

Above midrange the pressure controller is positioning the vent valve; below midrange the pressure controller is positioning the inert gas valve. Some like to view this as follows:

- When increasing its output, the controller first closes the inert gas valve and then opens the vent valve.
- When decreasing its output, the controller first closes the vent valve and then opens the inert gas valve.

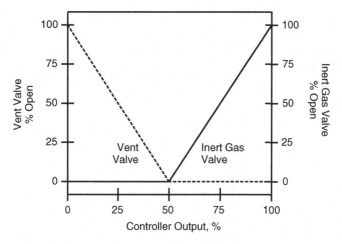

Figure 3.7   Ideal split range logic.

With the logic as in Figure 3.7, the tank pressure controller must be direct acting. On an increase in tank pressure, the controller should increase its output. This logic is referred to as split range because the controller output is positioning one valve when the controller output is below midrange and the other valve when the controller output is above midrange. Actually, the split does not have to occur at midrange, but only an occasional implementation will use something other than midrange.

This book follows the current trend with digital systems to express all valve outputs in terms of percent open (as if the valve is fail-closed). If the output is to a fail-open valve, the "inversion" to percent closed occurs in the valve block or its equivalent.

Implementation of Split-Range Logic.   The transition from conventional controls to digital controls has affected how split-range logic is implemented:

Conventional controls.   The split-range logic was always implemented at the control valves by specifying the appropriate directionality and range. A single output signal is used to drive both valves. The following specifications implement the logic in Figure 3.7:

Valve	Directionality	Range
Inert gas	Air-to-close (fails open)	0% (open) to 50% (closed)
Vent	Air-to-open (fails closed)	50% (closed) to 100% (open)

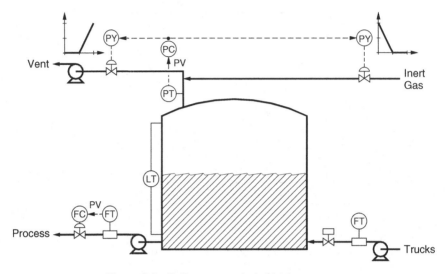

Figure 3.8   Split-range control of tank pressure.

On loss of power, the inert gas valve fails open and the vent valve fails closed, which pressurizes the storage tank to the inert gas supply pressure (which must not overpressurize the tank).

Digital controls. The split-range logic is implemented within the digital controls. In Figure 3.8, characterization functions designated by the "PY" elements implement the split-range logic. The nature of each characterization function is also indicated in Figure 3.8. This requires individual outputs to each final control element.

Ideal split-range logic as in Figure 3.7 can easily be expressed by equations. But incorporating practical issues (to be explained shortly) increases the complexity of the equations. Herein, split-range logic will always be implemented using characterization functions within the controls. Use of a fieldbus also permits the characterization function to be moved to a smart valve.

The logic in Figure 3.7 can be "reversed" in the sense that the inert gas valve opens above midrange and the vent valve opens below midrange. The following considerations apply:

Controller action. Reversing the logic affects the action of the pressure controller. On an increase in tank pressure, the controller must decrease its output, which means a reverse-acting controller.

Failure states. In conventional controls, reversing the logic also reverses the failure states (inert gas valve fails closed and vent valve fails open). But in digital controls, the failure states are determined by the specifications in the valve block or its equivalent. Choosing the split-range logic in Figure 3.7 or its reverse is a matter of personal preference.

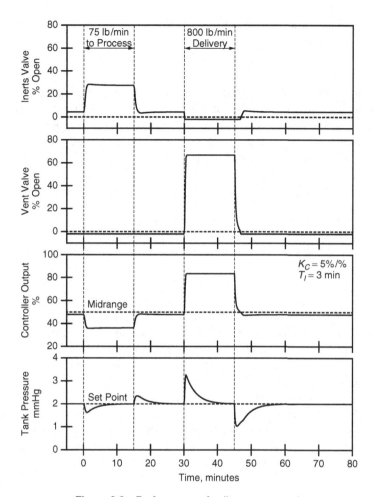

Figure 3.9 Performance of split-range control.

Split Range for the Storage Tank. Figure 3.9 presents the performance of the split-range configuration illustrated in Figure 3.8. The material flows are as follows:

- $t = 0$ *min*. A process flow of 75 lb/min occurs for 15 min.
- $t = 30$ *min*. A delivery of 800 lb/min occurs for 15 min.

The storage tank is 80% full of liquid. The pressure controller is tuned with a gain of 5%/% and a reset time of 3 min. The tank pressure measurement range is 0 to 20 mmHg.

The largest upset to the tank pressure is associated with a material delivery. In Figure 3.9, the pressure increases by more than 1 mmHg above the set point at the beginning of a delivery and drops by approximately 1 mmHg below the set point at the end of a delivery.

Controller Tuning. When separate controllers are provided, each controller can be adjusted to match the process characteristics with which it must contend. But with split range, there is only one controller. Can the same tuning be used for both venting and adding inert gas? The "one tuning fits all" approach is the simplest, but its success is not assured.

For controlling the storage tank using two controllers, the gains were different but the reset times were the same:

Controller	Controller Gain	Reset Time
Adding inert gas	10.0%/%	3.0 min
Venting	5.0%/%	3.0 min

Split range has no effect on the reset time but does affect the controller gain. The split-range logic in Figure 3.7 translates a change of 1% in the pressure controller output to a change of 2% in the valve opening. This inserts a gain of 2 into the loop. To compensate, the controller gain should be reduced by a factor of 2: that is, 5.0%/% for adding inert gas and 2.5%/% for venting. The larger of these gains was used for responses in Figure 3.8.

If the process characteristics differ significantly between one side of the split range and the other, conservative tuning is often the result. This is not necessarily bad unless it leads to actions that affect plant operations negatively, such as reducing the flow rate during a delivery or increasing the pressure set point.

There are a couple of possibilities for reducing the difference in the process sensitivity:

1. *Change the valve capacity.* There are two possibilities:
   - Change the C_V of the control valve.
   - Change the upstream or downstream pressure. For the storage tank, the capacity of the inert gas valve is affected by the inert gas supply pressure.
2. *Divide the split-range logic other than at midrange.* When split at midrange, the split-range logic increases the sensitivity for a factor of 2. The consequences of splitting at 33% instead of 50% are as follows:
   - Between 0 and 33%, a change of 1% in the pressure controller output produces a change of 3% in the valve opening, which inserts a gain of 3.
   - Between 33 and 100%, a change of 1% in the pressure controller output produces a change of 1.33% in the valve opening, which inserts a gain of 1.33.

In digital systems, scheduled tuning is a viable option that essentially permits different controller tuning to be used on each side of the split range. For the storage tank, a tuning table can be created as follows:

- Tune the controller while adding inert gas.
- Tune the controller while venting.

The results are stored in a tuning table. When the controller output is greater than 50%, one set of tuning coefficients is used. When the controller output is less than 50%, the other set is used. This approach can provide different values for the reset time as well as for the controller gain.

Practical Split-Range Logic. The logic in Figure 3.7 suggests that the transition from controlling with one valve and controlling at the other occurs exactly at midrange. Characterization functions can do this precisely, but how the valves actually behave depends on the accuracy of the span adjustments for the final control elements. Even with smart valves, perfection is unrealistic.

Using the storage tank as the example, the instrument technicians could assure that the inert gas valve and the vent valve would never be open at the same time by setting the range adjustments as follows:

- *Upper range value for the inert gas valve:* Slightly below midrange.
- *Lower range value for the vent valve:* Slightly above midrange.

The corresponding approach can be pursued with the characterization functions that implement the split-range logic in digital systems, the advantages being as follows:

- Quantitative values are available for the settings.
- The settings are easily modified.

But in each case, a deadband is inserted at midrange.

Any deadband has the potential of introducing a limit cycle into the controlled variable, especially if a controller output near midrange is required to maintain the controlled variable at its set point. In the responses in Figure 3.9, the inert gas valve must be 4.4% open to compensate for leaks from the storage tank. Should the controls cycle between venting and adding inert gas, the effective result is adding inert gas during one part of the cycle only to vent the gas during the next part of the cycle. Should this arise, adjustments must be made in the characterization functions that implement the split-range logic, the objective being to "tune out" the cycle.

In practical split-range logic, overrange is used to assure that valves are fully closed or fully open, the former being more crucial than the latter. For the vent valve for the storage tank, the characterization function in Figure 3.10 provides for the following:

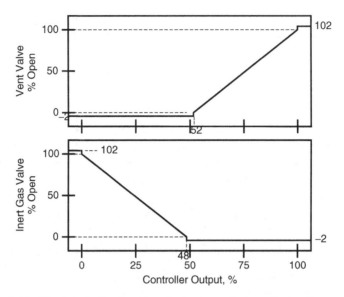

Figure 3.10 Characterization functions for implementing practical split-range logic.

- If the pressure controller output is less than 52%, the output to the vent valve is −2%. This assures that the vent valve is fully closed.
- From a pressure controller output of 52% up to a pressure controller output of 100%, the output to the vent valve varies linearly from 0 to 100%.
- If the pressure controller output is greater than 100%, the output to the vent valve is 102%. This assures that the vent valve is fully open.

The characterization function in Figure 3.10 for the inert gas valve is the mirror image of the characterization function for the vent valve.

Split Range vs. Two Controllers. The choice is often dictated by the following statements:

- If it is necessary to control to the same set point at all times, split range must be used. This is often the case for critical temperatures, such as reactor temperatures.
- In order to use two controllers, their set points must be separated. Consequently, the process operating conditions will differ depending on which controller is currently active. This eliminates applications such as reactor temperature control.

Although either can be applied to control the storage tank pressure, the consumption of inert gas will be less when two pressure controllers are used. Most storage tanks experience temperature variations between night and day. To maintain a constant tank pressure (that is, use the same set point for venting and for adding inert gas), the controls must add inert gas during the evening to compensate for the reduced gas volume at the lower temperature, and then must vent gas during the day to compensate for the increased gas volume at the higher temperature. When two pressure controllers are installed as in Figure 3.2, the set points can be separated sufficiently so that gas is not vented, due to the higher temperatures during the day.

3.3. TEMPERATURE CONTROL USING LIQUID BYPASS

Figure 3.11 illustrates a steam-heated exchanger where the liquid outlet temperature is controlled using a liquid bypass. There are two possibilities:

- *One control valve, usually in the bypass.* All liquid can be forced through the exchanger, but even with the bypass valve completely open, some liquid flows through the exchanger. The minimum possible heat transfer rate is not zero.
- *Two control valves, one in the bypass and one in series with the exchanger.* By closing the bypass valve, all liquid can be forced through the exchanger. By closing the valve in series with the exchanger, all liquid can be forced through the bypass, resulting in a heat transfer rate of zero. However, the liquid within the exchanger is heated to the steam supply temperature.

The exchanger illustrated in Figure 3.11 utilizes two control valves, and only that configuration is considered herein. The major advantages of this configuration are as follows:

- *Minimum heat transfer rate is zero.* With two control valves, all of the liquid can be bypassed, making a heat transfer rate of zero attainable.
- *Condensate return.* The shell pressure is the steam supply pressure, so condensate return is not a problem.
- *Fast response.* Changing the fraction of the liquid that flows through the bypass has a very rapid effect on the liquid outlet temperature.

Control Valve Openings. For the previous example of controlling the storage tank pressure, it made no sense for both valves to be open at the same time; or, said another way, one valve must be fully closed at all times. For the exchanger, it makes no sense for both valves to be fully closed at the same time—this would stop the liquid flow.

Steam-heated exchangers are sometimes used to provide the heating mode in recirculation loops for vessel jackets. The recirculation flow should be as large as possible at all times, so the valve openings specified by the controls should give

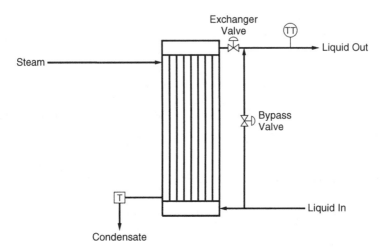

Figure 3.11   Controlling liquid outlet temperature using a liquid bypass.

the minimum restriction to the flow subject to the requirement that temperature control also be provided. This is normally achieved when one of the control valves is fully open and the other is partially open.

Perhaps the best way to think of the effect of the valve openings on heat transfer is as follows:

Heat Transfer	Exchanger Valve % Open	Bypass Valve % Open
None	0	100
Intermediate	100	100
Maximum	100	0

With the exchanger valve fully closed and the bypass valve fully open, no heat is transferred to the liquid. To increase the heat transfer rate, the controls should first open the exchanger valve, and only after it is fully open should the controls begin to close the bypass valve. The maximum heat transfer is attained when the bypass valve is fully closed and the exchanger valve is fully open.

Two Controllers. The two-controller configuration in Figure 3.12 is analogous to the configuration in Figure 3.1 for the storage tank. The set points for the two controllers must be separated sufficiently so that each controller does not start to close its valve until the other valve is fully open. Thus, the temperature is maintained at one set point for low heat transfer rates and a different set point for high heat transfer rates.

The two controllers in the configuration in Figure 3.12 are differentiated by the valve to which they output:

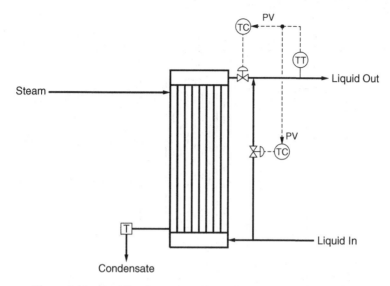

Figure 3.12 Liquid outlet temperature control using two controllers.

Exchanger valve. If the liquid outlet temperature is increasing, the controller must reduce the exchanger valve opening, that is, the controller
must be reverse acting. If the liquid outlet temperature is below its set
point, this controller will increase the opening of the exchanger valve. This
controller must have the higher set point. For heat transfer rates between
zero (exchanger valve fully closed) and the intermediate rate (when both
valves are open), this controller maintains the liquid outlet temperature at
the higher set point.

Bypass valve. If the liquid outlet temperature is increasing, the controller
must increase the bypass valve opening; that is, the controller must be direct
acting. If the liquid outlet temperature is above its set point, this controller
will increase the opening of the bypass valve. This controller must have the
lower set point. For heat transfer rates between the intermediate rate and
the maximum rate (bypass valve fully closed), this controller maintains the
liquid outlet temperature at the lower set point.

If the liquid outlet temperature is between the two set points, both controllers
drive their respective valves fully open. The two controllers in the configuration
in Figure 3.12 will maintain the liquid outlet temperature at their respective set
points. However, controlling to a different temperature set point depending on the
heat transfer rate is unacceptable in most temperature control applications. Later
in this chapter, a steam-heated exchanger with a liquid bypass is used to heat the
jacket of a vessel with a recirculation system. Such jackets must be maintained
at a constant temperature, so the configuration in Figure 3.12 is unacceptable.

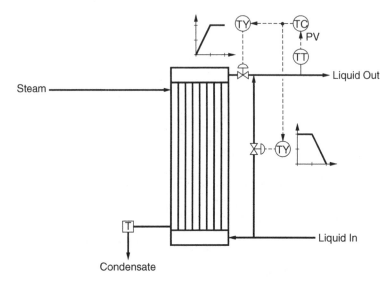

Figure 3.13   Split-range configuration for liquid outlet temperature control.

Split-Range Configuration. Figure 3.13 presents the following split-range configuration:

> *Below midrange.* The bypass valve is fully open. The exchanger valve is fully open at midrange and closes below midrange.
>
> *At midrange.* Both valves are fully open.
>
> *Above midrange.* The exchanger valve is fully open. The bypass valve is fully open at midrange and closes above midrange.

Increasing the output of the controller always increases the heat input to the exchanger. If the temperature is rising, the controller must reduce its output, so the controller must be reverse acting. As there is a single controller, there is only one temperature set point. The control configuration in Figure 3.13 attempts to maintain the liquid outlet temperature at that set point regardless of the heat transfer rate. Figure 3.13 provides a characterization function (the "TY" element) for each valve. The ideal nature of each characterization function is also presented in Figure 3.13.

Controller Tuning. One of the major advantages of the bypass configuration is the very rapid response. A valve movement is quickly translated into a change in the liquid outlet temperature. Consequently, the reset time for both controllers will be very short, and will not be significantly different.

The process gain or sensitivity is far more complex. There are two components of this relationship:

Sensible heat. The liquid outlet temperature is the result of mixing a hot stream and a cold stream. The relationship between this temperature and the mass flow of each stream is relatively simple.

Installed valve characteristics. The relationship between the opening of each control valve and the flow through that valve is rather complex, depending on the following:

- Valve flow coefficient C_V
- The inherent characteristics of each valve
- The exchanger pressure drop as a function of flow

These relationships are so complex that the only way to analyze them reliably is via a flow simulation.

Fluid Mixing. The exchanger exit temperature T_E is the temperature of the fluid at the exit of the exchanger. The liquid outlet temperature T_{out} is the result of mixing this fluid with the bypass flow, whose temperature is the liquid inlet temperature T_{in}. The liquid outlet temperature is described by an energy balance whose terms are the sensible heat of the respective streams. The equation is as follows:

$$F_E c_P (T_E - T_R) + F_B c_P (T_{in} - T_R) = F c_P (T_{out} - T_R)$$

where

c_P = liquid heat capacity (Btu/lb-°F)
$F = F_E + F_B$ total liquid flow (lb/min)
F_B = liquid flow through bypass (lb/min)
F_E = liquid flow through exchanger (lb/min)
T_E = temperature at exchanger exit (°F)
T_{in} = liquid inlet temperature (°F)
T_{out} = liquid outlet temperature (°F)
T_R = reference temperature for computing enthalpy (°F)

For a constant liquid heat capacity c_P, the energy balance simplifies to

$$F_E T_E + F_B T_{in} = F T_{out}$$

The following assumptions will be made:

- The exchanger exit temperature T_E is not a function of the exchanger flow F_E. For exchangers that are oversized, the exchanger exit temperature T_E will approach the temperature of the condensing steam. Otherwise, this will occur only at low values of the exchanger flow F_E.

- Changes in the exchanger flow F_E have little effect on the bypass flow F_B. This will be true if the pressure drop across the exchanger–bypass combination is constant. With this assumption,

$$\frac{\partial F}{\partial F_E} = \frac{\partial (F_E + F_B)}{\partial F_E} = 1$$

Sensitivity of Liquid Outlet Temperature to the Flows Through the Control Valves. Below midrange, the outlet temperature is controlled by manipulating the exchanger valve. The sensitivity with which the controller must contend is expressed by the partial derivative

$$\frac{\partial T_{\text{out}}}{\partial F_E}$$

Taking the partial derivative of the energy balance with respect to the exchanger flow F_E gives the following expression:

$$T_E = F\frac{\partial T_{\text{out}}}{\partial F_E} + T_{\text{out}}$$

Solving for the partial derivative gives the expression for the process sensitivity:

$$\frac{\partial T_{\text{out}}}{\partial F_E} = \frac{T_E - T_{\text{out}}}{F}$$

The process sensitivity is not constant, depending on the total liquid flow F and the liquid outlet temperature T_{out}.

When controlling the liquid outlet temperature using the bypass valve (the temperature controller output is above midrange), the process sensitivity is the partial derivative of the liquid outlet temperature T_{out} with respect to the bypass flow F_B. A similar approach yields the following expression for this partial derivative:

$$\frac{\partial T_{\text{out}}}{\partial F_B} = \frac{T_{\text{in}} - T_{\text{out}}}{F}$$

Since $T_{\text{in}} < T_{\text{out}}$, this process sensitivity is negative, as indeed it should be: Increasing the bypass flow reduces the liquid outlet temperature. But provided that the controller action is specified properly, only the magnitude of the sensitivity is of concern.

Sensitivity as a Function of Valve Opening. Suppose that with both valves fully open, the flow is split evenly between the exchanger and the bypass. Then as the respective valve moves from fully closed to fully open, the following occur:

1. Total liquid flow F increases by a factor of 2.
2. The respective process sensitivity decreases by a factor of 2.

Since the exchanger outlet temperature responds very rapidly to changes in either flow, the controller could most likely be tuned in a sufficiently conservative manner that a change in sensitivity by a factor of 2 would not cause problems.

Sensitivity as a Function of Liquid Outlet Temperature. The liquid outlet temperature T_{out} must be between the liquid inlet temperature T_{in} and the exchanger exit temperature T_E. The sensitivity of the outlet temperature to each stream is proportional to the difference between the temperature of that stream and the liquid outlet temperature. The following observations are based on the equations presented previously for the two respective sensitivities:

- If the liquid outlet temperature T_{out} is midway between the liquid inlet temperature T_{in} and the exchanger exit temperature T_E, the two sensitivities are equal.
- If the liquid outlet temperature T_{out} is nearer the exchanger exit temperature T_E, the sensitivity of the liquid outlet temperature to the exchanger flow is less than its sensitivity to the bypass flow.
- If the liquid outlet temperature T_{out} is nearer the liquid inlet temperature T_{in}, the sensitivity of the liquid outlet temperature to the exchanger flow is greater than its sensitivity to the bypass flow.

In continuous processes, changes to the target for the liquid outlet temperature are usually small and infrequent. In batch processes larger changes are likely, and each sensitivity can easily change by a factor of 10. The impact of such a change on controller performance would be noticeable even in a very fast loop.

Installed Valve Characteristics. This issue pertains to the following:

- How the exchanger flow F_E varies with the exchanger valve opening. This relationship affects the performance of the temperature controller only when the controller output is below midrange. Above midrange, the exchanger valve is fully open.
- How the bypass flow F_B varies with the bypass valve opening. This relationship affects the performance of the temperature controller only when the controller output is above midrange. Below midrange, the bypass valve is fully open.

This relationship is affected by several factors:

- The valve flow coefficient C_V of the respective valve.
- The inherent characteristics of the respective valve.

- The pressure drop-to-flow relationship for the exchanger. Usually, this can be approximated adequately by an orifice coefficient for the exchanger, which basically assumes that the pressure drop across the exchanger varies with the square of the flow. From the process design data, the exchanger pressure drop for the design flow should be available, which permits the orifice coefficient to be computed.
- The pressure drop across the exchanger–bypass combination. The actual situation is usually between the following two extremes:
  - *Constant pressure drop.* Changing the valve openings affects the total liquid flow but not the pressure drop across the exchanger–bypass combination.
  - *Constant total liquid flow.* Changing the valve openings affects the pressure drop across the exchanger–bypass combination but not the total liquid flow.

Flow simulations are the only way to analyze processes with either or both of the following:

1. Parallel flow paths
2. Two or more valves

The exchanger–bypass combination contains both. The flow simulation may have to encompass more than the exchanger–bypass combination. For example, if the exchanger is providing the heat for a vessel with a recirculating jacket, the flow simulation may have to encompass all components of the jacket recirculation loop.

A pure flow simulation analyzes only how the various flows are affected by the various valve openings. But for this example, the interest is how the liquid outlet temperature is affected by the various valve openings. To obtain this information, the simulation must include the following:

- Sensible heat equations for mixing the bypass flow and the exchanger flow
- Heat transfer equations for the exchanger, which includes the effect of the exchanger flow F_E on the exchanger exit temperature T_E

In some applications, a measurement of the total liquid flow is available, but measurements for the exchanger flow and the bypass flow are unusual.

The sizing of the control valves affects the relationship between the liquid outlet temperature and the respective valve openings. For this example, the valves are sized to give the following results when both valves are fully open:

- The flow through the exchanger is the same as the flow through the bypass.
- The exchanger valve is taking 25% of the total pressure drop across the exchanger and the exchanger valve.

The total pressure drop across the exchanger-bypass combination is assumed to be constant.

As the pressure drop across the bypass valve is constant, the guidelines for selecting the characteristics for control valves suggest that the bypass valve should have linear characteristics. But given the dominant practice of installing equal-percentage valves, in this example both will be equal-percentage.

The process operating line is a plot of the liquid outlet temperature as a function of the temperature controller output. Figure 3.14 presents the operating line for equal-percentage valves. The analysis of the operating line considers three issues:

1. *Regions of high sensitivity* (the operating line is nearly vertical). The operating line in Figure 3.14 does not exhibit this behavior.
2. *Regions of low sensitivity* (the operating line is flat). The operating line in Figure 3.14 exhibits such a region from a controller output of approximately 40% (exchanger valve 80% open) to midrange (exchanger valve fully open). This is basically a "dead zone" where changes in the controller output have no effect on the controlled variable.
3. *Significant change in sensitivity*. Except for the flat region, the operating line in Figure 3.14 exhibits relatively modest changes in the slope.

How can the dead zone in the operating line in Figure 3.14 be eliminated? With the split-range logic implemented in software, the easiest approach is usually to modify the split-range logic. There are two possibilities:

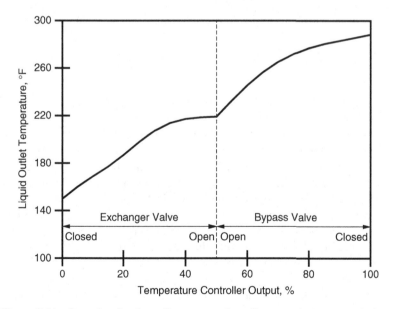

Figure 3.14 Operating line for split-range configuration, equal-percentage valves.

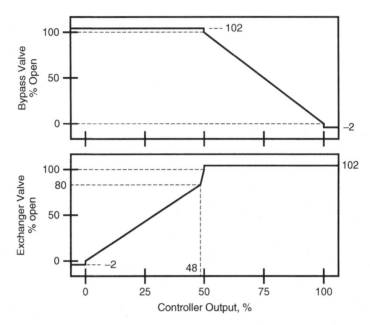

Figure 3.15   Modified split-range logic to avoid flat region on operating line.

1. *Figure 3.15*. The characterization for the exchanger valve opening as a function of controller output is as follows:

Controller Output	Exchanger Valve Opening
0%	0%
48%	80%
50%	100%

   Up to an opening of 80%, the exchanger valve is used to control the liquid outlet temperature. But once the exchanger valve is 80% open, it is quickly driven to 100% open.

2. *Figure 3.16*. Once the controller output attains 40%, the exchanger valve has no effect on the liquid outlet temperature. To reduce the dead zone in the operating line, the split-range logic in Figure 3.16 begins to close the bypass valve at a controller output of 40% instead of 50%. For controller outputs between 40 and 50%, neither valve is fully open, which reduces the total liquid flow. However, eliminating the dead zone in the operating line is more important than maintaining the maximum possible flow.

When the characterization functions for the split-range logic are implemented within a digital system, changes such as these are easily incorporated.

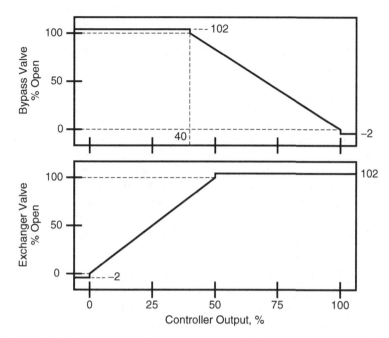

Figure 3.16   Overlapping split-range logic to avoid flat region on operating line.

3.4. RECIRCULATING JACKET WITH HEAT AND COOL MODES

To maintain the desired temperature in some process vessels, heating is required at times, but cooling at other times. The interest here is those applications where the transition occurs frequently (such as in every product batch), and its time of occurrence cannot be predicted precisely. For such applications, the controls must initiate the transition depending on whether heating or cooling is required to maintain the desired temperature within the vessel. Furthermore, a smooth transition is required, that is, with as little disturbance to the vessel temperature as possible.

For all examples herein, the jacketed reactor described in Chapter 2 is extended to provide heating as well as cooling. In the examples the heating medium will be hot water or steam; other possibilities include hot oil, direct fired, and others. The cooling medium will be tower cooling water; other possibilities include chilled water, glycol, refrigerant, air cooler, and others. As the focus of the examples is on temperature control, all feed and discharge streams will be omitted to keep the P&I diagrams as simple as possible.

Split-Range Logic. Most such applications require split-range control logic. Herein, the split-range logic will be configured as follows:

Controller Output	Cooling	Heating
0%	Maximum	Off
50%	Off	Off
100%	Off	Maximum

On increasing its output, the controller first reduces the cooling and then increases the heating. If the vessel temperature is increasing, the controller must reduce its output, which means a reverse-acting controller. The temperature control configuration is the temperature-to-temperature cascade commonly provided when the jacket is equipped with a recirculation system. The reactor temperature controller provides the set point to the jacket temperature controller. Split-range logic is required within the inner loop to manipulate the proper valve, depending on whether heating or cooling is required to maintain the desired jacket temperature.

Cooling Water and Hot Water. The configuration in Figure 3.17 provides heating using hot water and cooling using cooling water. The control valves are installed on the respective supplies. A pressure regulator may be required on the return to maintain the required pressure in the jacket.

For the configuration in Figure 3.17, the jacket temperature responds quickly to changes in either the cooling water flow or the hot water flow. The dynamics

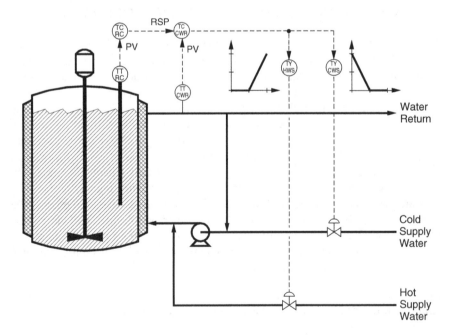

Figure 3.17   Hot water and cold water.

will be similar, but the sensitivities can be significantly different, due to the following factors:

Capacity of the respective control valve. The larger the valve, the larger the sensitivity of the jacket temperature to the valve opening. Oversizing the valves on the cooling and/or heating media supply is common, the consequence being that at large openings the control valve is offering so little resistance to fluid flow that changes in its opening have little effect on the media flow rate.

Jacket temperature relative to the media supply temperature. The larger the difference between the media supply temperature and the jacket tempera-ture, the larger the sensitivity of the jacket temperature to changes in the flow of that medium. Especially in batch reactors, this difference can change significantly during the batch.

As the split-range logic is within a fast inner loop of a cascade, the tuning of the jacket temperature controller can be relaxed to cope with these factors without degrading the performance of the reactor temperature loop.

The cooling water supply temperature (nominally, $68°F$) and the hot water supply temperature ($190°F$ or less) restrict the range of jacket temperatures that can be attained. Sometimes chilled water is provided either in addition to or in lieu of the cooling water, to attain lower temperatures. One approach for attaining lower temperatures is to use a cold and a hot glycol supply. To attain higher temperatures, a similar configuration using a separate hot oil and cold oil supply is a viable option. The control issues are basically the same as for the configuration in Figure 3.17.

Direct Steam Injection. In most plants, low-pressure steam is readily avail-able, making it an attractive medium for heating. The approach illustrated in Figure 3.18 is to inject the steam directly into the recirculation loop. The steam that is injected is not returned as condensate and must be replenished as boiler feedwater. Consequently, the configuration in Figure 3.18 is normally installed when the consumption of steam is small (or infrequent). The control valve for the cooling must be located on the cooling water supply. A pressure regulator may be required to maintain a pressure on the jacket, but this pressure cannot exceed the pressure of the steam supply.

The configuration in Figure 3.18 provides a flow controller for the steam. The flow controller is not mandatory; the split-range logic can output directly to the steam control valve. But without the flow controller, the steam flow would be affected by pressure changes both in the steam supply and within the recirculation jacket. Especially when the difference between the steam supply pressure and the jacket pressure is small, even small changes in either pressure would significantly affect the steam flow and, consequently, the jacket temperature.

When the split-range logic is in the heating mode, the control configuration in Figure 3.18 becomes a temperature-to-temperature-to-flow cascade. The steam

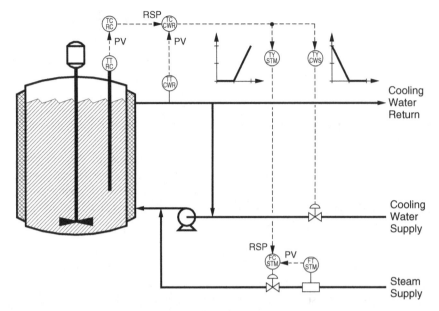

Figure 3.18 Direct steam injection.

flow loop will respond more rapidly than the jacket temperature loop, providing the desired factor of 5 for separation of the dynamics. In Chapter 2 we examined the potential for windup between the jacket temperature loop and the reactor temperature loop. When in the heating mode, a similar consideration arises between the steam flow loop and the jacket temperature loop. Of specific concern is the possibility of the jacket temperature controller providing a steam flow set point that cannot be attained. That is, the steam flow controller drives the steam control valve fully open, but the resulting steam flow is still below the steam flow set point.

The possibilities for implementing windup protection are the same as presented in Chapter 2:

- Integral tracking
- External reset
- Inhibit increase/inhibit decrease

The following issues arise when windup protection is provided for the jacket temperature controller:

- Two of these require that split-range logic be "inverted," that is, the jacket temperature output determined from the current steam flow and cooling water valve position.

• When windup protection is invoked to prevent windup in the jacket temperature loop, the potential for windup arises in the reactor temperature loop.

Inverting the Split-Range Logic. Two mechanisms, specifically integral tracking and external reset, for preventing windup in the jacket temperature controller require that the split-range logic for the steam flow be "inverted." Element TYSTM in Figure 3.18 computes the steam flow set point from the jacket temperature controller output. In using these mechanisms to provide windup protection, the jacket temperature controller output must be computed from the current value of the steam flow.

For the ideal form of the split-range logic, the relationship to convert steam flow to jacket temperature controller output is based on the following two points:

Steam Flow	Controller Output
Lower-range value or less	50%
Upper-range value or greater	100%

For the ideal split-range logic, a simple equation can be developed to relate the jacket temperature controller output to the current value of the steam flow. However, a characterization function can better accommodate the overrange, deadband, and other aspects of a practical split-range relationship.

The discussion that follows assumes that the characterization function FYSTMI is configured as basically the "inverse" of characterization function FTSTM in Figure 3.18. FTSTM converts the temperature controller output to a steam flow set point; FTSTMI converts the steam flow to a temperature controller output.

Integral Tracking. For the configuration in Figure 3.18, integral tracking is invoked in the jacket temperature controller when the steam flow controller has driven its output to the upper output limit. Consequently, only split-range logic for the heating mode must be "inverted." Integral tracking will never be activated during the cooling mode.

The inputs to the jacket temperature controller to implement integral tracking are configured as follows:

Input MRI. The appropriate value for input MRI is the output FYSTMI.Y of the characterization function described previously for inverting the split-range logic for the heating mode.

Input TRKMR. Integral tracking must be activated when the steam control valve is driven to its upper output limit (output QH from the steam flow controller is true).

The logic for integral tracking for the jacket temperature controller is expressed as follows:

```
TCJKT.TRKMR = FCSTM.QH
TCJKT.MRI = TYSTMI.Y
```

When windup protection is invoked in the jacket temperature controller, changes in the jacket temperature set point no longer have any effect on the jacket temperature, which is the condition for windup to occur in the reactor temperature controller. The inputs to the reactor temperature controller to implement integral tracking are configured as follows:

Input MRI. This input must be the measured value of the jacket temperature.

Input TRKMR. Integral tracking must be initiated in the reactor temperature controller when any of the following conditions are true:

- Output QH from the steam flow controller is true. The steam flow controller has driven is output to the upper output limit.
- Output QH from the jacket temperature controller is true. The jacket temperature controller has driven its output to its upper output limit. This could occur before the steam flow controller has driven its output to its upper output limit.
- Output QL from the jacket temperature controller is true. The jacket temperature controller has driven its output to its lower output limit.

The logic for output tracking and for integral tracking for the reactor temperature controller is expressed as follows:

```
TCRC.TRKMN = !TCJKT.RMT
TCRC.MNI = TCJKT.SP
TCRC.TRKMR = FCSTM.QH | TCJKT.QH | TCJKT.QL
TCRC.MRI = TTJKT.PV
```

External Reset. The objective for configuring external reset is to prevent windup should the steam flow controller drive its output to its upper output limit. This can only occur during the heating mode. But unlike integral tracking, input XRS is always used in performing the PID computations. Specifically, the value of input XRS is used in the cooling mode as well as the heating mode.

The value of input XRS must be as follows:

Cooling mode. (jacket temperature controller output below midrange). Windup protection is only required should the jacket temperature controller drive its output to the lower output limit. The standard windup protection provided by the PID controller is invoked should this occur. Consequently, the output of the jacket temperature controller can be used for input XRS, the result being "internal reset."

Heating mode. (jacket temperature controller output above midrange). Input XRS must be the output of the characterization function TYSTMI that inverts the split-range logic for the heating mode..

The logic for external reset for the jacket temperature controller is expressed as follows:

```
if (TCJKT.MN <= 50.0)
   TCJKT.XRS = TCJKT.MN
else
   TCJKT.XRS = TYSTMI.Y
```

For the reactor temperature controller, the measured value of the jacket temperature controller is configured for input XRS. The logic for output tracking and for external reset for the reactor temperature controller is expressed as follows:

```
TCRC.TRKMN = !TCJKT.RMT
TCRC.MNI = TCJKT.SP
TCRC.XRS = TTJKT.PV
```

Inhibit Increase/Inhibit Decrease. When contrasting the logic to be presented below for windup protection via inhibit increase/inhibit decrease to that presented for the temperature-to-temperature cascade in Chapter 2, there is one difference that must be taken into consideration. The action of the jacket temperature controller is not the same:

Without the split-range logic (Figure 2.3). The output of the jacket temperature controller is the opening of the cooling water valve. On an increase in jacket temperature, the jacket temperature controller must increase the opening of this valve, which requires a direct-acting controller.

With the split-range logic (Figure 3.18). Increasing the output of the jacket temperature controller first reduces the cooling and then increases the heating. If the jacket temperature is increasing, the jacket temperature controller must reduce its output, which requires a reverse-acting controller.

This difference has no impact on integral tracking and external reset, but for inhibit increase/inhibit decrease, it reverses the configuration for the NOINC and NODEC inputs.

To prevent windup in the jacket temperature controller when the steam control valve is driven to its upper output limit, the jacket temperature controller must not be permitted to further increase its output, which is the steam flow set point.

Input NOINC for the jacket temperature controller must be true when output QH from the steam flow controller is true. The logic for inhibit increase/inhibit decrease for the jacket temperature controller is expressed as follows:

```
TCJKT.NOINC = FCSTM.QH
```

To prevent windup in the reactor temperature controller should the maximum heating be attained, the input NOINC must be true if either of the following is true:

> *Output QH from the steam flow controller.* This covers the possibility that the steam control valve is driven fully open before the jacket temperature controller has driven its output to the upper output limit (which corresponds to the upper range value of the steam flow set point).
>
> *Output QH from the jacket temperature controller.* This covers the possibility that the jacket temperature controller drives its output to the upper output limit before the steam control valve is driven fully open.

To cover the possibility that the jacket temperature controller drives its output to the lower output limit, the input NODEC should be true if the output QL of the jacket temperature controller is true.

The logic for output tracking and for inhibit increase/inhibit decrease for the reactor temperature controller is expressed as follows:

```
TCRC.TRKMN = !TCJKT.RMT
TCRC.MNI = TCJKT.SP
TCRC.NOINC = FCSTM.QH | TCJKT.QH
TCRC.NODEC = TCJKT.QL
```

Steam-Heated Exchanger. One approach to provide heating is to insert a steam-heated exchanger into the recirculation loop. The cooling and heating modes are provided as follows:

> *Cooling mode.* Cooling is by cooling water. Water flows from the cooling water supply into the recirculation loop, with the excess flowing into the cooling water return. The control valve can be located on the return to provide the highest possible pressure within the recirculation loop.
>
> *Heating mode.* Three approaches can vary the heat transfer rate to the cooling water flowing through the exchanger:

- Control valve on the steam supply
- Control valve on the condensate
- Liquid bypass

Each is discussed herein.

Control Valve on the Steam Supply. A fully open steam valve gives the maximum condensing steam pressure. Closing the control valve reduces the pressure of the condensing steam, which has the following consequences:

- Reduces the condensing steam temperature and the driving force for heat transfer. The controls rely on this effect to vary the heat transfer rate.
- Reduces the driving force for condensate to flow through the steam trap into either the condensate return or a steam drain. The consequences of this effect lead to problems.

Figure 3.19 presents the split-range control configuration. As in previous configurations, the cooling water valve opens below midrange and the steam valve opens above midrange.

To achieve a smooth transition either from heating to cooling or from cooling to heating, there will be a period of time when low heat transfer rates are required. Consequently, the exchanger must deliver heat transfer rates from zero (or at least close to zero) up to the maximum possible heat transfer rate. With the control valve on the steam supply, there will be a minimum heat transfer rate that the exchanger can deliver without cycling.

With the control valve on the steam supply, the minimum pressure of the condensing steam depends on the condensate discharge:

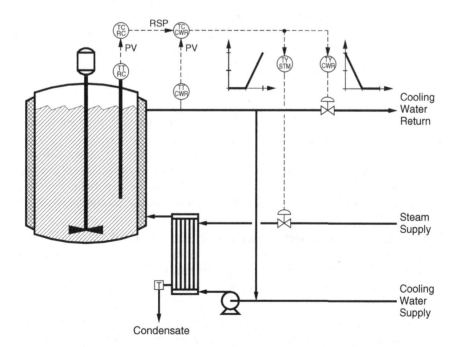

Figure 3.19   Steam-heated exchanger, control valve on steam supply.

- *To drain.* Minimum pressure is atmospheric pressure, giving a condensing steam temperature of 212°F.
- *To condensate return.* Minimum pressure is condensate return pressure, which will be above atmospheric pressure. The condensing steam temperature is above 212°F.

Either temperature is greater than the temperature of the jacket water in the recirculation loop. This temperature difference determines the minimum heat transfer rate that can be sustained without cycling.

When the pressure of the condensing steam drops to the pressure required to force the condensate out of the exchanger, the following cycle ensues:

- The exchanger begins to fill with condensate.
- This reduces the effective heat transfer area, which decreases the heat transfer rate.
- The reduced heat transfer rate causes the jacket temperature to drop.
- When the jacket temperature drops below its set point, the controller increases the steam control valve opening.
- This eventually increases the condensing steam pressure to the point where all of the condensate is forced out of the exchanger.
- With all surface area available for heat transfer, this increases the heat transfer rate and the jacket temperature.
- As the jacket temperature rises above its set point, the controller decreases the steam control valve opening.
- The pressure of the condensing steam drops, again causing the exchanger to fill with condensate.

The result is a limit cycle in the condensing steam pressure, the heat transfer rate, and the jacket temperature.

With the control valve on the steam supply, the exchanger cannot deliver a constant heat transfer rate that is below the minimum. If one averages the heat transfer rate over the cycle, the average will be below the minimum. However, such cycles are not desirable in vessels where good temperature control is crucial.

Control Valve on the Condensate. Figure 3.20 presents the split-range configuration that uses the control valve on the cooling water return for cooling and the control valve on the condensate return for heating.

With the control valve on the condensate, the following observations apply:

- The pressure of the condensing steam within the exchanger is close to the steam supply pressure at all times. The steam supply pressure provides the driving force for condensate flow.

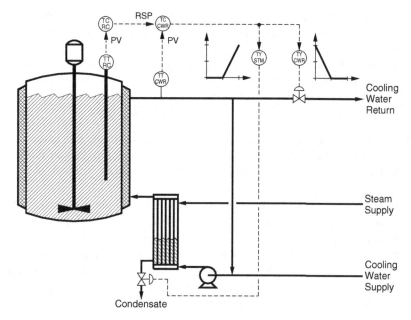

Figure 3.20   Steam-heated exchanger, control valve on condensate.

- With a constant pressure for the condensing steam, the temperature of the condensing steam will also be constant. This provides a constant temperature driving force for heat transfer.
- With the control valve on the condensate, the exchanger must be partially filled with condensate at all times. Otherwise, steam will flow through the control valve into the condensate return. More on this shortly.
- With the exchanger partially filled with condensate, the effective area for heat transfer is the surface area exposed to the condensing steam. The submerged heat transfer area subcools the condensate, but this contribution to the total heat transfer rate is small.

The minimum heat transfer rate occurs when the condensate valve is closed completely. The exchanger fills with condensate, giving a heat transfer rate of essentially zero. As the condensate valve opens, the condensate level within the exchanger drops, which increases the heat transfer area and the heat transfer rate.

Installing the control valve on the condensate adversely affects the dynamics. The volumetric flow of condensate (a liquid) is far smaller than the volumetric flow of steam (a gas), the consequences being:

Control valve on steam supply   (Figure 3.19). A change in the steam flow affects the condensing steam pressure (and consequently, the heat transfer rate) rapidly.

Control valve on condensate   (Figure 3.20). A change in the condensate flow affects the condensate level (and consequently, the heat transfer rate) much more slowly.

For production-scale vessels, the jacket temperature controller can be tuned to provide adequate performance. However, different tuning coefficients are likely to be required for cooling vs. heating.

The maximum heat transfer rate occurs when the exchanger is completely empty of condensate, which is likely to occur before the condensate valve is fully open. Should this occur, steam flows into the condensate return, a phenomenon known as *blowing steam*. There are several possible solutions:

- Install a steam trap upstream of the control valve on the condensate. This is a simple "fix" but with the side effect that windup begins in the jacket temperature controller as soon as the steam trap starts to block the steam flow.
- Install a condensate pot between the exchanger and the condensate return. This is an effective solution where one condensate pot can be shared between multiple exchangers.
- Measure the condensate level within the exchanger and provide a level override. Basically, when the exchanger level drops to the minimum allowable value, the override begins to adjust the condensate valve opening to maintain the condensate level at the minimum.

The level override configuration for a steam-heated exchanger is examined in Chapter 4. Its incorporation into the split-range configuration in Figure 3.20 is relatively straightforward.

Liquid Bypass.   Figure 3.21 presents a liquid bypass arrangement for the steam-heated exchanger in the recirculation loop for the jacket. This configuration provides two control valves:

1. *Bypass valve:*   Installed in the bypass around the exchanger.
2. *Exchanger valve:*   Installed on the exchanger exit.

In the cooling mode, all flow must bypass the exchanger, resulting in no heat transfer. The maximum heat transfer rate is when all recirculation flow passes through the exchanger. In addition to the minimum heat transfer rate being zero, the exchanger with bypass responds very rapidly and contributes nothing of significance to the dynamics of the jacket temperature loop. However, only a flow simulation of the complete jacket recirculation loop can assure that the valves are sized properly and that there are no dead zones.

The configuration in Figure 3.21 is essentially two split-range systems in series:

- The heating and cooling modes are implemented by applying split-range logic to the output of the jacket temperature controller. This split-range logic is the same as presented for previous examples in this section.
- In the heating mode, split-range logic is applied to obtain the openings for the exchanger valve and the bypass valve. This split range is the same as in Figure 3.13.

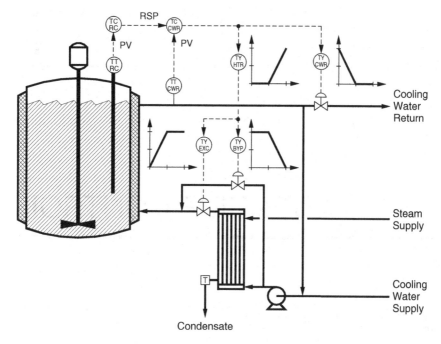

Figure 3.21 Steam-heated exchanger, liquid bypass.

Figure 3.21 uses four characterization functions to implement this logic, but the logic can be implemented with only three configured as follows:

- *Input:* The jacket temperature controller output.
- *Output:* Opening for one of the control valves.

The ideal split-range logic is as follows:

Jacket TC Output:	0%	50%	75%	100%
Mode	Cooling	All off	Heating	Heating
Cooling water valve opening	100%	0%	0%	0%
Exchanger valve opening	0%	0%	100%	100%
Bypass valve opening	100%	100%	100%	0%

From a controller output of 0 to 50%, jacket temperature is controlled with the cooling water valve; from 50 to 75%, the jacket temperature is controlled with the exchanger valve; and from 75 to 100%, the jacket temperature is controlled with the bypass valve.

4

OVERRIDE CONTROL

Constraints are limits on process operations. Such limits may arise in several different ways:

Limits on the controller output. These limits are usually imposed via the controller output limits.

Limits on the process variable. In most applications, these limits can be adequately imposed by specifying limits on the controller set point.

Limits on a dependent variable. When a controller output significantly affects the dependent variable, an override configuration is one approach to enforcing such limits.

Herein several examples will be used to explain the various aspects of override controls. The explanations also include the following:

- Override controls are implemented using the selector block (sometimes referred to as an *auctioneer*). This block was described in detail in Chapter 1.

- Windup is always a possibility for override controls. Approaches to preventing this windup are examined.

4.1. LIMIT ON THE COOLING WATER RETURN TEMPERATURE

Figure 4.1 presents a reactor with a once-through jacket. The focus herein is on temperature control, so the feed and discharge streams are not shown. Once-through means that the cooling water enters the jacket, makes one pass through

Advanced Process Control: Beyond Single-Loop Control   By Cecil L. Smith
Copyright © 2010 John Wiley & Sons, Inc.

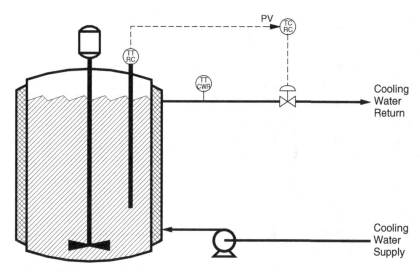

Figure 4.1 Reactor temperature control for a reactor with once-through jacket.

the jacket, and exits to the cooling water return. There is no recirculation of cooling water from jacket outlet to jacket inlet.

Cooling water always contains minerals whose solubility decreases with temperature. The conditioning of the cooling water determines the degree to which these minerals are removed, which imposes an upper limit on the cooling water return temperature. Should these minerals precipitate within the jacket, scale forms on the heat transfer surfaces and degrades the ability to transfer heat.

Figure 4.1 presents a simple feedback configuration for reactor temperature control. If the rate of heat generation within the reactor is low, the controller will reduce the cooling water flow, which causes the cooling water return temperature to approach the reactor temperature. This usually leads to two requirements:

Equipment protection. Appropriate action must be taken should the cooling water return temperature exceed the upper limit.

Control system. The process controls should never take actions that would necessitate a response from the equipment protection logic. Even during normal operations, the configuration in Figure 4.1 could reduce the cooling water flow to a point where the cooling water return temperature exceeds the upper limit.

Herein only the latter will receive attention.

Cooling Water Return Temperature Control. Figure 4.2 presents a configuration that controls the cooling water return temperature using the control valve on the cooling water return. This loop can potentially be used in the following configurations:

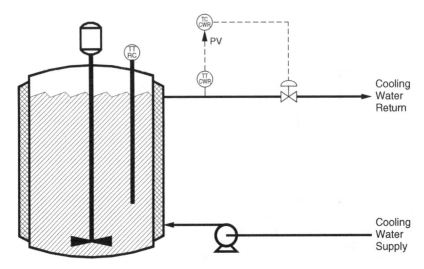

Figure 4.2   Cooling water return temperature control for a reactor with once-through jacket.

Cascade. The loop in Figure 4.2 is the inner loop; the outer loop is a reactor temperature loop that provides the set point for the cooling water return temperature. This configuration is analogous to the configuration in Figure 2.3 for a reactor with a recirculating jacket.

Override. An override configuration uses the loop in Figure 4.2 to avoid the high cooling water return temperatures that occur at low cooling water flows. At other times, the override configuration controls the reactor temperature using the configuration in Figure 4.1.

At intermediate heat transfer rates, the loop in Figure 4.2 will function properly. But issues arise at the extremes:

- *Very low cooling water flow rates.* The cooling water return temperature is very close to the reactor temperature. Changes in the cooling water flow rate have little influence on the cooling water return temperature.
- *Very high cooling water flow rates.* The cooling water return temperature is very close to the cooling water supply temperature. Again, changes in the cooling water flow rate have little influence on the cooling water return temperature.

Cascade Configuration. The cascade configuration in Figure 4.3 is used in practice with some degree of success. The justification of cascade is often based on its improved response to disturbances such as cooling water supply temperature and pressure. In this respect, cascade is equally effective for once-through and recirculating jackets. But in practice, cascade control of a once-through jacket is not generally as successful as cascade control of a recirculating jacket. For a

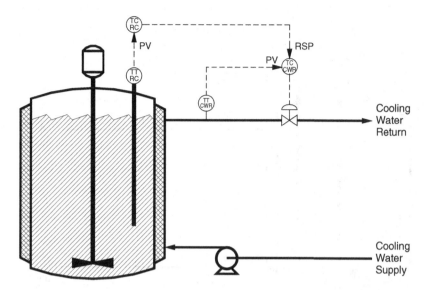

Figure 4.3 Temperature-to-temperature cascade for a reactor with a once-through jacket.

recirculating jacket, the inner loop completely removes the heat transfer non-linearities from the relationship between the reactor temperature and the jacket temperature, which significantly improves the performance of the reactor temperature controller. The inner loop does not achieve this for a once-through jacket.

If the cascade configuration in Figure 4.3 is installed, the upper limit on the cooling water return temperature can easily be imposed. These limits can be imposed as either or both of the following:

- *As an upper output limit for the reactor temperature controller.* Initiating windup protection at the upper output limit is a standard feature of a PID block.

- *As an upper set point limit for the cooling water return temperature controller.* Windup protection in the reactor temperature controller must be invoked should this limit be attained. For this to occur, additional configuration may be required.

A cascade configuration is often a viable alternative to an override configuration. However, the cascade configuration must be viable, and the inner loop of the cascade must function effectively under all regions of process operation.

Override Configuration. Figure 4.4 presents the override configuration for controlling the reactor temperature subject to an upper limit on the cooling water return temperature. The configuration in Figure 4.4 requires three control blocks:

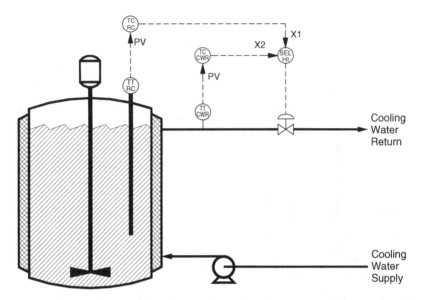

Figure 4.4 Reactor temperature control with cooling water return temperature override.

Reactor temperature controller (TCRC). The set point for this controller is the desired reactor temperature.

Cooling water return temperature controller (TCCWR). The set point for this controller is the upper limit on the cooling water return temperature.

Select block (SEL). For this application, a high select is required.

At a given instant, only one of the controllers in Figure 4.4 will be effectively in operation (in the sense that its output is affecting the cooling water valve opening). There are two possible situations:

1. *Reactor temperature at set point; cooling water return temperature below set point.* At high heat transfer rates, the reactor temperature controller is "in control" and is specifying the opening of the cooling water control valve. As the cooling water return temperature is below its set point, the cooling water return temperature controller is decreasing its output. Windup protection is required to prevent the cooling water return temperature controller from driving its output to the lower output limit.

2. *Cooling water return temperature at set point; reactor temperature below set point.* At low heat transfer rates, the cooling water return temperature controller is "in control" and is specifying the opening of the cooling water control valve. As the reactor temperature is below its set point, the reactor temperature controller is decreasing its output. Windup protection is required to prevent the reactor temperature controller from driving its output to the lower output limit.

The control configuration in Figure 4.4 would normally be described as follows:

Reactor temperature control with cooling water return temperature override.

The main objective is to control the reactor temperature, so this is a reactor temperature control configuration. But because of the upper limit imposed on the cooling water return temperature, under some conditions the cooling water return temperature (the override) must be controlled instead of the reactor temperature. Although the terminology perhaps implies otherwise, the high select in Figure 4.4 does not in any way give priority to the reactor temperature controller. It merely examines the values of the two inputs and sets its output to the larger of the two.

The following description could also be applied to the control configuration in Figure 4.4:

Cooling water return temperature control with reactor temperature override.

This statement implies that the primary objective is to control the cooling water return temperature, which is not the case for the process in Figure 4.4.

Valve Failure Mode. As for other configurations within this book, the output of the override configuration in Figure 4.4 is the control valve opening (as if the valve is fail-closed). For a fail-open valve, the conversion from % open to % closed is performed by the valve block or its equivalent.

Conventional systems did not provide this flexibility. For a failed-closed valve, the output of each controller had to be % closed. This necessitated "inverting" the following:

Controller action. Both controllers in Figure 4.4 must be reverse acting instead of direct acting.

Selector option. A low select is required instead of a high select.

Although this approach works, most find working with control valve opening more convenient.

4.2. EXAMPLE WITHOUT WINDUP PROTECTION

The performance of the override control configuration in Figure 4.4 will be examined for the following two cases:

1. *A slow decrease in the rate of heat generation within the reactor.* This necessitates a switch from reactor temperature control to cooling water return temperature control.

2. *A slow increase in the rate of heat generation within the reactor.* This necessitates a switch from cooling water return temperature control to reactor temperature control.

The first case is typical for many batch reactors. The rate of reaction is highest (and thus is releasing the most heat) during the early portion of the batch. As the reactions are driven to completion, the rate of heat generation slowly decreases. To maintain a constant reactor temperature, the controls must reduce the flow of cooling water. The cooling water return temperature increases, and can potentially exceed the upper limit.

In all examples the parameters for the controllers are as follows:

	Reactor Temperature Controller	Cooling Water Return Temperature Controller
Set point	150°F	130°F (upper limit)
Measurement range	0 to 300°F	0 to 200°F
Measurement resolution	0.1°F	0.1°F
Action	Direct ($E = PV - SP$)	Direct ($E = PV - SP$)
K_C	8.0%/%	1.0%/%
$K_{C,EU}$	2.67%/°F	0.5%/°F
T_I	10.0 min	1.0 min
T_D	0.0 min	0.0 min
Lower output limit	−2%	−2%

Switch from Reactor Temperature Control to Cooling Water Return Temperature Control. The responses in Figure 4.5 are to a slow decrease in the rate of heat generation beginning at $t = 0$. The starting conditions are as follows:

- The reactor temperature is 150°F (the set point).
- A cooling water valve opening of 27.3% is required to give a reactor temperature of 150°F.
- The cooling water return temperature is 113.4°F, which is below its set point (and upper limit) of 130°F.
- With no windup protection, the cooling water return temperature controller has driven its output to the lower output limit (−2%).

From $t = 0$ to $t \cong 120$ min, the controls behave as follows:

Reactor temperature controller. This controller is "in control." The rate of heat generation is decreasing, so:
- The reactor temperature drops slightly below its set point of 150°F.
- The controller decreases the cooling water valve opening.

Cooling water return temperature controller. The cooling water return temperature is increasing but is below its set point of 130°F. With no windup

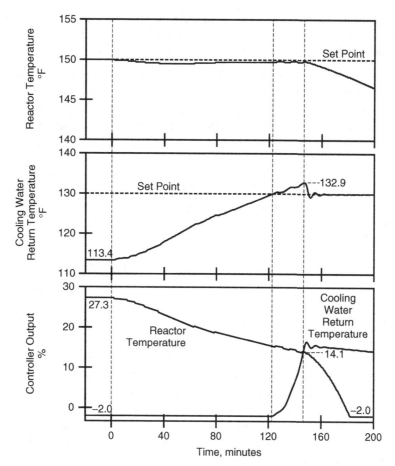

Figure 4.5 Switching from reactor temperature control to cooling water return temperature control, no windup protection.

protection, the reset mode drives the controller output to the lower output limit (−2%).

High select. Input X1 (output of the reactor temperature controller) is greater than input X2 (output of the cooling water return temperature controller). The block selects input X1.

With windup protection, the behavior will be the same, but with one exception: the reset mode in the cooling water return temperature controller is not permitted to drive the controller output to its lower output limit (−2%). At $t \cong 120$ min in Figure 4.5, the cooling water return temperature crosses its set point of 130°F. On or about this time, control of the cooling water return temperature should start. The cooling water return temperature controller properly begins to increase its output. But the increase starts from the lower output limit of -2%, so the reactor temperature controller remains "in control."

From $t \cong 120$ min to $t \cong 145$ min in Figure 4.5, the following occurs:

Reactor temperature controller. To maintain the reactor temperature near its set point of 150°F, the controller decreases the cooling water valve opening.

Cooling water return temperature controller. The cooling water return temperature overshoots its set point. The greater the overshoot, the more rapidly the controller increases its output.

At $t \cong 145$ min, the selector switches from reactor temperature control to cooling water return temperature control. Without windup protection, the switch occurs approximately 25 min later than appropriate, the consequence being an overshoot of 2.9°F in the cooling water return temperature.

Once the transients associated with the switch have passed, the controls behave as follows:

Cooling water return temperature controller. This controller is "in control." The rate of heat generation is decreasing, so:
- The cooling water return temperature drops slightly below its set point of 130°F.
- The controller decreases the cooling water valve opening.

Reactor temperature controller. The reactor temperature decreases below its set point of 150°F. With no windup protection, the reset mode drives the controller output to the lower output limit (-2%).

High select. Input X1 (output of the reactor temperature controller) is less than input X2 (output of the cooling water return temperature controller). The block selects input X2.

Windup is occurring when the reset mode drives the reactor temperature controller output to its lower output limit of -2% (as in Figure 4.5). The applicable statement is:

Reset windup occurs in a controller when changes in the controller output have no effect on the process variable.

The high select is selecting the output of the cooling water return temperature controller, so changes in the output of the reactor temperature controller have no effect on the reactor temperature.

Switch from Cooling Water Return Temperature Control to Reactor Temperature Control. The responses in Figure 4.6 are to a slow increase in the rate of heat generation in the reactor beginning at $t = 0$. The starting conditions are as follows:

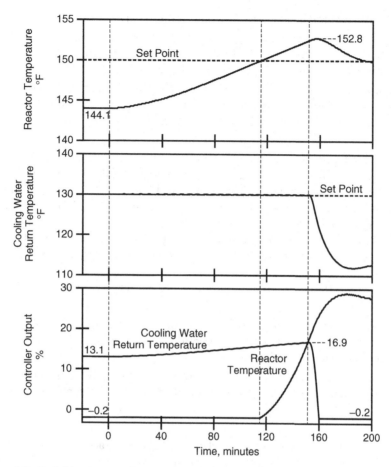

Figure 4.6   Switching from cooling water return temperature control to reactor temperature control, no windup protection.

- The cooling water return temperature is 130°F (the set point).
- A cooling water valve opening of 13.1% is required to give a cooling water return temperature of 130°F.
- The reactor temperature is 144.1°F, which is less than its set point of 150°F.
- With no windup protection, the reactor temperature controller has driven its output to the lower output limit (−2%).

From $t = 0$ to $t \cong 120$ min, the controls behave as follows:

Cooling water return temperature controller. This controller is "in control." The rate of heat generation is increasing, so:

- The cooling water return temperature rises slightly above its set point of 130°F.

- The controller increases the cooling water valve opening.

Reactor temperature controller. The reactor temperature is increasing but is below its set point of 150°F. With no windup protection, the reset mode drives the controller output to the lower output limit (−2%).

High select. Input X1 (output MN of the reactor temperature controller) is less than input X2 (output MN of the cooling water return temperature controller). The high select block selects input X2.

With windup protection, the behavior will be the same, but with one exception: The reset mode in the reactor temperature controller is not permitted to drive the controller output to its lower output limit (−2%). At $t \cong 120$ min in Figure 4.6, the reactor temperature crosses its set point of 150°F. On or about this time, control of the reactor temperature should begin. The reactor temperature controller properly begins to increase its output. But the increase starts from the lower output limit of −2%, so the cooling water return temperature controller remains "in control."

From $t \cong 120$ min to $t \cong 150$ min in Figure 4.6, the following occurs:

Cooling water return temperature controller. To maintain the cooling water return temperature near its set point of 130°F, the controller increases the cooling water valve opening.

Reactor temperature controller. The reactor temperature overshoots its set point. The greater the overshoot, the more rapidly the controller increases its output.

At $t \cong 150$ min, the selector switches from cooling water return temperature control to reactor temperature control. Without windup protection, the switch occurs approximately 30 min later than appropriate, the consequence being an overshoot of 2.9°F in the reactor temperature.

Once the transients associated with the switch have passed, the controls behave as follows:

Reactor temperature controller. This controller is "in control." The rate of heat generation is increasing, so:
- The reactor temperature rises slightly above its set point of 150°F.
- The controller increases the cooling water valve opening.

Cooling water return temperature controller. The cooling water return temperature decreases below its set point of 130°F. With no windup protection, the reset mode drives the controller output to the lower output limit (−2%).

High select. Input X1 (output of the reactor temperature controller) is greater than input X2 (output MN of the cooling water return temperature controller). The high select block selects input X1.

Windup is occurring when the reset mode drives the cooling water return temperature controller output to its lower output limit of -2% (as in Figure 4.6). The applicable statement is:

Reset windup occurs in a controller when changes in the controller output have no effect on the process variable.

The high select is selecting the output of the reactor temperature controller, so changes in the output of the cooling water return temperature controller have no effect on the cooling water return temperature.

4.3. INTEGRAL TRACKING

As described in Chapter 1, the PID block takes the following actions when integral tracking is active:

1. The controller output bias M_R is set equal to the value of input MRI (the usual reset mode calculations are not performed).
2. The output of the controller is computed by the proportional-plus-bias equation: $M = K_C E + M_R$

Reactor Temperature Controller. Integral tracking is to be active if the output of the reactor temperature controller (input X1 to the high select) is not selected. Inputs MRI and TRKMR must be configured as follows:

Input MRI (value for integral tracking): Output Y of the high select.

Input TRKMR (condition for integral tracking): Inverse (logical NOT) of output Q1 of the high select.

When there are only two inputs to the high select, input TRKMR could be output Q2 of the high select. However, the inverse of output Q1 can be used regardless of the number of inputs to the high select.

The logic for integral tracking in the reactor temperature controller is expressed as follows:

```
TCRC.TRKMR = !SEL.Q1
TCRC.MRI = SEL.Y
```

Cooling Water Return Temperature Controller. Integral tracking is to be active if the output of the cooling water return temperature controller (input X2 to the high select) is not selected:

MRI (value for integral tracking): Output Y of the high select.

TRKMR (condition for integral tracking): Inverse (logical NOT) of output Q2 of the high select.

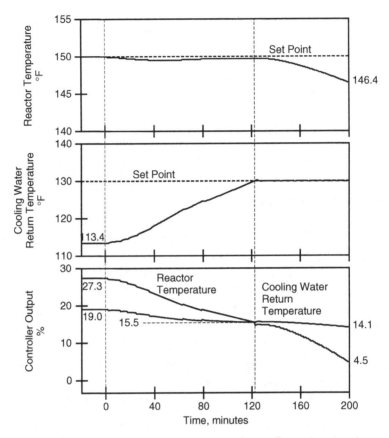

Figure 4.7   Switching from reactor temperature control to cooling water return temperature control, integral tracking for windup protection.

The logic for integral tracking in the cooling water return temperature controller is expressed as follows:

```
TCCWR.TRKMR = !SEL.Q2
TCCWR.MRI = SEL.Y
```

Switch from Reactor Temperature Control to Cooling Water Return Temperature Control. The responses in Figure 4.7 are to a slow decrease in the rate of heat generation beginning at $t = 0$. Prior to the switch at $t \cong 120$ min, the trends in Figure 4.7 (integral tracking) are identical to those in Figure 4.5 (no windup protection), the exception being the output of the cooling water return temperature controller.

The starting conditions are as follows:

• Reactor temperature is $150°F$ (the set point).

- A cooling water valve opening of 27.3% is required to give a reactor temperature of 150°F.
- The cooling water return temperature is 113.4°F, which is below its set point (and upper limit) of 130°F.
- Integral tracking is active in the cooling water return temperature controller.
- The cooling water return temperature controller output is calculated as follows:

$$SP = 130.0°F$$

$$PV = 113.4°F$$

$$E = PV - SP = 113.4°F - 130.0°F = -16.6°F$$

$$M_R = 27.3\% \text{ (output of reactor temperature controller)}$$

$$K_{C,EU} = 0.5\%/°F$$

$$M = K_{C,EU}E + M_R$$
$$= (0.5\%/°F) \times (-16.6°F) + 27.3\%$$
$$= 19.0\%$$

From $t = 0$ to $t \cong 120$ min, the controls behave as follows:

Cooling water return temperature controller:

input TRKMR = true (integral tracking is active)

$$M_R = \text{input MRI} = \text{reactor temperature controller output}$$

$$E = PV - SP < 0 \text{ (controller is direct acting)}$$

$$M = K_C E + M_R$$
$$= K_C E + \text{reactor temperature controller output}$$
$$< \text{reactor temperature controller output}$$

High select:

input X1 = reactor temperature controller output

input X2 = cooling water return temperature controller output

$$< \text{input X1}$$

output Y = input X1

output Q1 = true; integral tracking is not active in the
     reactor temperature controller

output Q2 = false; integral tracking is active in the cooling water
     return temperature controller

The switch occurs when the cooling water return temperature crosses its set point (the control error E changes sign), which is at $t \cong 120$ min in the trends in Figure 4.7. The subsequent overshoot in the cooling water return temperature is minimal.

At the time of the switch at $t \cong 120$, the trend for the reactor temperature controller output in Figure 4.7 exhibits an abrupt decrease (this change does not affect the cooling water valve opening; the selector has switched to the output of the cooling water return temperature controller). This decrease is the result of an abrupt change in the controller output bias M_R that occurs the instant integral tracking is activated in the reactor temperature controller:

Control error E (controller is direct acting):

$$PV = 149.7°F$$

$$E = PV - SP = 149.7°F - 150.0°F = -0.3°F$$

Immediately prior to the switch (integral tracking not active):

$$M_R = 16.3\% \text{ (result of the reset mode calculations)}$$

$$M = K_{C,EU}E + M_R = (2.67\%/°F) \times (-0.3°F) + 16.3\% = 15.5\%$$

$$\cong \text{output of cooling water return temperature controller}$$

Immediately after the switch (integral tracking is active):

$$M_R = 15.5\%. \text{ (output of cooling water return temperature controller)}$$

$$M = K_{C,EU}E + M_R = (2.67\%/°F) \times (-0.3°F) + 15.5\% = 14.7\%$$

At the instant integral tracking is activated in the reactor temperature controller, the controller output bias M_R changes from 16.3% to 15.5%, which causes the controller output to change from 15.5% to 14.7%. The magnitude of the change is the product of the control error E and the controller gain K_C.

Subsequent to the switch at $t \cong 120$ min in Figure 4.7, the continued decrease in the rate of heat generation has the following impact:

Cooling water return temperature controller. To maintain the cooling water return temperature close to its set point of 130°F, the controller decreases its output only slightly below the value of 15.5% at the time the switch occurs.

Reactor temperature controller. The lower the rate of heat generation, the further the reactor temperature drops below its set point of 150°F. The controller output is decreasing rapidly, and its behavior has some similarities to the behavior of windup. The conditions at $t = 200$ min in Figure 4.7 are as follows:

$$M_R = 14.1\% \text{ (cooling water return temperature controller output)}$$

$$E = PV - SP = 146.4°F - 150.0°F = -3.6°F$$

$$M = K_{C,\text{EU}}E + M_R$$
$$= (2.67\%/^{\circ}\text{F}) \times (-3.6^{\circ}\text{F}) + 14.1\%$$
$$= 4.5\%$$

If the reactor temperature continues to decrease, the controller output will eventually attain the lower output limit (-2%). However, the proportional term ($K_C E$) of the proportional-plus-bias equation is largely responsible.

Just because a controller has driven its output to an output limit does not necessarily mean that windup has occurred. Windup is only associated with the reset mode. If the controller output is at an output limit because of the proportional term ($K_C E$), windup has not occurred.

Switch from Cooling Water Return Temperature Control to Reactor Temperature Control.

The responses in Figure 4.8 are to a slow increase in the rate of heat generation in the reactor beginning at $t = 0$. Prior to the switch at $t \cong 120$ min, the trends in Figure 4.8 (integral tracking) are identical to those in Figure 4.6 (no windup protection), the exception being the output of the reactor temperature controller.

The starting conditions are as follows:

- The cooling water return temperature is 130°F (the set point).
- A cooling water valve opening of 13.1% is required to give a cooling water return temperature of 130.0°F.
- The reactor temperature is 144.1°F, which is below its set point of 150°F.
- Integral tracking is active in the reactor temperature controller.
- The reactor temperature controller output is calculated as follows:

$$\text{SP} = 150.0^{\circ}\text{F}$$
$$\text{PV} = 144.1^{\circ}\text{F}$$
$$E = \text{PV} - \text{SP} = 144.1^{\circ}\text{F} - 150.0^{\circ}\text{F} = -5.9^{\circ}\text{F}$$
$$M_R = 13.1\% \text{ (output of cooling water return temperature controller)}$$
$$K_{C,\text{EU}} = 2.67\%/^{\circ}\text{F}$$
$$M = K_{C,\text{EU}}E + M_R$$
$$= (2.67\%/^{\circ}\text{F}) \times (-5.9^{\circ}\text{F}) + 13.1\%$$
$$= -2.7\% \text{ (less than lower output limit of } -2\%)$$

From $t = 0$ to $t \cong 120$ min, the controls behave as follows:

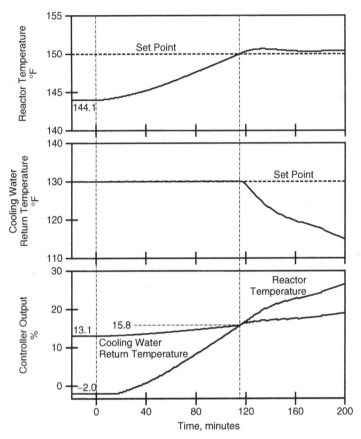

Figure 4.8   Switching from cooling water return temperature control to reactor temperature control, integral tracking for windup protection.

Reactor temperature controller:

input TRKMR = true (integral tracking is active)

M_R = input MRI = cooling water return temperature controller output

$E = PV - SP < 0$ (controller is direct acting)

$M = K_C E + M_R$

= $K_C E$ + cooling water return temperature controller output

< cooling water return temperature controller output

High select:

> input X1 = output of the reactor temperature controller
>
> < input X2
>
> input X2 = output of the cooling water return
> temperature controller
>
> output Y = input X2
>
> output Q1 = false; integral tracking is active in
> the reactor temperature controller
>
> output Q2 = true; integral tracking is not active in
> the cooling water return temperature controller

The switch occurs when the reactor temperature crosses its set point (the control error E changes sign), which is at $t \cong 120$ min in the trends in Figure 4.8. The subsequent overshoot in the reactor temperature is less than $1°F$.

Activating integral tracking in the cooling water return temperature controller causes an abrupt increase in the controller output that is the product of the following:

- Control error E ($0.1°F$ at the time of the switch)
- Controller gain $K_{C,EU}$ ($0.5\%/°F$)

The change in the controller output is 0.05%, which is too small to be apparent in Figure 4.8.

4.4. EXTERNAL RESET

As explained in the description of the PID block in Chapter 1, external reset relies on the feedback form of the PID control equation as presented in Figure 1.5. The controller output bias M_R is the output of a first-order lag:

- The time constant is the reset time T_I.
- The input is the value of input XRS, or if input XRS is not configured, the input is the controller output.

The controller output is computed by the usual proportional-plus-bias control equation:

$$M = K_C E + M_R$$

Unlike integral tracking, external reset is not activated or inactivated (there is no counterpart to input TRKMR). If the controller is in automatic and output tracking is not active, the reset mode calculations are always performed using the value of input XRS.

Reactor Temperature Controller. External reset is configured as follows:

Input XRS (value for external reset): Output Y of the high selector.

The logic for external reset in the reactor temperature controller is expressed as follows:

```
TCRC.XRS = SEL.Y
```

Cooling Water Return Temperature Controller. External reset is configured as follows:

Input XRS (value for external reset): Output Y of the high selector.

The logic for external reset in the cooling water return temperature controller is expressed as follows:

```
TCCWR.XRS = SEL.Y
```

Switch from Reactor Temperature Control to Cooling Water Return Temperature Control. The responses in Figure 4.9 are to a slow decrease in the rate of heat generation beginning at $t = 0$. Prior to the switch at $t \cong 120$ min, the trends in Figure 4.9 (external reset) are identical to those in Figure 4.5 (no windup protection), the exception being the output of the cooling water return temperature controller. The trends in Figure 4.9 (external reset) are very similar to those in Figure 4.7 (integral tracking).

The starting conditions are as follows:

- Reactor temperature is at its set point of $150°F$.
- A cooling water valve opening of 27.3% is required to give a reactor temperature of $150°F$.
- The cooling water return temperature is $113.4°F$, which is below its set point (and upper limit) of $130°F$.
- The cooling water return temperature controller output is calculated as follows:

$$SP = 130.0°F$$
$$PV = 113.4°F$$
$$E = PV - SP = 113.4°F - 130.0°F = -16.6°F$$

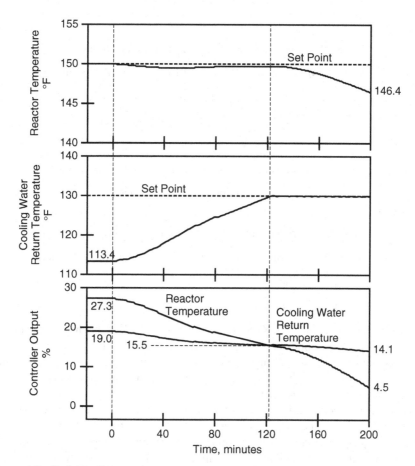

Figure 4.9 Switching from reactor temperature control to cooling water return temperature control, external reset for windup protection.

input $XRS = 27.3\%$ (output of reactor temperature controller)

$M_R = 27.3\%$ (at steady-state the output of a lag equals the input to the lag)

$K_{C,EU} = 0.5\%/°F$

$$M = K_{C,EU}E + M_R$$
$$= (0.5\%/°F) \times (-16.6°F) + 27.3\%$$
$$= 19.0\%$$

From $t = 0$ to $t \cong 120$ min, the controls behave as follows:

Cooling water return temperature controller:

input XRS = reactor temperature controller output

M_R = input XRS lagged by 1.0 min (the reset time)

$\cong$ reactor temperature controller output

$E = PV - SP < 0$ (controller is direct acting)

$M = K_C E + M_R$

$\cong K_C E$ + reactor temperature controller output

< reactor temperature controller output

High select:

Input X1 = reactor temperature controller output

Input X2 = cooling water return temperature controller output

< input X1

output Y = input X1

During this interval, the XRS input to both controllers is the output of the reactor temperature controller. Consequently, the reset mode in the reactor temperature controller is integrating the control error. But in the cooling water return temperature controller, the reset mode is not integrating the control error, which prevents windup from occurring.

The switch occurs just prior to the time that the cooling water return temperature crosses its set point, which is at $t \cong 120$ min in the trends in Figure 4.9. The subsequent overshoot in the cooling water return temperature is minimal.

Subsequent to the switch at $t \cong 120$ min in Figure 4.9, the continued decrease in the rate of heat generation has the following impact:

Cooling water return temperature controller. Input XRS is the controller output. Consequently, the reset mode is integrating the cooling water return temperature control error. To maintain the cooling water return temperature at its set point, the controller is slowly decreasing its output.

Reactor temperature controller. Input XRS is the output of the cooling water return temperature controller. Consequently, the reset mode is not integrating the reactor temperature control error, which prevents windup from occurring in the controller. However, the controller output is decreasing rapidly, and its behavior has some similarities to the behavior of windup. The conditions at $t = 200$ min in Figure 4.9 are as follows:

input XRS = 14.1% (cooling water return temperature controller output)

M_R = 14.1% lagged by 10.0 min (the reset time)

$\cong 14.1\%$

$$E = PV - SP = 146.4°F - 150.0°F = -3.6°F$$

$$M = K_{C,EU}E + M_R$$

$$\cong (2.67\%/°F) \times (-3.6°F) + 14.1\%$$

$$\cong 4.5\%$$

The rapid decrease in the controller output in Figure 4.9 is due largely to the proportional term $(K_C E)$, so windup is not occurring.

The computation above is very similar to the one made for integral tracking and in most cases, the performance of external reset and integral tracking are very similar. However, abrupt changes in the controller output bias never occur when external reset is providing the windup protection. On the switch from reactor temperature control to cooling water return temperature control, the trends in Figure 4.7 (integral tracking) show an abrupt change in the reactor temperature controller output; no such change occurs in the trends in Figure 4.9 (external reset). Otherwise, these trends are very similar.

Switch from Cooling Water Return Temperature Control to Reactor Temperature Control. The responses in Figure 4.10 are to a slow increase in the rate of heat generation in the reactor beginning at $t = 0$. Prior to the switch at $t \cong 120$ min the trends in Figure 4.10 (external reset) are identical to those in Figure 4.6 (no windup, protection), the exception being the output of the cooling water return temperature controller. The trends in Figure 4.10 (external reset) are very similar to those in Figure 4.8 (integral tracking).

The starting conditions are as follows:

- The cooling water return temperature is at its set point of 130°F.
- A cooling water valve opening of 13.1% is required to give a cooling water return temperature of 130°F.
- The reactor temperature is 144.1°F, which is below its set point of 150°F.
- The reactor temperature controller output is calculated as follows:

$$SP = 150.0°F$$

$$PV = 144.1°F$$

$$E = PV - SP = 144.1°F - 150.0°F = -5.9°F$$

$$\text{input XRS} = 13.1\% \text{ (output of cooling water return}$$
$$\text{temperature controller)}$$

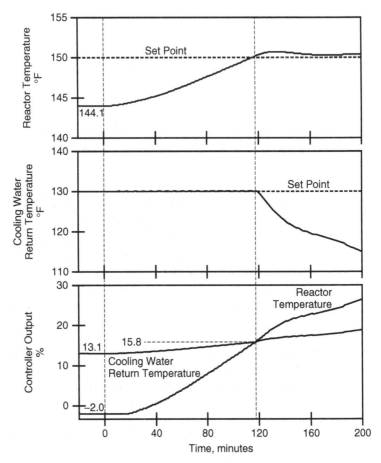

Figure 4.10 Switching from cooling water return temperature control to reactor temperature control, external reset for windup protection.

$$M_R = 13.1\% \text{ (at steady-state the output of a lag equals the}$$
$$\text{input to the lag)}$$

$$M = K_{C,\text{EU}}E + M_R$$
$$= (2.67\%/^\circ\text{F}) \times (-5.9^\circ\text{F}) + 13.1\%$$
$$= -2.7\% \text{ (less than lower output limit of} -2\%)$$

The output is at its lower output limit because of the reactor temperature control error (-5.9°F or -1.93%) and the high controller gain ($8.0\%/\%$), not because of windup.

From $t = 0$ to $t \cong 120$ min, the controls behave as follows:

Reactor temperature controller:

input XRS $=$ cooling water return temperature controller output

$M_R =$ input XRS lagged by 10.0 min (the reset time)

$\cong$ cooling water return temperature controller output

$E = \text{PV} - \text{SP} < 0$ (controller is direct acting)

$M = K_C E + M_R$

$\cong K_C E +$ cooling water return temperature controller output

$<$ cooling water return temperature controller output

High select:

input X1 $=$ reactor temperature controller output

$<$ input X2

input X2 $=$ cooling water return temperature controller output

output Y $=$ input X2

During this interval, the XRS input to both controllers is the output of the cooling water return temperature controller. Consequently, the reset mode in the cooling water return temperature controller is integrating the control error. But in the reactor temperature controller, the reset mode is not integrating the control error, which prevents windup from occurring.

The switch occurs just prior to the time that the reactor temperature crosses its set point, which is at $t \cong 120$ min in the trends in Figure 4.10. The subsequent overshoot in the reactor temperature is less than $1°F$.

4.5. INHIBIT INCREASE/INHIBIT DECREASE

The reactor temperature control with cooling water return temperature override in Figure 4.4 includes a high select to auctioneer the outputs of the two controllers. In any override configuration with a high select, the integral mode in the controller whose output is not selected will decrease its output to the maximum extent possible. When no windup protection is provided, the output will be driven to the lower output limit. A simple concept for preventing this windup is as follows:

For the controller whose output is not selected, do not permit the controller to decrease its output.

This is the purpose of input NODEC to the PID block described in Table 1.1. For low selectors, input NOINC must be used to prevent the controller whose output is not selected from increasing its output.

Reactor Temperature Controller. As the output of this controller is input X1 to the high select, input NODEC must be the inverse (logical NOT) of output Q1 from the high select.

The logic for inhibit increase/inhibit decrease in the reactor temperature controller is expressed as follows:

```
TCRC.NODEC = !SEL.Q1
```

Cooling Water Return Temperature Controller. As the output of this controller is input X2 to the high select, input NODEC must be the inverse (logical NOT) of output Q2 from the high select.

The logic for inhibit increase/inhibit decrease in the cooling water return temperature controller is expressed as follows:

```
TCCWR.NODEC = !SEL.Q2
```

Switch from Reactor Temperature Control to Cooling Water Return Temperature Control. The responses in Figure 4.11 are to a slow decrease in the rate of heat generation beginning at $t = 0$. Prior to the switch at $t \cong 110$ min, the trends in Figure 4.11 (inhibit increase/inhibit decrease) are identical to those in Figure 4.5 (no windup protection), the exception being the output of the cooling water return temperature controller. The overall performance of inhibit increase/inhibit decrease is comparable to that of integral tracking (Figure 4.7) and external reset (Figure 4.9).

The starting conditions are as follows:

- Reactor temperature is at its set point of 150°F.
- A cooling water valve opening of 27.3% is required to give a reactor temperature of 150°F.
- The cooling water return temperature is 113.4°F, which is below its set point (and upper limit) of 130°F.
- Input NODEC to the cooling water return temperature controller is true, which prevents windup in the cooling water return temperature controller.
- Unlike integral tracking and external reset, it is not possible to compute a value for the output of the cooling water return temperature controller. In Figure 4.11 the cooling water return temperature controller output is 15.5%. Often, this will be the value of the controller output on the most recent occasion that input NODEC changed from false to true. However, this is not assured, as the controller is permitted to increase its output when input NODEC is true.

In Figure 4.11, the switch from reactor temperature control to cooling water temperature control occurs initially at $t \cong 110$ min. Prior to the switch, the reactor temperature controller is "in control" and the cooling water return temperature

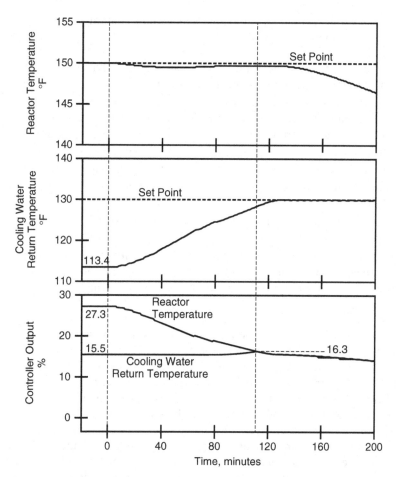

Figure 4.11 Switching from reactor temperature control to cooling water return temperature control, inhibit increase/inhibit decrease for windup protection.

controller is not permitted to decrease its output (input NODEC is true). Subsequent to the switch, the expectation is that the cooling water return temperature controller is "in control" and the reactor temperature controller is not permitted to decrease its output (input NODEC is true).

But subsequent to the switch, the trends in Figure 4.11 indicate that both controllers reduce their output. At the time of the switch, the output of each controller is 16.3%. But if input NODEC is continuously true for the reactor temperature controller, the controller cannot reduce its output. For the behavior in Figure 4.11 to occur, input NODEC to the reactor temperature controller must be false on some executions of the PID block.

To obtain the behavior in Figure 4.11, the controllers alternately decrease their outputs. Starting with the output of the reactor temperature controller less than the output of the cooling water return temperature controller, the sequence of events is as follows:

- Input NODEC is true for the reactor temperature controller but is false for the cooling water return temperature controller.
- The decrease in the rate of heat generation in the reactor causes the cooling water return temperature to rise slightly above its set point, causing the cooling water return temperature controller to decrease its output.
- This continues until the output of the cooling water return temperature controller is less than the output of the reactor temperature controller.
- Input NODEC is false for the reactor temperature controller but is true for the cooling water return temperature controller.
- The reactor temperature is below its set point, so the controller quickly reduces its output.
- In most cases, only one execution is required for the output of the reactor temperature controller to be less than the output of the cooling water temperature controller. This is the starting point, so the sequence of events repeats.

The usual situation is that the input NODEC is true for several executions of the cooling water return temperature controller, and is then true for one execution of the reactor temperature controller. Consequently, the switch from one controller to the other is not "clean." In essence, "chatter" accompanies the switch. Such chatter is often associated with noise. But when inhibit increase/inhibit decrease is providing windup protection, the chatter will occur in a noise-free environment.

In conventional controls chatter associated with a switch is undesirable. Chatter in a hardware component usually leads to wear or other degradation of the hardware. However, the chatter in the control configuration in Figure 4.4 is entirely in software. As for its effect on performance, compare the trends in Figure 4.11 (inhibit increase/inhibit decrease) to the trends in Figure 4.7 (integral tracking) and in Figure 4.9 (external reset). The difference in performance is insignificant.

Switch from Cooling Water Return Temperature Control to Reactor Temperature Control. The responses in Figure 4.12 are to a slow increase in the rate of heat generation in the reactor beginning at $t = 0$. Prior to the switch at $t \cong 110$ min, the trends in Figure 4.12 (inhibit increase/inhibit decrease) are identical to those in Figure 4.6 (no windup protection), the exception being the output of the reactor temperature controller. The performance of inhibit increase/inhibit decrease is comparable to that of integral tracking (Figure 4.8) and external reset (Figure 4.10).

The starting conditions are as follows:

- The cooling water return temperature is $130°F$ (the set point).

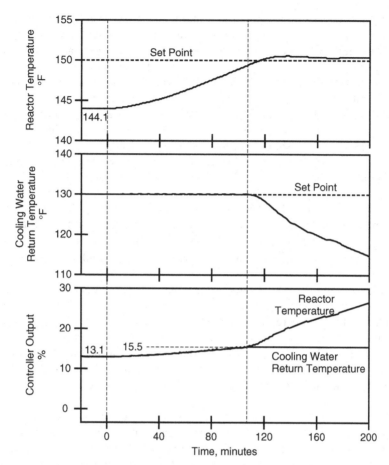

Figure 4.12 Switching from cooling water return temperature control to reactor temperature control, inhibit increase/inhibit decrease for windup protection.

- A cooling water valve opening of 13.1% is required to give a cooling water return temperature of 130.0°F.
- The reactor temperature is 144.1°F, which is below its set point of 150°F.
- Input NODEC to the reactor temperature controller is true, which prevents windup in the reactor temperature controller.
- Unlike integral tracking and external reset, it is not possible to compute a value for the output of the reactor temperature controller. In Figure 4.11 the reactor temperature controller output is 13.1%.

In Figure 4.12, the switch from reactor temperature control to cooling water temperature control occurs at $t \cong 110$ min. Following the switch, the reactor temperature controller increases its output, which is logical since the rate of heat generation is increasing. The cooling water temperature drops below its set point of 130°F, so the cooling water temperature controller would normally decrease its output. But since input NODEC to the cooling water temperature controller is true, these decreases are not allowed and the controller output remains at 15.5%.

Prior to the switch, the outputs of both controllers are slowly increasing. The rate of heat generation is increasing, so the cooling water flow must increase in order to maintain the cooling water return temperature at its set point of 130°F. But with the reactor temperature several degrees below its set point, the reactor temperature controller should not be increasing its output.

Prior to the switch, the control modes in the reactor temperature controller are performing as follows:

Proportional mode. The reactor temperature is increasing, so the proportional mode is increasing the controller output (controller is direct acting).

Integral mode. The control error is negative, so the integral mode is decreasing the controller output bias and consequently the controller output.

In the trend in Figure 4.12, the starting value of the control error is -5.9°F. For this control error, the integral mode dominates, causing the result of the control calculations to be a decrease in the controller output. But since input NODEC is true, this decrease is not permitted, and the controller output remains fixed at its initial value of 13.1%. This should continue for most of the time between $t = 0$ and at $t \cong 110$ min. Only when the reactor temperature approaches its set point should an increase in the controller output be observed (increases are permitted regardless of the state of input NODEC).

The behavior expected is that the reactor temperature controller output should remain constant at 13.1% until the reactor temperature approaches its set point, and then should increase slightly. That the reactor temperature controller output is slowly increasing is a consequence of the finite resolution of 0.1°F in the temperature measurements. If both temperature measurements have infinite resolution, the output of the reactor temperature controller behaves in the manner expected.

The trends in Figure 4.13 provide higher resolution over the interval 54 to 72 min following the start of the slow increase in the rate of heat generation. The 0.1°F resolution in the temperature measurements is clearly visible in these trends. The trend in Figure 4.13 for the cooling water return temperature controller output will be examined first. The cooling water return temperature switches between 130.0 and 130.1°F. As the controller gain is 0.5%/°F, the following abrupt changes can be seen in the cooling water return temperature controller output:

• When the cooling water return temperature changes from 130.0°F to 130.1°F, the controller output increases by 0.05%.

• When the cooling water return temperature changes from 130.1°F to 130.0°F, the controller output decreases by 0.05%.

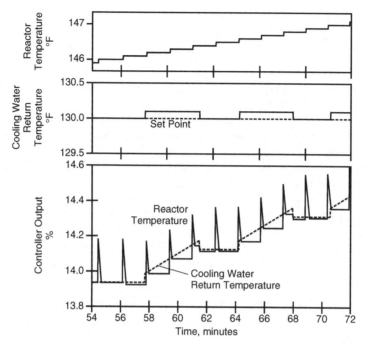

Figure 4.13 Impact of 0.1°F resolution in temperature measurements.

The effect of the reset mode on the controller output depends on the cooling water return temperature:

- When the cooling water return temperature is 130.0°F, the control error is zero and the controller output does not change.
- When the cooling water return temperature is 130.1°F, the control error is 0.1°F and the cooling water return temperature controller output increases in a ramp fashion. The slope of the ramp in controller output is the same as the slope of the ramp in the controller output bias:

$$\frac{dM_R}{dt} = \frac{K_{C,EU}E}{T_I} = \frac{(0.5\%/°F)(0.1°F)}{1.0 \text{ min}} = 0.05\%/\text{min}$$

Next, the trend for the reactor temperature controller output will be examined. The reactor temperature changes in 0.1°F increments. As the controller gain is 2.67%/°F, each 0.1°F increase in reactor temperature causes an increase of 0.267% in the controller output. These are clearly visible in the trend in Figure 4.13.

Each increase in the reactor temperature controller output is followed by a rapid decrease in the controller output. This decrease is the result of the reset mode. For a reactor temperature of 146.0°F, the control error is −4.0°F and the rate of change is as follows:

$$\frac{dM_R}{dt} = \frac{K_{C,EU}E}{T_I} = \frac{(2.67\%/^\circ F)(-4.0^\circ F)}{10.0 \text{ min}} = -1.07\%/\text{min}$$

When the reactor temperature is $147.0^\circ F$, the control error is $-3.0^\circ F$ and the rate of change is $-0.80\%/\text{min}$.

The increase of 0.267% in the controller output occurs on each $0.1^\circ F$ increase in the reactor temperature. Only input NODEC is configured; increases in the controller output occur regardless of the state of input NODEC. Following each increase of 0.267% in the controller output, the reactor temperature controller output is greater than the cooling water return temperature controller output. Consequently, input NODEC is false, which permits the reset mode to decrease the controller output. However, once the reactor temperature controller output is less than the cooling water return temperature controller output, input NODEC is true and further decreases are inhibited. Consequently, the nature of the trend in the reactor temperature controller output on each $0.1^\circ F$ increase in reactor temperature is as follows:

- The controller output increases by 0.267%, due to the proportional action.
- The controller output then decreases in a ramp fashion, due to the reset action.
- Once the controller output becomes less than the cooling water temperature controller output, input NODEC is true and no further decreases are allowed.

The long-term result is that the reactor temperature controller output increases along with the cooling water return temperature controller output. This is evident in both the high resolution trend in Figure 4.13 and the low resolution trend in Figure 4.12. The $0.1^\circ F$ resolution in the temperature measurements has little effect on the behavior of integral tracking and external reset. But in the case of inhibit increase/inhibit decrease, it can cause the behavior of the output of a controller to be different than suggested by an analysis of the control equations. However, such aberrant behavior rarely has a significant impact on the performance. Basically, the trends in both Figures 4.11 and 4.12 for inhibit increase/inhibit decrease compare very favorably to the corresponding trends for integral tracking (Figures 4.7 and 4.8) and external reset (Figures 4.9 and 4.10).

4.6. LIMITS ON HEAT TRANSFER

Limits on process operations can be classified as follows:

Hard limits. A common example is the control valve fully closed or fully open. It is not possible to violate such constraints. However, the control configuration must cope with consequences such as windup should such a constraint be encountered.

Soft limits. An example is the upper limit on the cooling water return temperature. Violating this limit is undesirable, but it is possible. The controls must contain logic to prevent this limit from being violated, along with logic to cope with consequences such as windup.

Fuzzy limits. These generally arise on transitions from one limiting mechanism to another. The transition is rarely sharp, but occurs over a range. This can occur in oversized control valves. The hard limit is valve fully open. But if a control valve is oversized by a factor of 4, valve openings beyond 50% have little effect on the flow through the valve. Beyond half-open, the valve it taking such a small percentage of the total pressure drop that it has little influence on the flow. The hard limit is still valve fully open; however, consequences such as reset windup begin at the fuzzy limit, that is, valve half-open.

Such a fuzzy limit arises in a once-through jacket. Jackets share one aspect in common with an oversized valve—routine practice is to oversize the ability to pump cooling water through the jacket.

Heat Transfer Mechanisms. For the reactor with a once-through jacket illustrated in Figure 4.1, the heat removal is basically a two-step process:

1. Heat is first transferred from the vessel contents to the cooling water. This is described by the heat transfer equation:

$$Q = U A \Delta T_{\mathrm{LM}}$$

where

Q = heat transfer rate (Btu/min)
U = heat transfer coefficient (Btu/hr-ft^2-$^\circ$F)
A = heat transfer area (ft^2)
ΔT_{LM} = logarithmic mean temperature difference ($^\circ$F)
     = $(\Delta T_{\mathrm{in}} - \Delta T_{\mathrm{out}})/ \ln (\Delta T_{\mathrm{in}}/\Delta T_{\mathrm{out}})$
$\Delta T_{\mathrm{in}} = T - T_{\mathrm{CWS}}$ = temperature difference at jacket inlet ($^\circ$F)
$\Delta T_{\mathrm{out}} = T - T_{\mathrm{CWR}}$ = temperature difference at jacket outlet ($^\circ$F)
T = reactor temperature ($^\circ$F)
T_{CWS} = cooling water supply temperature ($^\circ$F)
T_{CWR} = cooling water return temperature ($^\circ$F)

2. Heat is removed from the jacket by the cooling water in the form of sensible heat. This is described by the sensible heat equation for the cooling water:

$$Q = W c_P (T_{\mathrm{CWR}} - T_{\mathrm{CWS}})$$

where

Q = rate of heat removal by cooling water (Btu/min)
W = cooling water flow (lb/min)
c_P = cooling water heat capacity (Btu/lb-$^\circ$F)

At steady-state conditions, these two must be equal. That is, the heat transferred from the vessel contents to the jacket must equal the heat being removed from the jacket by the cooling water in the form of sensible heat.

Maximum Heat Transfer Rate. For a once-through jacket, there is a maximum or limiting heat transfer rate. The cooling water return temperature is a function of cooling water flow. As the cooling water flow increases, the cooling water return temperature approaches the cooling water supply temperature. At very high water flows, the temperature rise from jacket inlet to jacket outlet is very small, and the exit temperature is essentially equal to the inlet temperature. Thus, the maximum driving force for heat transfer is the difference between the vessel temperature and the cooling water supply temperature. The maximum heat transfer rate Q_{max} is

$$Q_{max} = UA(T - T_{CWS})$$

Heat Transfer as a Function of Cooling Water Flow. Figure 4.14 presents the cooling water return temperature and the heat transfer rate as functions of cooling water flow. As the cooling water flow rate increases from zero, the heat transfer rate initially increases with water flow. But for large cooling water flows, the cooling water return temperature approaches the cooling water supply temperature. The heat transfer rate approaches the maximum possible heat transfer rate, and the sensitivity approaches zero. For very large water flows, the cooling water flow has little effect on the heat transfer rate. Those with little understanding of heat transfer sometimes believe that if the cooling water flow is half the maximum, only half of the available heat transfer is being used. This is definitely not the case!

Heat Transfer Limited. At high water flows, heat transfer from vessel contents to jacket is the limiting mechanism for heat transfer. The cooling water flow affects the heat transfer by altering the temperature difference for heat transfer. But at high water flows, the temperature rise from jacket inlet to jacket outlet is very small. If the cooling water return temperature is already close to the cooling water supply temperature, further increases in the cooling water flow will reduce the temperature rise from jacket inlet to jacket outlet but will have little effect on the temperature difference for heat transfer.

A numerical example for a heat transfer process with the following characteristics will illustrate this effect:

$$A = 100.0 \text{ ft}^2$$

$$c_P = 1.0 \text{ Btu/lb-}^\circ\text{F}$$

$$T_{CWS} = 62^\circ\text{F}$$

$$T = 150^\circ\text{F}$$

$$U = 60.0 \text{ Btu/hr-ft}^2\text{-}^\circ\text{F}$$

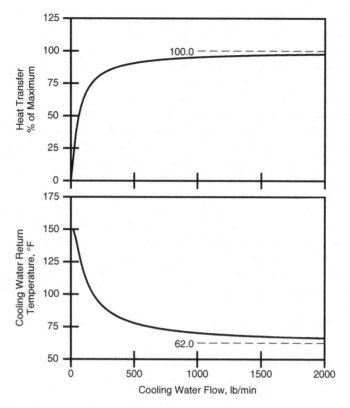

Figure 4.14   Effect of cooling water flow on heat transfer rate for a once-through jacket.

For these parameters, the maximum possible heat transfer is

$$Q_{\max} = UA(T - T_{\text{CWS}}) = 8800 \text{ Btu/min}$$

The following numerical example computes the heat transfer rates for two different water flow rates:

Cooling Water Flow:	500 lb/min	1000 lb/min
Vessel temperature	150.0°F	150.0°F
Cooling water supply temperature	62.0°F	62.0°F
Cooling water return temperature	78.0°F	70.4°F
Temperature rise	16.0°F	8.4°F
ΔT_{LM}	79.7°F	83.7°F
Heat transfer rate Q	7980 Btu/min	8370 Btu/min
$Q/Q_{\max}$	0.907	0.951

Doubling the cooling water flow only increases the heat transfer rate about 5%.

Effect on Process Gain. At high water flows, the heat transfer is the controlling mechanism for removing heat from the vessel. With heat transfer in control, the cooling water flow has little effect on the heat transfer rate, and consequently little effect on vessel temperature. That is, at high cooling water flow rates, the sensitivity of vessel temperature to cooling water flow approaches zero. This spells trouble for the temperature controller.

This problem is commonly encountered in practice. Cooling water is an inexpensive commodity. Designers see no need to conserve it. So when it comes to sizing piping, pumps, and other parts of the jacket, oversizing is common. But realize that the oversizing is only in regard to the ability to pump cooling water through the jacket. Increasing the heat transfer area would be productive, but increasing the ability to flow cooling water through the jacket is not.

The ability to pump excess water through the jacket has undesirable consequences on the vessel temperature control loop. At high cooling water flows, the sensitivity of vessel temperature to cooling water flow approaches zero. This means that the cooling water valve opening has little effect on the vessel temperature. Controllers cannot function if the process gain is zero. The process gain never goes completely to zero, but it gets so close to zero that the controller cannot accomplish its task.

What is the consequence on the control loop? The condition for reset windup was stated in Chapter 1 as follows:

Reset windup occurs in a controller when changes in the controller output have no effect on the process variable.

Under heat transfer limited conditions, the output of the controller has such a small effect on the reactor temperature that this statement is effectively true. If the vessel temperature is above its set point, the controller responds by opening the cooling water valve. The flow increases, but this has little effect on the vessel temperature. So the controller continues to open the valve. The result is reset windup.

The Onset of Heat Transfer Limited Conditions. At the onset of heat transfer limited conditions, there is a significant decrease in the slope of the graph in Figure 4.14 for the heat transfer as a function of cooling water flow. As the slope changes gradually, the "line in the sand" marking the onset of heat transfer limited conditions is somewhat fuzzy. For water flows less than about 250 lb/min, the process is not heat transfer limited. For water flows greater than 500 lb/min, the process is definitely heat transfer limited.

How can one detect the onset of heat transfer limited conditions? Consider using the following two temperature differences:

- $T_{CWR} - T_{CWS}$: The temperature rise from jacket inlet to jacket outlet.
- $T - T_{CWS}$: The maximum temperature difference for heat transfer, which is the reactor temperature less the cooling water supply temperature.

The onset of heat transfer limited conditions can be determined by examining the following ratio:

$$\frac{\text{temperature rise of the cooling water}}{\text{maximum } \Delta T \text{ for heat transfer}} = \frac{T_{CWR} - T_{CWS}}{T - T_{CWS}}$$

For the heat transfer process for the preceding numerical example, the effect of cooling water flow on the cooling water return temperature, the heat transfer rate, and the ratio is as follows:

W (lb/min)	T_{CWR} (°F)	$T_{CWR} - T_{CWS}$ (°F)	$T - T_{CWS}$ (°F)	Q (Btu/min)	$\dfrac{Q}{Q_{max}}$	$\dfrac{T_{CWR} - T_{CWS}}{T - T_{CWS}}$
125	110.5	48.5	88.0	6057	0.688	0.551
250	91.0	29.0	88.0	7253	0.842	0.330
500	78.0	16.0	88.0	7976	0.906	0.181
1000	70.4	8.4	88.0	8374	0.952	0.095
2000	66.3	4.3	88.0	8584	0.975	0.049
∞	62.0	0	88.0	8800	1.000	0.000

This suggests the following criteria for the onset of heat transfer limited:

$$\frac{T_{CWR} - T_{CWS}}{T - T_{CWS}} < 0.2$$

One could argue that the coefficient should be 0.25. The demarcation between sensible heat limited and heat transfer limited is fuzzy, so slightly different values or the onset of heat transfer limited can be expected.

Controlling Cooling Water Temperature Rise. The control configuration in Figure 4.15 controls the temperature rise of the cooling water by manipulating the cooling water valve opening. If the temperature rise is increasing, the controller must increase its output. In Figure 4.15 the cooling water temperature rise is computed by subtracting the cooling water supply temperature from the cooling water return temperature. This involves subtracting two large numbers to obtain a small number. Numerically, this is not a good practice, as errors in either of the large numbers are amplified in the small number.

A preferred approach is to measure the temperature difference directly. Many modern temperature transmitters can sense the difference in temperature between two RTDs or two thermocouples. If values for both the cooling water supply temperature and the cooling water return temperature are required, either:

- Measure the cooling water supply temperature; measure the cooling water temperature rise; compute the cooling water return temperature.

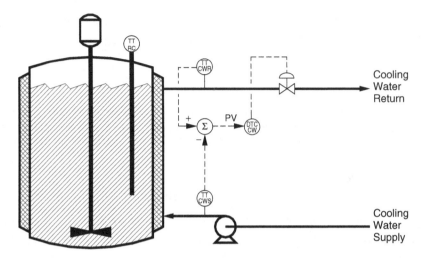

Figure 4.15 Controlling cooling water temperature rise for a reactor with once-through jacket.

- Measure the cooling water return temperature; measure the cooling water temperature rise; compute the cooling water supply temperature.

Given the fuzzy line for the onset of heat transfer limited conditions, the cooling water temperature rise set point can usually be computed from normal operating conditions and used under other conditions. Only when large changes occur in the reactor temperature must a computation be provided for the cooling water temperature rise set point.

Reactor Temperature Control with Cooling Water Temperature Rise Override. Figure 4.16 presents an override configuration that consists of the following components:

Reactor temperature controller. The corresponding simple feedback configuration is presented in Figure 4.1.

Cooling water rise temperature controller. The corresponding simple feedback configuration is presented in Figure 4.15.

Selector. A low select "auctioneers" the output of the two controllers.

As the primary requirement is to control the reactor temperature, this configuration is referred to as reactor temperature control with a cooling water temperature rise override. This configuration requires windup protection in the same manner as required for the configuration in Figure 4.4 for reactor temperature control with cooling water return temperature override. The options for windup protection are integral tracking, external reset, and inhibit increase/inhibit decrease.

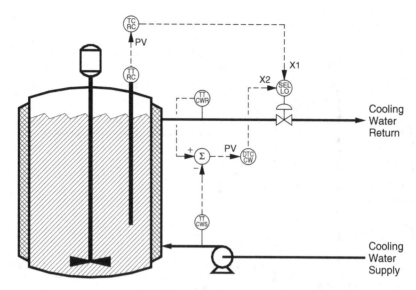

Figure 4.16   Reactor temperature control with cooling water temperature rise override.

Reactor Temperature Control with Cooling Water Return Temperature Override and Cooling Water Temperature Rise Override.

A control configuration can provide any number of overrides. The reactor temperature control configuration in Figure 4.17 provides two overrides:

> *Override on cooling water return temperature.* The configuration in Figure 4.4 provides only this override.
>
> *Override on cooling water temperature rise.* The configuration in Figure 4.16 provides only this override.

The override on cooling water return temperature requires a high select; the override on cooling water temperature rise requires a low select. In the configuration in Figure 4.17, the high select is followed by the low select. However, the order can be reversed if desired. Where both overrides require the same type of select (that is, both require a high select or both require a low select), a single selector can normally be used, provided that the selector block provides for three or more inputs (as most do). For each of the three controllers in Figure 4.17, windup occurs if the output of that controller is not driving the control valve on the cooling water.

Integral Tracking. For each controller, integral tracking must be active whenever the controller is not driving the control valve. For all three controllers, input MRI is the output from the last selector. Input TRKMR is configured as follows:

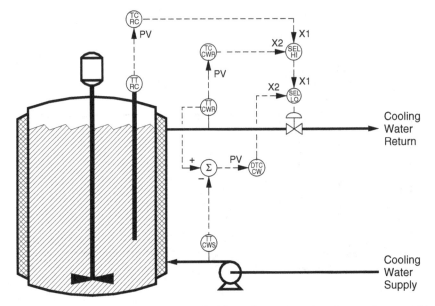

Figure 4.17 Reactor temperature control with cooling water return temperature override and cooling water temperature rise override.

Reactor temperature controller. The output of this controller is positioning the valve provided that:

- Output Q1 of the high select is true.
- Output Q1 of the low select is true.

Input TRKMR is the inverse (logical NOT) of the logical AND of these two outputs. The logic for integral tracking for the reactor temperature controller is expressed as follows:

```
TCRC.TRKMR = !(SELHI.Q1 & SELLO.Q1)
TCRC.MRI = SELLO.Y
```

Cooling water return temperature controller. The output of this controller is positioning the valve provided that:

- Output Q2 of the high select is true.
- Output Q1 of the low select is true.

Input TRKMR is the inverse (logical NOT) of the logical AND of these two outputs. The logic for integral tracking for the cooling water return temperature controller is expressed as follows:

```
TCCWR.TRKMR = !(SELHI.Q2 & SELLO.Q1)
TCCWR.MRI = SELLO.Y
```

Cooling water temperature rise controller. The output of this controller is positioning the valve provided output Q2 of the low select is true. Input TRKMR is the inverse (logical NOT) of output Q2 of the low select. The logic for integral tracking for the cooling water temperature rise controller is expressed as follows:

```
DTCCW.TRKMR = !SELLO.Q2
DTCCW.MRI = SELLO.Y
```

External Reset. For this example (and frequently for others), the external reset approach to windup protection is very easy to configure. For each controller, the XRS input must be configured as follows:

Input XRS (value for external reset): Output of the last selector (the order of the selectors is immaterial).

The logic for external reset is as follows:

Reactor temperature controller.

```
TCRC.XRS = SELLO.Y
```

Cooling water return temperature controller.

```
TCCWR.XRS = SELLO.Y
```

Cooling water temperature rise controller.

```
DTCCW.XRS = SELLO.Y
```

Inhibit Increase/Inhibit Decrease. The logic for windup protection using inhibit increase/inhibit decrease is developed by examining each controller's behavior when it is "not in control." For controller X, the procedure is as follows:

- Assume that one of the other controllers is "in control."
- Determine the direction that controller X would change its output, if permitted.
- If controller X attempts to increase its output, input NOINC to controller X must be true. If controller X attempts to decrease its output, input NODEC to controller X must be true.

This must be repeated for each of the other controllers.

Before developing the logic, the conditions for each controller to be "in control" are as follows:

Reactor temperature controller. The reactor temperature controller is "in control" when both of the following are true:

- Output Q1 from the high select is true (input X1 from the reactor temperature controller is selected).
- Output Q1 from the low select is true (input X1 from the high select is selected).

Cooling water return temperature controller. The cooling water return temperature controller is "in control" when both of the following are true:

- Output Q2 from the high select is true (input X2 from the cooling water return temperature controller is selected).
- Output Q1 from the low select is true (input X1 from the high select is selected).

Cooling water temperature rise controller. The cooling water temperature rise controller is "in control" when the following is true:

- Output Q2 from the low select is true (input X2 from the cooling water temperature rise controller is selected).

The reactor temperature controller is "not in control" if either the cooling water return temperature controller is "in control" or the cooling water temperature rise controller is "in control":

Cooling water return temperature controller. When this controller is "in control," the behavior of the reactor temperature controller is as follows:

- The reactor temperature will be below its set point.
- If permitted, the reactor temperature controller would decrease its output.

Input NODEC to the reactor temperature controller must be true when the cooling water return temperature controller is "in control."

Cooling water temperature rise controller. When this controller is "in control," the behavior of the reactor temperature controller is as follows:

- The reactor temperature will be above its set point.
- If permitted, the reactor temperature controller would increase its output.

Input NOINC to the reactor temperature controller must be true when the cooling water temperature rise controller is "in control."

The logic for inhibit increase/inhibit decrease for the reactor temperature controller is as follows:

```
TCRC.NOINC = SELLO.Q2
TCRC.NODEC = SELHI.Q2 & SELLO.Q1
```

The cooling water return temperature controller is "not in control" if either the reactor temperature controller is "in control" or the cooling water temperature rise controller is "in control":

Reactor temperature controller. When this controller is "in control," the behavior of the cooling water return temperature controller is as follows:

- The cooling water return temperature will be below its set point.
- If permitted, the cooling water return temperature controller would decrease its output.

Input NODEC to the cooling water return temperature controller must be true when the reactor temperature controller is "in control."

Cooling water temperature rise controller. When this controller is "in control," the behavior of the cooling water return temperature controller is as follows:

- The cooling water return temperature will be below its set point.
- If permitted, the cooling water return temperature controller would decrease its output.

Input NODEC to the cooling water return temperature controller must be true when the cooling water temperature rise controller is "in control."

There are no conditions for which input NOINC must be true, so this input need not be configured. Input NODEC must be true when either the reactor temperature controller is "in control" or the cooling water temperature rise controller is "in control." The logic for input NODEC to the cooling water return temperature controller could be expressed as follows:

```
TCCWR.NODEC = (SELHI.Q1 & SELLO.Q1) | SELLO.Q2
```

An equivalent statement is that input NODEC must be true when the cooling water return temperature controller is "not in control." An alternative expression of the logic for inhibit increase/inhibit decrease for the cooling water return temperature controller is as follows:

```
TCCWR.NODEC = !(SELHI.Q2 & SELLO.Q1)
```

The cooling water temperature rise controller is "not in control" if either the reactor temperature controller is "in control" or the cooling water return temperature controller is "in control":

Reactor temperature controller. When this controller is "in control," the behavior of the cooling water temperature rise controller is as follows:

- The cooling water temperature rise will be above its set point.
- If permitted, the cooling water temperature rise controller would increase its output.

Input NOINC to the cooling water temperature rise controller must be true when the reactor temperature controller is "in control."

Cooling water return temperature controller. When this controller is "in control," the behavior of the cooling water temperature rise controller is as follows:

- The cooling water temperature rise will be above its set point.

- If permitted, the cooling water temperature rise controller would increase its output.

  Input NOINC to the cooling water temperature rise controller must be true when the cooling water return temperature controller is "in control."

There are no conditions for which input NODEC must be true, so this input need not be configured. Input NOINC must be true when either the reactor temperature controller is "in control" or the cooling water return temperature controller is "in control." The logic for input NOINC to the cooling water temperature rise controller could be expressed as follows:

```
DTCCW.NOINC = (SELHI.Q1 & SELLO.Q1) | (SELHI.Q2 & SELLO.Q1)
```

An equivalent statement is that input NOINC must be true when the cooling water temperature rise controller is "not in control." An alternative expression of the logic for input NOINC to the cooling water temperature rise controller is as follows:

```
TCCWR.NOINC = !SELLO.Q2
```

4.7. OTHER EXAMPLES

Override control configurations are relatively common in the process industries. This section provides a few additional examples. Only the P&I diagrams will be presented, but windup protection must be provided for all.

Reactions with One Reactant Being a Gas. For some liquid- or solid-phase reactions, one of the reactants is a gas. A pressurized vessel is used as a batch reactor, with this reactant being totally charged at the beginning of the batch. Assume that there is no liquid phase for this reactant. The pressure–temperature relationship is expressed by the equations of state, not by vapor–liquid equilibrium relationships. For a given temperature, the reactor pressure will be the highest when the reactor is initially charged. Being a pressurized vessel, a relief device must be provided for equipment protection, which means that the controls must not heat the reactor to a temperature that would cause the pressure to exceed the pressure setting on the relief device.

This could lead to the following strategy for the batch reaction:

- Begin by adjusting the heat input to the reactor to control the reactor pressure. The reactor pressure set point must be safely below the pressure setting on the relief device.
- As the reaction proceeds, gas is consumed. To maintain constant pressure, the reactor temperature must increase.

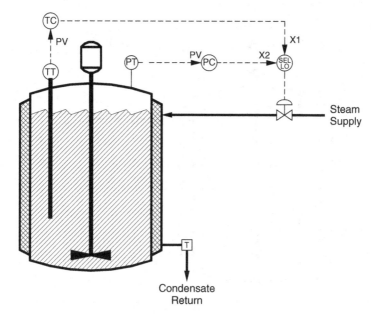

Figure 4.18 Reactor temperature control with pressure override.

- Once the reactor temperature attains the desired value for the reaction, change to controlling the reactor temperature by adjusting the heat input to the reactor.
- As the reaction continues to progress, the reactor pressure will drop.

The control configuration in Figure 4.18 is reactor temperature control with a reactor pressure override. This configuration requires three components:

1. Reactor temperature controller
2. Reactor pressure controller
3. Low select

A cascade control configuration is a possible alternative to an override configuration. Figure 4.19 presents a temperature-to-pressure cascade control configuration for the reactor. The limit on the reactor pressure would be imposed by specifying an upper limit for the pressure set point.

Although this approach would effectively limit the reactor pressure, there is a problem with the cascade configuration. For acceptable performance, the pressure loop (the inner loop) must be faster than the temperature loop (the outer loop), preferably by a factor of 5. This will not be the case. Changes in the heat input to the reactor will affect the temperature and the pressure with approximately the same dynamics (the temperature and pressure are related by the equation of state).

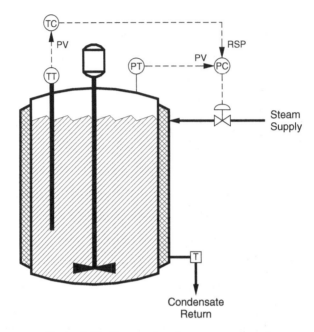

Figure 4.19 Temperature-to-pressure cascade.

Column Flooding. For distillation columns such as those illustrated in Figure 4.20, the driving force for liquid to flow down the tower is the gravity head of the liquid. This is opposed by a pressure drop due to the vapor flowing up the tower. If the pressure drop from the vapor flow ever exceeds the gravity head, liquid cannot flow down the tower and the tower fills with liquid, resulting in a phenomenon known as *flooding*. The onset of flooding can be detected from differential pressure measurements. Figure 4.20 illustrates a differential pressure measurement over the lower section only. The measurement should be across the section where flooding is most likely, although sometimes the differential pressure across the entire column is sensed.

The heat input to the reboiler determines the vapor flow up the tower. The heat input can be used to control either:

- Bottoms composition, as in Figure 4.20. The temperature of a stage in the lower tower section is often used in lieu of the composition measurement.
- Bottoms level.

In either case, the controller could potentially increase the heat input and consequently the vapor flow sufficiently to exceed the pressure drop for the onset of flooding.

Figure 4.20 provides a steam flow measurement and a steam flow controller, which permits the heat input to be changed via the steam flow set point. When

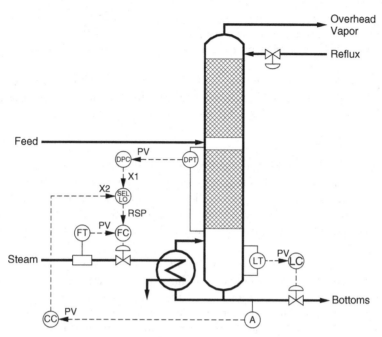

Figure 4.20 Composition-to-flow cascade with differential pressure override on the steam flow set point.

the heat input is used to control the composition or stage temperature, the flow controller is normally recommended. However, the flow controller is optional, and the heat input can be varied by changing the steam valve position.

The configuration in Figure 4.20 provides two controllers:

Bottoms composition controller. Increasing the heat input drives more of the low-boiling components from the bottoms of the tower. This increases the stage temperatures in the lower packed section.

Lower section differential pressure controller. Increasing the heat input increases the vapor flow throughout the tower, which in turn increases the differential pressure across each packed section.

The output of the low select is the smaller of the outputs of these two controllers. This output is the set point for the steam flow controller. The result will always be a cascade configuration:

- When the input from the composition controller is selected, the resulting configuration is a composition-to-flow cascade.
- When the input from the differential pressure controller is selected, the resulting configuration is a differential pressure-to-flow cascade.

Both configurations are viable cascade configurations in that the dynamics of the flow loop are faster than the dynamics of either the composition loop or the differential pressure loop. Consequently, the configuration in Figure 4.20 is a viable configuration.

The cascade configuration, that is, the flow controller as an inner loop, must be justified by one of the control loops, which for this example is the composition loop. With regard to the other loop, there are two possibilities:

1. *The cascade configuration will function properly for the other loop.* For this example, the issue is the viability of the differential pressure-to-flow cascade. Although the differential pressure responds fairly rapidly to changes in the heat input, the flow controller is much faster, making the differential pressure-to-flow cascade a viable cascade. However, using a flow controller as an inner loop of the differential pressure-to-flow cascade offers few benefits.

2. *The cascade configuration will not function properly for the other loop.* Normally, the problem is an inadequate separation of dynamics. To eliminate the inner loop for the differential pressure controller, Figure 4.21 inserts the selector between the flow controller and the steam control valve. This configuration eliminates the flow controller as the inner loop for the differential pressure controller.

In this example, the flow controller as an inner loop for the differential pressure controller is unlikely to contribute much to performance. But since it does no harm, most would prefer to retain the inner loop and use the configuration in Figure 4.20.

As noted previously, cascade is a possible alternative to an override configuration. Figure 4.22 presents a composition-to-differential pressure-to-flow cascade (if desired, the steam flow controller could be omitted to give a composition-to-differential pressure cascade). To avoid flooding, an upper limit can be imposed on the differential pressure set point. However, this requires that the cascade configuration be viable.

There is no problem with respect to the dynamics. The flow responds more rapidly than the differential pressure, which in turn responds more rapidly than the composition (or stage temperature). However, there is a problem with respect to the differential pressure, either with or without a steam flow controller. The differential pressure varies approximately with the square of the vapor flow. At the high vapor flows encountered near the flooding limit, the differential pressure is very sensitive to changes in the heat input. In the override configuration in either Figure 4.20 or Figure 4.21, the differential pressure loop would only be used at vapor flows near the flooding limit. But at lower vapor flows, the sensitivity of the differential pressure to the vapor flow is much lower. Consequently, the performance of the differential pressure loop degrades as the vapor flow decreases. The problem is similar to the problem with head-type flow meters (such as the orifice

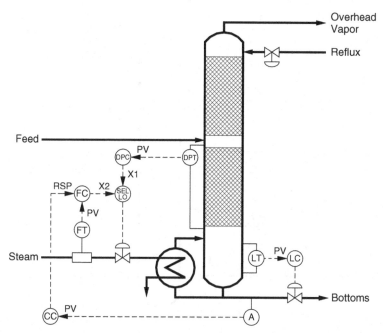

Figure 4.21 Composition-to-flow cascade with differential pressure override on the steam valve position.

meter) when the flow rate is low. Consequently, the configuration in Figure 4.22, either with or without the flow controller, is not recommended.

Steam-Heated Exchanger. For the steam-heated exchanger in Figure 4.23, the liquid outlet temperature is controlled via a control valve on the condensate. A major advantage of this configuration is that the full steam supply pressure is available as the driving force for condensate return. For this approach, the heat transfer is varied through the area for heat transfer. The effective area for heat transfer is the area exposed to condensing steam. The submerged heat transfer area can only subcool the condensate, with little contribution to the total heat transfer. This leads to the following limits on heat transfer:

Minimum limit. This occurs when the exchanger is completely filled with condensate, and is essentially zero.
Maximum limit. This occurs when the exchanger is completely drained of condensate, exposing the total heat transfer area to the condensing steam.

Given the tradition in this industry of oversizing control valves, the condensate will fully drain from the exchanger at a valve opening less than fully open. Further opening the control valve leads to the undesirable consequence known as *blowing*

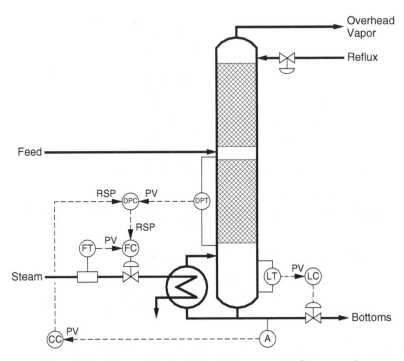

Figure 4.22 Composition-to-differential-pressure-to-flow cascade.

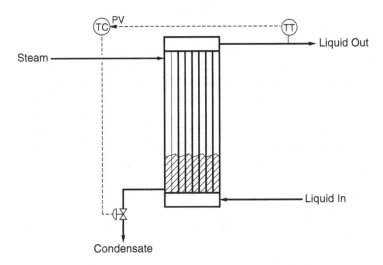

Figure 4.23 Liquid outlet temperature control for a steam-heated exchanger.

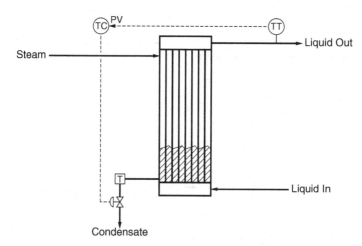

Figure 4.24 Liquid outlet temperature control of a steam-heated exchanger with steam trap.

steam. The condensate level drops completely out of the exchanger, permitting steam to flow into the condensate return system. This is a major drawback of the configuration in Figure 4.23 and must be addressed.

A simple way to prevent the exchanger from blowing steam is to insert a steam trap upstream of the condensate control valve, as illustrated in Figure 4.24. However, there is a side effect. Once the control valve has opened sufficiently for the steam trap to be blocking the steam flow, the exchanger is completely drained of condensate and the maximum heat transfer rate has been attained. Opening the steam valve further has no effect on the heat transfer, the consequence being reset windup in the liquid outlet temperature controller. Unfortunately, there is no indication that the steam trap is blocking the steam flow, which means that the controls have no basis for invoking windup protection. The only option is to provide a value for the upper output limit for the level controller that corresponds approximately to the valve opening at which the steam trap begins to block the steam flow. This will be imperfect at best.

Another alternative is the override configuration in Figure 4.25. The additional components are as follows:

Condensate level measurement. This is the only additional hardware component; all others are implemented in software.
Condensate level controller. The set point for the condensate level controller is the minimum condensate level to be permitted in the exchanger.
Low select. The output is the minimum of the output of the two controllers.

The resulting configuration is liquid outlet temperature control with condensate level override. If it makes one feel more comfortable, the steam trap can be

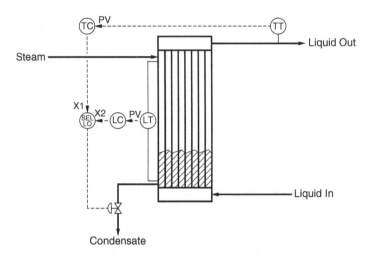

Figure 4.25   Liquid outlet temperature control with condensate level override.

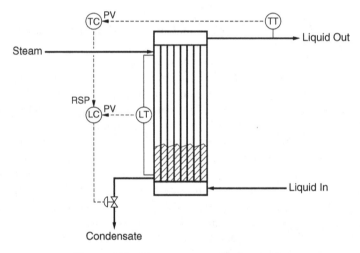

Figure 4.26   Temperature-to-level cascade.

retained. As long as the override configuration is in operation, the trap should be exposed only to condensate. The override configuration prevents the exchanger from blowing steam, but in a manner that avoids reset windup in the liquid outlet temperature controller.

The cascade control configuration in Figure 4.26 is a potential alternative to the override configuration. The liquid outlet temperature controller adjusts the set point to the condensate level controller. The minimum allowable condensate level is imposed by a low limit on the condensate level set point. Unfortunately,

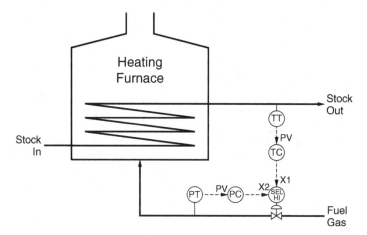

Figure 4.27 Liquid outlet temperature control with burner pressure override.

there is a problem with the dynamics. The level loop will be slower than the temperature loop. Exchanger liquid outlet temperatures respond very quickly. In most applications, the liquid outlet temperature loop in Figure 4.26 will be faster than the condensate level loop. The cascade configuration in Figure 4.26 will not provide adequate control of the liquid outlet temperature.

Fired Heater. All combustion processes must adhere to a minimum firing rate. The fired heater in Figure 4.27 is no exception. The burner designers will state the minimum firing rate, below which flame instabilities appear within the furnace. Where the fuel is a gas and the burners have fixed orifices, this can usually be translated into the minimum burner header pressure. The objective of the control system is to adjust the firing rate so as to attain the desired liquid outlet temperature. The firing rate is determined by a control valve on the fuel. As this control valve opens, the fuel flow and the burner header pressure increase. Closing the control valve has the opposite effect. However, the control valve must not be closed to a point where the firing rate is below the minimum.

With any combustion furnace, safety issues arise and must be addressed. These are not examined herein and are not included in the P&I diagrams. For the process controls, the primary objective is to maintain the liquid outlet temperature at or near its target. However, the process controls must do this without taking any action that would cause the safety system to react, usually in a manner that shuts down the furnace.

The override configuration in Figure 4.27 requires three components:

Liquid outlet temperature controller. The set point is the desired value for the liquid outlet temperature.

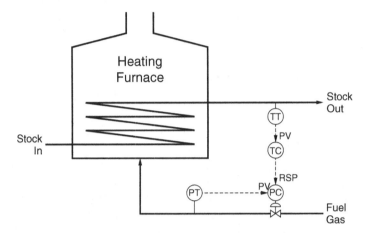

Figure 4.28   Temperature-to-pressure cascade.

Burner header pressure controller. The set point is a value slightly above the burner header pressure that corresponds to the minimum firing rate.

High select. The output is the maximum of the output of the two controllers.

This configuration provides liquid outlet temperature control with a burner header pressure override. Should the liquid outlet temperature controller reduce the firing rate to the minimum (or actually just above the minimum), the burner header pressure controller assumes control and maintains the firing rate just above the minimum.

Figure 4.28 provides a cascade control configuration that is an alternative to the override configuration. The liquid outlet temperature controller is the outer loop of the cascade and provides the set point for the burner header pressure controller. The minimum firing rate is imposed by a lower limit on the burner header pressure set point. The cascade configuration in Figure 4.28 is a viable cascade. Especially for heating furnaces, the liquid outlet temperature loop will be far slower than the burner header pressure loop.

In all previous examples, the override configuration was preferred over the cascade configuration. This is only because the subject of this chapter is override controls. Override configurations are not generally preferred over cascade configurations. Each approach must be evaluated on its own merits.

5

VALVE POSITION CONTROL

A valve position controller must not be confused with a valve positioner:

Valve positioner. A high-gain feedback loop that senses the actuator position and regulates the air to the actuator so as to drive the valve to the position corresponding to the control signal. A valve positioner is implemented within the final control element.

Valve position control. A PID block whose process variable is the position of a final control element. This is consistent with the customary nomenclature for controllers:

	Designation	Process Variable
Temperature controller	TC	Temperature
Pressure controller	PC	Pressure
Level controller	LC	Level
Flow controller	FC	Flow
Valve position controller	VPC	Valve position

In most valve position control applications the objective is process optimization through constraint control. Most often the optimum operating conditions are at constraints, and sometimes, one of the constraints is a control valve fully open or fully closed. But if a controller is permitted to drive its control valve fully open or fully closed, the controller ceases to function, which means that its process variable is not maintained at its set point. So that the controller will continue to function, the control valve must be driven to a target such as 10% open or 90% open.

Advanced Process Control: Beyond Single-Loop Control   By Cecil L. Smith
Copyright © 2010 John Wiley & Sons, Inc.

5.1. POLYMER PUMPING EXAMPLE

The usual justification for variable-speed pumping (a variable-speed drive coupled to a centrifugal pump) is energy savings—putting energy into a pump just to dissipate it across a control valve is not very productive. But occasionally, there are other justifications. For this example, the polymer being pumped is sensitive to shear; that is, excessive shear (as in a centrifugal pump) breaks some of the high-molecular-weight polymer molecules into lower-molecular-weight molecules. To minimize this, the pump should be operated at the minimum possible speed.

Figure 5.1 illustrates the polymer pumping system with a constant-speed pump. Polymer must be delivered to two destinations, referred to as destination A and destination B. For each destination, a flow controller is available. No assumptions can be made about the respective flow paths. Specifically:

- The two flow rates are not necessarily the same.
- The pressures at the two destinations are not necessarily the same.
- The resistance to fluid flow offered by the piping is not the same (distance is different, fittings are different, etc).

Valve Position Controller. The constant-speed drive on the pump will be replaced by a variable-speed drive equipped with a speed controller. A control configuration with the following objectives is required:

- Maintain the polymer flow to each destination at or near its set point. The flow controllers will achieve this objective, provided that neither control valve is driven fully open.
- Operate the pump at the minimum possible speed, subject to the requirement that neither control valve be fully open.

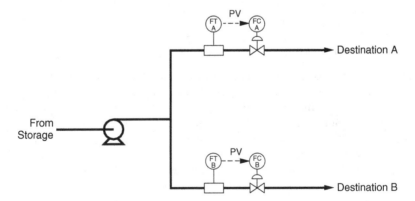

Figure 5.1 Polymer pumping system, constant-speed pump.

If one starts with the variable-speed drive operating at the same speed as the constant-speed drive and then gradually reduces the pump speed, what will be observed?

- Each polymer flow will drop slightly below its set point. But since the flow controllers respond rather rapidly, the drop will be small.
- Each flow controller increases the respective control valve opening so as to maintain the flow at or near its set point.

This can continue until one of the valves has been driven fully open. Thereafter, continuing to decrease the pump speed will cause the flow through that valve to drop below its set point.

Suppose that the flow to destination A is always the most demanding. The basis for the configuration in Figure 5.2 is the following:

- Fully open the valve to destination A.
- Maintain the flow to destination A by adjusting the pump speed.

But if the flow to destination B is ever the most demanding, the configuration in Figure 5.2 will not maintain the flow to destination B at its set point.

At any point in time, which flow is most demanding? It will be the flow through the most open control valve. This is the purpose of the high select in Figure 5.3. The output of the high select is the opening of the control valve for the most demanding flow. The minimum pump speed is the speed at which the control valve for the most demanding flow is fully open. But if a valve is fully open, the flow through that valve could be any flow less than the set point. If the flow is below the set point, the flow controller needs to open the control valve. But once the valve is fully open, no control action is available to increase the flow. To be certain that a PID controller is maintaining the process variable at

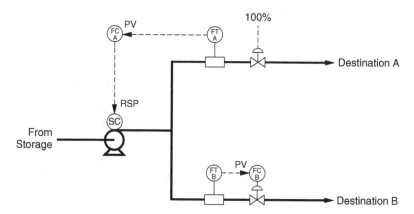

Figure 5.2 Adjusting pump speed to control one of the polymer flows.

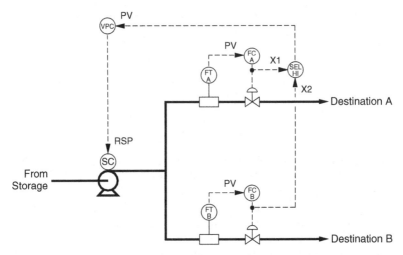

Figure 5.3 Valve position control configuration for polymer pumping system with a variable-speed pump.

the set point, the control valve must not be driven to either fully open or fully closed.

Consider adjusting the pump speed so that the most open control valve is 90% open. Since neither valve is driven fully open, each flow must be at or near its set point. A simple way to achieve this objective is to configure a PID block as follows:

- *Process variable:* opening of the most open control valve
- *Output:* set point for the speed controller of the variable-speed drive

The result is the valve position controller illustrated in Figure 5.3. If the opening of the most open valve is increasing, the controller must increase the pump speed (the controller is direct acting). If the opening of the most open valve is less than 90%, the pump speed can be reduced. If the opening of the most open valve is greater than 90%, the pump speed must be increased.

Relationship of Valve Position Controller to Flow Controllers. If both flow controllers in Figure 5.3 are on manual, the valve position controller cannot function. If the opening of the most open valve is less than the set point for the valve position controller, the valve position controller will decrease the pump speed. The expected result is that reducing the pump speed will force the flow controllers to open their valves. But with both flow controllers on manual, this does not happen. Changes in the output of the valve position controller have no effect on its measured variable, which is the condition for windup.

In a sense, a similar statement can be made for a cascade control configuration. If the inner loop is on manual, changes in its set point (the output of the outer loop) have no effect on the measured variable for either the inner loop or the outer loop. To avoid windup, output tracking must be active in the outer loop when the inner loop is not on remote.

The configuration in Figure 5.3 is not a cascade configuration as the term is normally used. That is, the valve position controller is not providing the set point to either flow controller. But if both flow controllers are on manual, the valve position controller cannot function properly. Although the control configuration is different, the valve position controller is dependent on the flow controllers in exactly the same manner as the outer loop of a cascade is dependent on the inner loop. Therefore, the statements made for cascade can be translated to corresponding statements for the valve position control configuration:

- The valve position controller cannot function if both flow controllers are in manual. Actually, it cannot function if the flow controller that outputs to the most open valve is on manual.
- The dynamics of the valve position loop must be slower than the dynamics of the flow loops, preferably by a factor of 5.

Issues Pertaining to Operational Modes. For the valve position controller in Figure 5.3 to function properly, the flow controller that outputs to the most open valve must be on automatic. Suppose that the most open valve is 70%, the corresponding flow controller is on manual, and the set point to the valve position controller is 90%. The reset mode in the valve position controller will drive its output to the lower output limit.

An equivalent requirement for the valve position controller to function properly is that the flow controller providing the input currently selected by the high select must be on automatic. This is the basis for the following discrete logic:

1. For each input to the high select, compute the logical AND of the following:
   - The corresponding output that indicates that this input is currently selected
   - Output AUTO of the flow controller that outputs to this input of the high select

   The result can be true only for the input that is currently selected, and will be true only if the respective flow controller is in automatic.
2. Compute the logical OR of the two results. A result of true means that the valve position controller can function.

If the result is false, action is required to prevent the valve position controller from driving its output to an extreme. The possibilities are as follows:

1. Force the valve position controller to manual. The appropriate statement is the following:

```
VPC.FMANL = !((SELHI.Q1 & FCA.AUTO) | (SELHI.Q2 & FCB.AUTO))
```

When FMANL is true, the controller is forced to manual. But when FMANL is false, the controller is not automatically switched to auto.

2. Suspend the PID calculations and hold the output at its current value. Most PID blocks have a mechanism to do this. For the PID block for Table 1.1, this is accomplished by setting inputs NODEC and NOINC to true:

```
VPC.NODEC = !((SELHI.Q1 & FCA.AUTO) | (SELHI.Q2 & FCB.AUTO))
VPC.NOINC = !((SELHI.Q1 & FCA.AUTO) | (SELHI.Q2 & FCB.AUTO))
```

Controller Tuning. The valve position controller is implemented using the standard PID block. When the objective is to achieve more optimal process operating conditions, rapid response is not essential. The controller is normally tuned with a relatively low controller gain and a relatively long reset time. Derivative is not recommended for valve position controllers.

As noted above, the valve position controller is totally dependent on the flow controllers reacting to changes in the valve position controller output (the pump speed). Consequently, the valve position controller must respond more slowly than the flow controllers, preferably by a factor of 5. For polymer pumping applications, the flow controllers are fast, so the valve position controller can be tuned accordingly. But in most applications, valve position controllers respond very slowly.

Set Point for the Valve Position Controller. For the polymer pumping application a value of 90% was previously suggested for the valve position controller set point. Suppose that a set point of 95% is used instead of 90%. The consequences are:

- The shear applied by the pump to the polymer is reduced. The degree to which this is beneficial depends on the application.
- Should a flow controller need to increase its flow, less control action is available. With a set point of 90%, a flow controller can increase its valve opening an additional 10%. With a set point of 95%, a flow controller can increase its valve opening only an additional 5%. This degrades their ability to react to any disturbances that cause the polymer flow to decrease.

The primary concern is how often either flow controller drives its valve fully open and for how long. Once the flow controller has driven its valve fully open, the valve position controller will begin to increase the pump speed. However, valve position controllers respond relatively slowly. The two relevant factors are the following:

1. During routine process operations, what disturbances occur that cause either flow controller to drive its valve fully open?

2. What are the consequences of the polymer flow being less than its set point?

As the valve position controller set point is moved closer to 100%, the flow controllers will fully open their valves more frequently and for longer periods of time. The consequences of this must not offset the benefits of running the pump at a slightly slower speed.

The answer to both of the questions above depends very much on the application. In some, the flow must be closely controlled, and even brief dips below the set point are undesirable. But for most batch processes, the flow is totalized and it is the flow total that is most important; flows less than the set point for a short duration have no adverse consequences.

In practice, none of these issues can be easily quantified. The only viable approach is to adjust the set point of the valve position controller based on operational experience. During commissioning the application, a conservative value is advisable for the set point. Once production operations are running smoothly, examine the trend plot of the output of each polymer flow controller. Except during abnormal events (such as equipment failures), is the output of either flow controller driven fully open? If the answer is no, the valve position controller set point can be increased. If the answer is yes, how often and for how long? Then comes the hard part: Do the benefits of the increased valve position controller set point (running the pump at a slower speed and thus exposing the polymer to less shear) offset the consequences of the polymer flow occasionally being below its set point?

5.2. TERMINAL REHEAT SYSTEMS

When heating and/or cooling is required to control the temperature in each of several rooms, two approaches are possible:

Terminal reheat (Figure 5.4). For each room, a manually adjusted damper admits a constant flow of cold air from a common supply. The temperature is controlled by reheating the air as it enters each room. One option is a steam-heated exchanger, as shown in Figure 5.4.

Variable volume. Separate hot and cold air supplies are required. For each room, an automatic damper is required on each supply. Usually, a minimum airflow is required to each room, so at times hot and cold air mixing will be required.

Energywise, terminal reheat is inefficient, but the initial installation costs are less.

The terminal reheat system in Figure 5.4 will be used to illustrate an application of valve position control. To keep the example simple, the following aspects will be ignored:

• A minimum flow of fresh air into the recirculation system is usually required. This is not included in Figure 5.4.

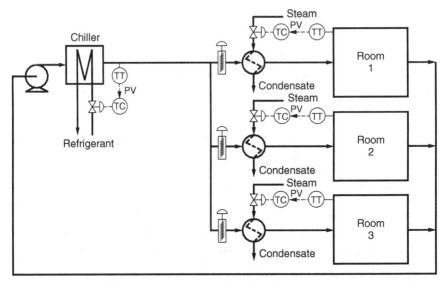

Figure 5.4 Terminal reheat.

- In high-humidity environments, a maximum cold air temperature is imposed to provide the necessary dehumidification.

The temperature controller provided for each room adjusts the opening of the control valve on the steam supply to the exchanger for that room. The temperature controller can only add heat. Consequently, the cold air supply temperature must be low enough that some reheat is required in every room. The maximum cooling that can be supplied to a room is when cold air temperature is the lowest that the chiller could attain. But the colder the air supply, the greater the reheat for each room.

Valve Position Controller. To operate as efficiently as possible, a control configuration is desired that will achieve the following objectives:

- Maintain the temperature in each room at or neat its set point. The individual room-temperature controllers will do this, provided that no steam supply valve is fully closed.
- Operate with as high a cold air temperature as possible, subject to no steam valve being fully closed. A valve position controller can be configured to achieve this objective.

Figure 5.5 presents the valve position control configuration. The components are as follows:

- The opening of each steam valve is an input to a low select. The output of the low select is the opening of the least open steam valve.

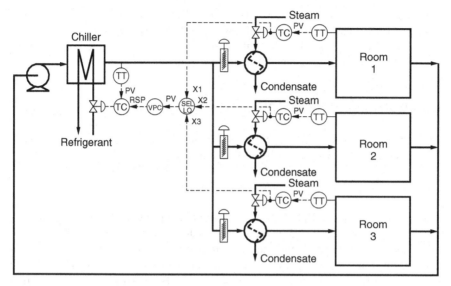

Figure 5.5   Valve position controller for terminal reheat.

- The opening of the least open steam valve is the process variable for a valve position controller. The output of this controller is the set point for the temperature of the cold air leaving the chiller.

If the opening of the least open steam valve is increasing, more heat is being added in the corresponding steam heater. Consequently, the set point for the cold air temperature can be increased. The valve position controller must be direct acting.

Configuration and Tuning. The issues associated with the valve position controller for the terminal reheat system are analogous to the issues for the polymer pumping application:

- For the valve position controller to function properly, the temperature controller that outputs to the least open steam valve must be in automatic. Discrete logic analogous to that for the polymer pumping application is required to detect when this is not the case and to suspend the control calculations in the valve position controller.
- The valve position controller must be tuned to respond more slowly than both of the following:
  - *Cold air temperature loop.* The relationship between the valve position controller and the cold air temperature controller is a true cascade.
  - *Room temperature loops.* The valve position controller depends on these controllers to respond to changes in the cold air temperature.

- The set point of the valve position controller is the desired opening of the least-open steam valve. How close this set point can be to fully closed depends on the disturbances to the room temperatures and the consequences of temperature excursions above the set point.

5.3. EQUILIBRIUM REACTION

Figure 5.6 illustrates a reactor in which the reaction is being driven to an equilibrium state. The reaction can be considered to be

$$A + B \leftrightarrows C + D$$

Product C is the desired product and is essentially nonvolatile at the conditions within the reactor. Product D happens to be water. Reactant B is volatile, but less so than water. To increase the yield of product C, the reaction must be shifted to the right. One way to achieve this is to remove water (product D) from the reactor by boiling off the water. But since reactant B is also volatile, a separation column is required to separate water from reactant B. The product draw from the separation column is water containing minute amounts of reactant B.

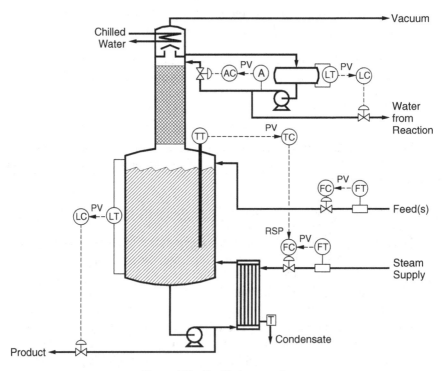

Figure 5.6 Equilibrium reaction.

The roles of the various control loops illustrated in Figure 5.6 are as follows:

Feed flow control. There are multiple feeds, but only one is shown in Figure 5.6. Flow measurement and control is provided for each feed, and additional controls are provided to maintain the feed rates in the proper ratio.

Reactor level control. To maintain material balance closure, the reactor level controller adjusts the opening of the control valve on the reactor discharge.

Reactor temperature control. The temperature in the reactor is maintained by adjusting the set point for the steam flow.

Composition of water leaving condenser. To avoid excess loss of reactant B, the composition of the column overhead is measured by an on-stream analyzer, and the reflux rate to the column adjusted to maintain this composition at the desired value.

Reflux drum level control. The reflux drum level controller adjusts the opening of the control valve on the product draw to maintain a constant level in the reflux drum.

Valve Position Controller. What ultimately determines the rate of heat input required for the reactor? It is the feed rate to the reactor. Only trace amounts of water remain in the reaction medium, so essentially all water produced by the reaction is removed. For each unit of feed, the amount of water to be removed is determined by the reaction stoichiometry. For each unit of water to be removed, a certain quantity of heat must be input to the reactor to provide the boil-up to the separation column.

To operate the reactor so as to produce as much product as possible, the feed rate to the reactor must be the maximum possible. This creates a role for the valve position controller illustrated in Figure 5.7. The process variable input is the opening of the steam control valve. The output of the valve position controller is the set point for the master feed flow controller.

For the valve position controller to function, the reactor temperature controller must be in automatic and the steam flow controller must be in remote. Discrete logic must be provided to suspend the PID calculations for the valve position controller if this is not the case.

Heat Transfer Limited. The transfer of heat from the steam supply to the reaction medium is a two-step process:

1. Steam must flow into the steam heater. This is a fluid-flow issue.
2. The heat from the condensing steam must be transferred to the reacting media. This is a heat transfer issue.

Each could be the limiting factor. But since valves are commonly oversized, heat transfer is very likely the limiting mechanism.

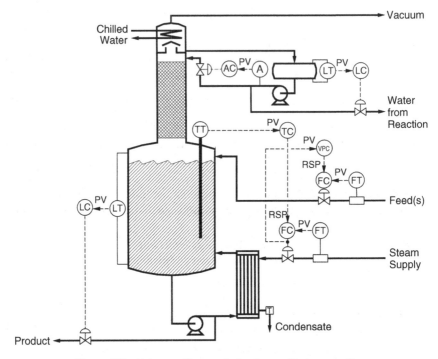

Figure 5.7 Valve position controller for equilibrium reaction.

If heat transfer is the limiting mechanism, the following will occur at large steam valve openings:

- The pressure of the condensing steam approaches the steam supply pressure.
- The temperature of the condensing steam approaches the saturation temperature at the steam supply pressure.
- The heat transfer rate approaches its maximum.
- The reactor temperature is little affected.

This creates a serious problem for the valve position controller configuration in Figure 5.7. Suppose that it is provided a set point of 90%. If the steam control valve is oversized to any degree (and it probably is), the onset of heat transfer limited conditions occurs at a lower valve opening. At 90% open, the steam control valve has little effect on the reactor temperature, which means that the reactor temperature controller will not perform properly. Control of reactor temperature is crucial, so this consequence is not acceptable.

The configuration in Figure 5.7 is appropriate when the heat input rate is limited by the rate at which steam can flow into the exchanger. When the heat input rate is limited by the heat transfer mechanisms, the process variable for the valve position controller must somehow indicate the current heat transfer rate with respect to the maximum heat transfer rate. The rate of heat transfer to the reacting medium is given by the following expression:

$$Q = UA \ (T_C - T)$$

where

A = heat transfer area
Q = heat transfer rate
T_C = temperature of condensing steam
T = reactor temperature
U = heat transfer coefficient

The upper limit for the condensing steam temperature T_C is the saturation temperature T_S for steam at the steam supply pressure. The maximum possible heat transfer rate is

$$Q_{max} = UA \ (T_S - T)$$

where

Q_{max} = maximum possible heat transfer rate
T_S = saturation temperature for steam at the steam supply pressure

The onset of heat transfer limited generally occurs when the heat transfer rate is approximately 90% of the maximum possible heat transfer rate. The ratio Q/Q_{max} is easily related to the temperatures:

$$\frac{Q}{Q_{max}} = \frac{T_C - T}{T_S - T}$$

To operate at 90% of the maximum possible rate, the temperature of the condensing steam can be computed as follows:

$$T_C = T + 0.9(T_S - T) = T_S - 0.1(T_S - T) = 0.9T_S + 0.1T$$

In the P&I diagram in Figure 5.8, the PV for the valve position controller is the condensing steam temperature. Technically, the controller is a temperature controller, but Figure 5.8 retains the term "valve position controller." The real purpose of this controller is to operate the process at the maximum possible heat transfer rate, which is at the transition from media limited to heat transfer limited. Hence, the term *valve position controller* more accurately conveys its purpose than *temperature controller*.

To implement the control configuration in Figure 5.8, a temperature transmitter is required for the condensing steam temperature. If the set point for the valve position controller must be computed, a value is also required for the saturation temperature of steam at the steam supply pressure. There are two possibilities:

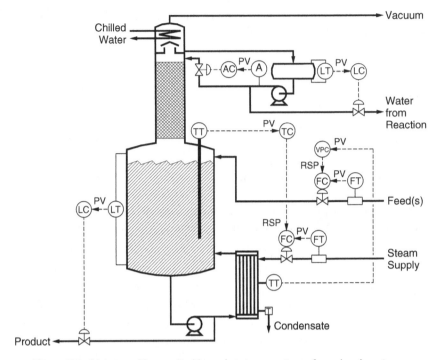

Figure 5.8 Valve position control based on temperature of condensing steam.

1. Assuming no superheat, the steam supply temperature could be measured directly.
2. The steam supply pressure could be measured and the saturation temperature computed by a characterization function. If the changes in the steam supply pressure are small, a linear approximation over the range of interest can be used in lieu of the characterization function.

5.4. REACTOR WITH A ONCE-THROUGH JACKET

Figure 5.9 illustrates a batch reactor with a once-through jacket for the removal of the heat generated within the reactor. A common procedure is to charge enough nonreacting material initially to provide a "heel" within the reactor so that the agitator can be started. This is followed by a period of time during which other materials are fed on a continuous basis. Although only one feed stream is illustrated in Figure 5.9, most reactions require multiple feed streams. So that precise ratios can be maintained, flow control is provided for all streams.

Often, the rate at which materials may be fed to the reactor is limited by the rate at which heat may be removed from the reactor. In the simplest approach, the heel is charged, the feeds are started slowly until the reaction is proceeding

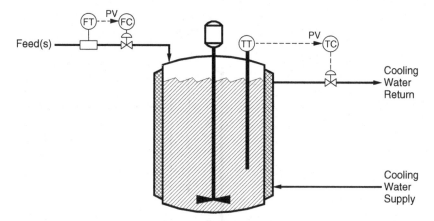

Figure 5.9   Reactor with once-through jacket, temperature control via cooling media.

as expected (some call this the initiation phase), the feed rates are ramped up to a specified target, and this flow is maintained until the necessary quantities have been fed to the reactor.

The target for the feed rates must not cause the reaction to generate more heat than can be removed via the jacket. In batch reactors, the conditions within the reactor vary throughout the batch, and consequently the rate at which heat can be removed also varies. With a constant feed rate, there are periods when the heat transfer capabilities are not being used to their fullest.

In the control configuration in Figure 5.9, the opening of the cooling water valve reflects the current heat transfer rate: The more open the valve, the higher the heat transfer rate. From a trend of the opening of the cooling water valve, one can assess the variability of the heat transfer rate during the batch and the extent to which the heat transfer capability is being utilized. To increase the production rate from a batch facility, ways must be found to shorten the batch cycle time. One possibility is to vary the feed rate in such a manner that the heat transfer capabilities are fully utilized throughout the continuous feed part of the batch.

Reactor Temperature Control by Adjusting the Feed Rates. For some reactions, the configuration in Figure 5.10 can be used during the time that materials are being fed to the reactor (but subsequent to the initiation phase). The cooling water valve is opened fully, and the temperature is controlled by adjusting the feed flow rates. Whether this works depends on the nature of the reaction:

- When one of the feeds is an activator, the reaction rate (and consequently the rate of heat generation within the reactor) is determined by the flow rate of the activator feed stream. For such reactions, the reactor temperature responds rapidly to a change in the feed rate. Controlling the reactor temperature by adjusting the feed rates usually causes the reactor to perform

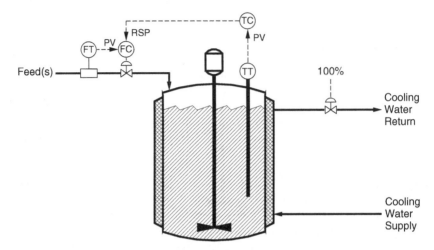

Figure 5.10 Reactor with once-through jacket, temperature control via reactor feed rate.

better than does controlling the reactor temperature by adjusting the cooling water valve opening.

• For most reactions carried out at a constant temperature, the reaction rate is determined by the concentration of the reactants within the reactor. Often, all but one of the reactants is in excess, so the reaction rate depends primarily on the concentration of the limiting reactant within the reactor. If the feed rate of this reactant is changed, this must first translate into a change in the concentration of the reactant within the reactor, and only then does it affect the reaction rate and the rate of heat generation. The reactor temperature responds slowly, which degrades the performance of the control configuration in Figure 5.10.

For all reactions, temperature control is crucial. If the temperature control configuration in Figure 5.9 provides the best control of the reactor temperature, it must be retained. This example assumes that this is the case.

Valve Position Controller. Figure 5.11 presents a control configuration that uses a valve position controller to adjust the feed rates. The process variable for the valve position controller is the opening of the cooling water valve. The set point for the valve position controller is the desired opening of the cooling water valve. If the current cooling water valve opening is below the target, the valve position controller can increase the feed rates to the reactor. This will generate more heat, forcing the reactor temperature controller to increase the cooling water valve opening. If the current cooling water valve opening is above the target, the valve position controller must decrease the feed rates to the reactor. If the opening of the cooling water valve is increasing, the valve position controller must decrease the feed rates. The valve position controller must be reverse acting.

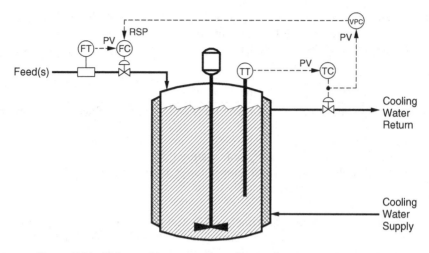

Figure 5.11   Valve position control based on cooling water valve opening.

For most polymerization reactions, the reacting medium has a low viscosity at the beginning of the batch, so the heat transfer capability is good. As the polymerization reaction proceeds, the viscosity increases. This degrades the heat transfer capability. In such reactors, the feed rates can be high at the beginning, but must be decreased as the batch proceeds. At the beginning of the continuous feed part of the batch, the valve position controller might increase the feed rates, but thereafter the valve position controller would be decreasing the feed rates.

Heat Transfer Limited. As discussed in Chapter 4, the removal of heat from the reacting medium is a two-step process:

Heat transfer. Heat is transferred from the reacting medium to the water in the jacket.
Sensible heat. Heat is removed from the jacket as sensible heat in the cooling water.

Either mechanism may be the limiting mechanism, but for most jackets, the heat transfer proves to be the limiting mechanism. The usual practice is to oversize the ability to flow cooling water through the jacket.

As the cooling water flow increases, the cooling water return temperature approaches the cooling water supply temperature, and the ΔT value for heat transfer approaches the difference between the reactor temperature and the cooling water supply temperature. Under these conditions, the heat transfer mechanisms become the limiting factor in determining the rate of heat removal from the reacting mass. When the heat transfer mechanisms are limiting, changes in the cooling water flow have little effect on ΔT for heat transfer, and consequently little effect on the reactor temperature. The temperature controller in Figure 5.11 cannot provide adequate performance under these conditions.

In Chapter 4 the following ratio was proposed for determining the onset of heat transfer limited conditions:

$$\frac{\text{temperature rise of the cooling water}}{\text{maximum } \Delta T \text{ for heat transfer}} = \frac{T_{CWR} - T_{CWS}}{T - T_{CWS}}$$

where

T_{CWS} = cooling water supply temperature (°F)
T_{CWR} = cooling water return temperature (°F)
T = reactor temperature (°F)

If this ratio is less than approximately 0.2, heat transfer mechanisms determine the rate of heat removal from the reacting mass. For the temperature controller in Figure 5.11 to function properly, this ratio must be greater than 0.2. Said another way, the temperature rise $T_{CWR} - T_{CWS}$ for the cooling water must be greater than 20% of $T - T_{CWS}$, which is the maximum ΔT for heat transfer. If the reactor temperature T and the cooling water supply temperature T_{CWS} do not change significantly, the onset of heat transfer limiting conditions can be ascertained from the temperature rise for the cooling water (this value is 20% of $T - T_{CWS}$).

This provides the basis for the valve position controller configuration in Figure 5.12. The cooling water supply temperature and the cooling water return temperature are measured. The cooling water temperature rise is the difference between these two temperatures (actually, a better approach is to obtain the temperature rise by measuring the temperature difference directly). The cooling water

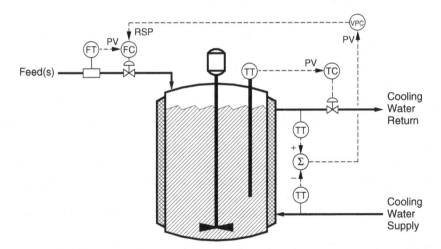

Figure 5.12 Valve position control based on cooling water temperature rise.

temperature rise is the process variable for the valve position controller. Technically, this controller is a differential temperature controller (DTC), but calling it
a valve position controller more clearly conveys its true purpose.

What if either the cooling water supply temperature T_{CWS} or the reactor
temperature T change significantly? The only recourse is to compute the ratio
$(T_{CWR} - T_{CWS})/(T - T_{CWS})$ and use this value as the process variable for the
valve position controller. This requires values for both T_{CWS} and T_{CWR}. They can
be measured directly as in Figure 5.11, but the following approach is preferable:

- Measure the temperature difference $T_{CWR} - T_{CWS}$ directly.
- Either measure T_{CWS} and add the temperature difference to obtain T_{CWR}, or
  measure T_{CWR} and subtract the temperature difference to obtain T_{CWS}.

6

RATIO AND FEEDFORWARD CONTROL

For most loops, simple feedback control provides adequate performance. One needs a measurement device, a final control element (usually, a control valve), and a PID controller. The upfront engineering work is minimal and requires only a qualitative understanding of process behavior. Tuning is often the biggest challenge, and this is left to the plant startup steam. It should be no surprise that engineering will specify simple feedback control with few exceptions, one of which being vessels with a recirculating jacket, where cascade is known to be superior.

In most plants, there are a handful of loops where enhanced performance translates into quantifiable improvements in plant operations. Many of these are temperature or composition loops, which also tend to be slow loops. In some, tools such as cascade can enhance the performance significantly. Another possibility is feedforward control, which is the subject of this chapter. Most ratio loops can be viewed as a simple manifestation of feedforward, where changes in one flow are translated into changes in another flow by maintaining the two flows in the proper ratio.

To implement feedforward control, three obstacles must be overcome:

- The behavior of the process must be understood quantitatively. This can be in the form of basic mechanism models (material balances, energy balances, etc), empirical relationships, characterization functions, or else. As long as the relationship between the important variables is quantified, any type of relationship is satisfactory. Obtaining such a relationship can be an obstacle, but it is abating as our modeling capabilities continue to improve. With few

Advanced Process Control: Beyond Single-Loop Control   By Cecil L. Smith
Copyright © 2010 John Wiley & Sons, Inc.

exceptions, the technical capability is now available to model any industrial process. Often, the obstacle is identifying sufficient benefits to justify the effort.

- Of the variables in the model, one will be controlled to a specified target. For feedforward to be most successful, measurement devices must be supplied for all other variables that significantly affect the variable being controlled. These measurement devices must be purchased, installed, and maintained. This involves both technical commitments and financial commitments. Measurement technology continues to improve, but our demands seem to increase at a comparable pace.

- The relationships embodied in the feedforward control formulation must be implemented within the controls. With conventional pneumatic and electronic analog controls, this meant special computing elements with numerous consequences. But with digital controls, this obstacle is history. The simpler feedforward formulations can be implemented using function blocks, but as the complexity increases, programmed implementations become more attractive.

Feedback control of any form (simple feedback, cascade, or else) is ultimately limited by the dynamics of the process. A control loop for a slowly responding process will respond accordingly. Feedforward is not limited in this manner. Even when dynamic compensation must be included in the feedforward formulation, the dynamic compensation must be consistent with the process dynamics, but does not limit the performance of the feedforward controls. As the demands on the controls continue to increase, at some point the feedback approaches will not be able to meet the requirements, leaving feedforward as the only recourse.

The initial attention must be directed to developing the feedforward formulation that will meet the control requirements. However, one must not stop there. The final formulation must include the appropriate initialization or tracking to achieve the following objectives:

Bumpless transfer from manual to automatic. As in cascade configurations, this need arises when a value computed by one part of the control configuration is not used by another part of the control configuration.

Prevent windup in a PID controller. In Chapter 1 we introduced the following simple statement for the condition for windup:

Reset windup occurs in a controller when changes in the controller output have no effect on the process variable.

Should any limiting condition be encountered, this statement is usually true, so windup protection of some form must be provided.

What are the consequences if these issues are ignored? When the process is operating in the normal manner, there are none. The consequences arise only under certain conditions. Sometimes the consequences are largely a nuisance. If so, the process operators can respond by switching appropriate controllers to either manual or local automatic, thus disabling appropriate parts of the control calculations (controllers on manual do not wind up). If the occurrence is infrequent and the consequences are minor, this may be an acceptable "solution." But if a nuisance occurs frequently, it is both a burden to the operators and a distraction that could have consequences. In some applications, the consequences are more serious, such as the safety system initiating a trip (a sudden shutdown of part or all of the process). The cost of one unnecessary trip can offset the benefits accrued for days or months.

The presentation herein will include considerable discussion of initialization and tracking. This adds complexity to the control formulation and can lead to what is best described as "creeping elegance." One must weight the trade-offs, which are, basically:

- The increased complexity within the controls has implications for commissioning and subsequent support.
- As operations staffing continues to be right-sized, the burden on the operators assumes increasing importance, especially the distractions.

In this chapter we pay special attention to such issues, explaining how limits, zero flow rates, and so on, can lead to problems that if not addressed within the controls become a burden to the operators. If this burden proves too great or if consequences such as an inappropriate trip arise only occasionally, even the best control formulation will be switched to manual or otherwise disabled.

6.1. SIMPLE RATIOS

The initial examples will implement ratio control using a flow-to-flow controller. In conventional pneumatic and electronic controls, such controllers were available as standard commercial products. Most digital systems implement ratio control in a slightly different way, which we present shortly.

A flow-to-flow controller accepts two measured variables, one designated as the wild flow and the other designated as the controlled flow:

Controlled flow. This is the flow through the control valve whose opening is at the discretion of the flow-to-flow controller. The output of the flow-to-flow controller determines the value of the controlled flow.

Wild flow. This is the measured value of some other flow within the process. This flow is "wild" in the sense that the flow-to-flow controller cannot affect its value but must respond to whatever its value happens to be. However, it is possible that the wild flow is being controlled by some other controller.

For the flow-to-flow controller, the process variable and set point are as follows:

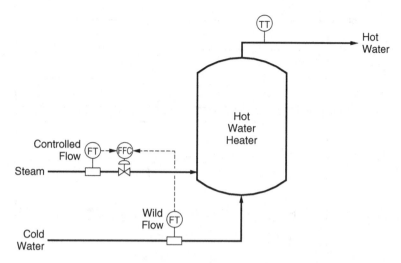

Figure 6.1 Flow-to-flow control for the hot water process.

Process variable: The current ratio of the controlled flow to the wild flow.

Set point: The desired value for the ratio of the controlled flow to the wild flow.

The vast majority of ratio applications involve two flows, and the customary terminology used for ratio control reflects this. But occasionally, ratio control is applied to maintain the ratio of two variables other than flows.

Ratio Control for a Hot Water Process. The process in Figure 6.1 produces hot water by mixing steam and cold water. There are no heat transfer surfaces—the condensed steam leaves with the hot water. This is a utility process that responds primarily to changes in the hot water demand; changes in the set point for the hot water temperature are very infrequent.

The hot water temperature is determined largely by the ratio of the steam flow to the cold water flow. Especially in applications where changes in hot water demand are frequent and large, significant improvement in hot water temperature control can be achieved by maintaining the proper ratio of steam flow to cold water flow. In the configuration in Figure 6.1, measurements of steam flow and cold water flow are inputs to a flow-to-flow controller (FFC) that adjusts the steam valve opening so as to maintain the desired ratio of steam flow to cold water flow.

To maintain the hot water temperature at 150°F with a hot water demand of 500 lb/min, the required steam-to-water ratio is 0.070 lb/lb. Using only the flow-to-flow controller in Figure 6.1, Figure 6.2 presents the response in hot water temperature to an increase in the hot water demand from 500 lb/min to 1000 lb/min and subsequently back to 500 lb/min. The departure of the hot water

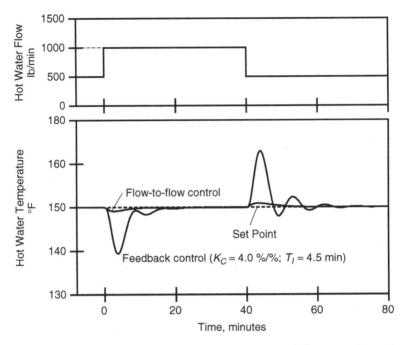

Figure 6.2 Response of flow-to-flow controller to change in hot water demand.

temperature from its set point is about 1°F. This is far superior to that of simple feedback control (also illustrated in Figure 6.2).

In the response in Figure 6.2, the hot water temperature is approximately 150°F when the hot water demand is 1000 lb/min. This is not normally the case, and given enough resolution, is not exactly the case in Figure 6.2. To present a more realistic situation, a small heat loss will be added. The heat loss is independent of throughput, which means that a slightly higher steam-to-water ratio is required at a hot water demand of 500 lb/min than at a hot water demand of 1000 lb/min.

Figure 6.3 illustrates the effect of the heat loss on the ratio control performance. The steam-to-water ratio is fixed at the value that gives a hot water temperature of 150°F at a hot water demand of 500 lb/min. When the hot water demand is 1000 lb/min, the hot water temperature lines out slightly above 150°F (the difference increases with the magnitude of the heat loss).

The set point to the flow-to-flow controller is the desired ratio of steam flow to cold water flow. The appropriate value for the steam-to-water ratio is influenced by the steam enthalpy, cold water temperature, heat losses, desired hot water temperature, and so on. To address the effect of such influences, the configuration in Figure 6.4 is a cascade configuration that includes a hot water temperature controller to adjust the set point of the flow-to-flow controller. If the hot water temperature is above its set point, the controller reduces its output, which is the set point for the steam-to-water ratio.

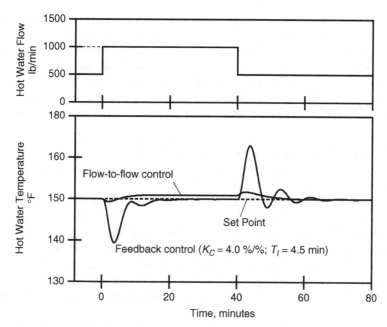

Figure 6.3 Response of flow-to-flow controller to change in hot water demand with a small heat loss from the process.

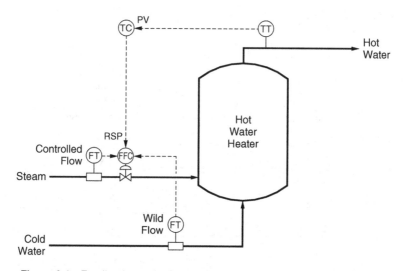

Figure 6.4 Feedback trim for flow-to-flow control of the hot water process.

The flow-to-flow controller provides the primary response to the major disturbance: specifically, a change in the hot water demand. The primary role of the hot water temperature controller in Figure 6.4 is to make small adjustments in the set point to the flow-to-flow controller to maintain the hot water temperature at its set point. This temperature controller is normally said to provide *feedback trim*. Such controllers are not normally tuned for aggressive response but, instead, make slow adjustments in the set point of the flow-to-flow controller.

Ratio Control for a Steam-Heated Exchanger. The process in Figure 6.5 is a steam-heated exchanger that heats a liquid stream to a specified outlet temperature. The liquid stream enters at 210°F and is heated with 75-psig steam. The desired liquid outlet temperature is 280°F. The liquid outlet temperature is determined largely by the ratio of the steam flow to the liquid flow. Especially in applications where changes in liquid flow are frequent and large, significant improvement in liquid outlet temperature control can be achieved by automatic control of the ratio of steam flow to liquid flow. In the configuration in Figure 6.5, measurements of steam flow and liquid flow are inputs to a flow-to-flow controller (FFC) that adjusts the steam valve opening so as to maintain the desired ratio of steam flow to liquid flow.

The performance of the flow-to-flow control configuration is illustrated in Figure 6.6, along with the performance of feedback control. The flow-to-flow control performance is superior to that of feedback control, but the improvement

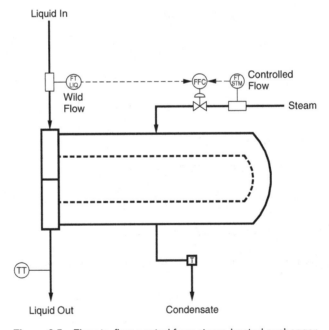

Figure 6.5 Flow-to-flow control for a steam-heated exchanger.

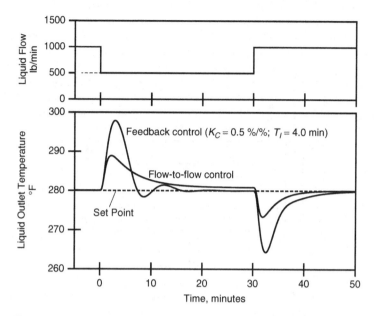

Figure 6.6 Response of flow-to-flow controller to change in liquid flow.

is not as dramatic as for the hot water process. Figure 6.6 illustrates two types of errors:

Dynamic. Each change in the liquid flow causes a significant excursion of the liquid temperature from its desired value. Dynamic compensation is required to address these errors.

Steady-state. Initially, the liquid flow is 1000 lb/min and the liquid outlet temperature is 280.0°F. For a liquid flow of 500 lb/min, the liquid outlet temperature does not line-out at exactly 280.0°F. The value required for the steam-to-liquid ratio varies slightly with the liquid flow, so maintaining a constant steam-to-liquid ratio will not maintain the same liquid outlet temperature. The liquid outlet temperature controller in Figure 6.7 provides the feedback trim to address this issue.

Tuning Flow-to-Flow Controllers. Tuning a flow-to-flow controller is substantially the same as tuning a flow controller. In most cases, these controllers are tuned with a low value of the controller gain and a short reset time. Any of the following tuning values usually provides acceptable performance:

K_C (%/%)	T_I (sec)
0.1	2.0
0.2	3.0
0.3	5.0

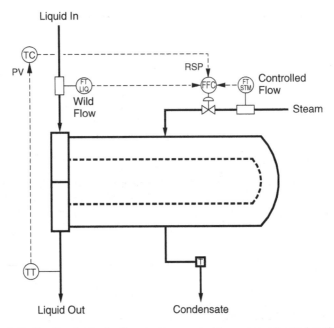

Figure 6.7 Feedback trim for flow-to-flow control for a steam-heated exchanger.

In all examples in this book, the flow-to-flow controller is tuned with a gain of 0.2%/% and a reset time of 3 sec (6.5 min). This is very conservative tuning. If necessary, the flow-to-flow controller could be tuned to respond more rapidly, but this rarely results in a significant improvement in process performance.

6.2. RATIO CONTROL IN DIGITAL SYSTEMS

Some digital systems provide a flow-to-flow controller as a standard function block, but many do not. When a standard function block is not available, additional function blocks, along with the PID block, must be configured to provide the equivalent functionality. There are two approaches.

Flow Ratio Computation. As illustrated in Figure 6.8, a divider block computes the flow ratio by dividing the steam flow (input x) by the liquid flow (input y). The output of this function block is the process variable input to a PID block. Since the process variable is a flow-to-flow ratio, the PID controller in this configuration is logically labeled FFC for "flow-to-flow controller." However, it should not be confused with the two-input FFC module that internally computes the flow ratio as well as providing PID control.

An issue that must be addressed with the configuration in Figure 6.8 is the possibility that either or both of the flows may be zero (or nearly zero). Suppose

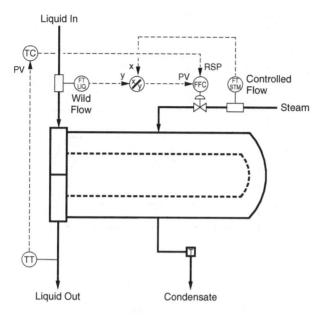

Figure 6.8 Implementation of flow-to-flow control by computing the PV.

that the process is shut down. Ideally, both the steam flow and the liquid flow would be exactly zero. The divider block would be asked to divide zero by zero, a mathematical operation that presents some difficulties.

In the real world, the flow measurements are usually nearly zero but not exactly zero. The divider is actually asked to divide one small number by another. The computation can be performed, but the result will be erratic, especially if there is any noise on either of the flow measurements. One approach is to configure output tracking to force the controller output to zero when the wild flow is less than the minimum flow that can be accurately measured.

Computed Flow Set Point. In the implementation in Figure 6.9, the set point for the steam flow controller is computed as the product of the temperature controller output and the measured value of the liquid flow. The temperature controller output is the desired ratio of steam flow to liquid flow. Superficially, division by zero does not appear to be a problem with this configuration. If the liquid flow is zero (or nearly zero), multiplying by the output of the temperature controller gives a steam flow set point that is zero (or nearly zero). As long as the steam flow controller is using its remote set point, this observation is true.

The configuration in Figure 6.9 is a temperature-to-flow cascade with a computation inserted between the output of the temperature controller and the flow controller. As for all cascades, output tracking is required in the temperature controller to provide bumpless transfer when the flow controller is switched from local to remote. As will be explained shortly in the discussion pertaining to the

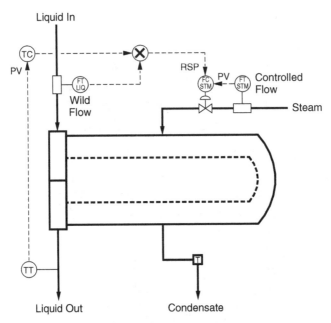

Figure 6.9 Implementation of flow-to-flow control by computing the set point for a flow controller.

feedback trim controller, the possibility of a division by zero (or a value that is nearly zero) arises in the computation of input MNI for output tracking.

The multiplier that computes the set point for the flow controller is sometimes designated as "RC" on a P&I diagram. In this context, RC could be understood to be a ratio controller. However, *ratio computation* is a more accurate understanding. The RC element is simply a multiplier, possibly including a bias in some applications. The output of the temperature controller is multiplied by the liquid flow (the wild flow) to obtain the set point for the steam flow controller.

Performance. While the performance of the two configurations for ratio control is not exactly the same, the difference rarely has a significant impact on process operations. In either case, the dynamics are that of a flow loop, and thus both are far faster than most other loops in the process.

For the exchanger, the trend in Figure 6.10 presents the response in the steam flow and hot water temperature for a liquid flow increase from 1000 lb/min to 1200 lb/min. The coefficients for the controllers are as follows:

Configuration	Controller	K_C	T_I	PV Measurement Range
Figure 6.8	FFC	0.2%/%	3 sec	0.0 to 0.1 lb/lb
Figure 6.9	FC	0.2%/%	3 sec	0 to 200 lb/min

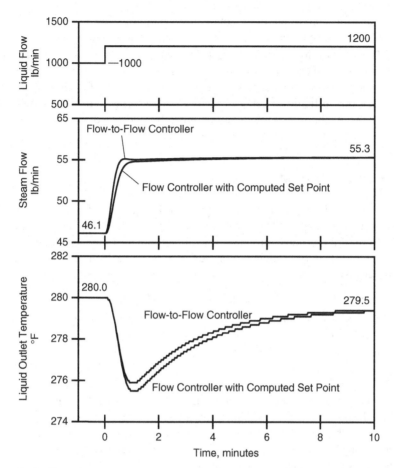

Figure 6.10 Response of liquid outlet temperature for the two options for implementing ratio control.

For this example, the response of the flow-to-flow controller configuration is slightly faster, but in the context of the temperature dynamics of the process, not significantly faster. However, the flow-to-flow controller is not always the faster of the two. The performance is affected by the controller gain and the measurement range of the process variable. If the controller gain for the flow controller is increased to 0.4%/% or the measurement range is changed to 0 to 100.0 lb/min, the flow controller configuration would be the faster of the two.

For both configurations the following characteristics are evident in the responses of the liquid outlet temperature in Figure 6.10:

1. The initial liquid outlet temperature is 280.0°F; the final liquid outlet temperature is 279.5°F. Maintaining a constant steam-to-liquid ratio does not

maintain a constant liquid outlet temperature, even at steady-state. The addition of the feedback trim is required to address this behavior.

2. The liquid flow increased by 20%, which caused both ratio configurations to increase the steam flow by 20%. But for both configurations, the liquid outlet temperature decreased to approximately 276°F before recovering. This dip in liquid outlet temperature is due to the dynamic characteristics of the exchanger. This behavior must be addressed by adding dynamic compensation to the control configuration.

6.3. FEEDBACK TRIM

Most, but not all, applications of ratio control require feedback trim, which is normally provided by a PID controller. Where required, feedback trim is normally in the form of a cascade configuration with the output of the outer loop providing the value for the ratio coefficient. For the steam-heated exchanger, the temperature controller provides this function in the following control configurations:

- Function block for flow-to-flow controller (Figure 6.2)
- PID controller function block with computed flow-to-flow as its PV (Figure 6.8)
- Flow controller with computed set point (Figure 6.9)

Maintaining a fixed ratio of steam flow to liquid flow provides a nearly constant value for the liquid outlet temperature. But as Figure 6.10 illustrated, increasing the liquid flow by 20% and maintaining a constant ratio of steam flow to liquid flow causes the equilibrium or steady-state value of the liquid outlet temperature to change from 280.0°F to 279.5°F. This is a relatively small change, but nevertheless is undesirable in most applications.

Performance. Normally, the feedback trim controller is tuned very conservatively. For the same case as Figure 6.10, Figure 6.11 presents the performance of the feedback trim controller for two sets of tuning coefficients:

Temperature at equilibrium. Given enough time, the liquid outlet temperature will line-out at its set point of 280.0°F.

Maximum departure from set point. For both responses the liquid outlet temperature dips to approximately 276°F. The feedback trim controller has little effect, as evidenced by the following:

- The improvement over the responses in Figure 6.10 (no feedback trim) is insignificant.
- Adjusting the tuning in the feedback trim controller has little effect.

The dip is due primarily to the dynamics of the exchanger and must be addressed by dynamic compensation.

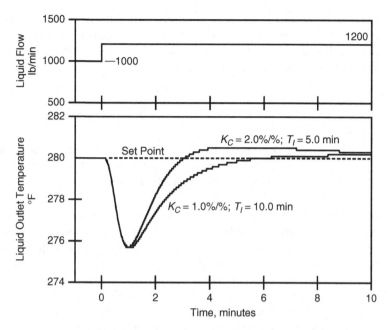

Figure 6.11 Addition of feedback trim to flow controller with computed set point.

Steam-to-Liquid Ratios. For the subsequent discussion pertaining to bumpless transfer and windup protection, three different steam-to-liquid ratios will arise:

Ratio based on current set point for steam flow. Output tracking uses the following ratio:

$$R_{SP} = \frac{F_{SP}}{W}$$

where
> R_{SP} = value for ratio coefficient for current steam flow set point
> F_{SP} = set point for the steam flow (the controlled flow)
> W = liquid flow (the wild flow)

Ratio based on current steam flow. Integral tracking and external reset use the following ratio:

$$R_{PV} = \frac{F}{W}$$

where
> R_{PV} = value for ratio coefficient for current steam flow
> F = steam flow (the controlled flow
> W = liquid flow (the wild flow)

Ratio based on normal operating conditions. When the liquid flow W is zero or nearly zero, neither R_{SP} nor R_{PV} can be computed. Under these conditions, the ratio must be based on either normal operating conditions or design conditions:

$$R_0 = \frac{F_0}{W_0}$$

where
R_0 = value for ratio coefficient for normal operating conditions
F_0 = steam flow at normal operating conditions
W_0 = liquid flow at normal operating conditions

Output Tracking. The configuration in Figure 6.9 is a temperature-to-flow cascade with a computation inserted between the output of the temperature controller and the flow controller. As for all cascades, output tracking is required in the temperature controller to provide bumpless transfer from local to remote.

To obtain a value for input MRI, the computation between the output of the outer loop (the temperature loop) and the set point of the inner loop (the flow loop) must be inverted. For the configuration in Figure 6.9, the inputs to the liquid outlet temperature controller must be configured as follows:

Input TRKMN: Inverse (logical NOT) of output RMT of the flow controller.
Input MNI: Target for the steam-to-liquid flow ratio (ratio R_{SP}, which is the steam flow set point divided by the liquid flow).

The logic is expressed as follows:

```
TC.TRKMN = !FCSTM.RMT
TC.MNI = FCSTM.SP / FTLIQ.PV
```

In the computation for input MNI, division by zero (or by a small number) occurs when the wild flow is stopped. This deserves further attention.

Let FTLIQMIN be the minimum value of the liquid flow that can be measured accurately. Division by zero can be avoided by using the following expression for input MNI to the liquid outlet temperature controller:

```
TC.MNI = FCSTM.SP / max(FTLIQ.PV, FTLIQMIN)
```

Although division by zero cannot occur, other issues should be considered. When the wild flow is zero.

Zero Liquid Flow. The discussion below is intended for those applications where on occasions the wild flow is stopped briefly (minutes or hours) and then resumed. The frequency of stopping the wild flow is such that requiring operator intervention becomes a nuisance and a distraction. Various issues must be addressed for the controls to recover smoothly when the wild flow resumes.

For the configuration in Figure 6.9 (flow controller with a computed set point), the following consequences with regard to stopping the wild flow will be considered:

- The measured value of the wild flow may be a small positive or negative value instead of exactly zero.
- In some applications (including the steam-heated exchanger), the measured variable for the feedback trim controller is not meaningful when the wild flow is stopped.
- Should tracking begin when the wild flow is stopped, the initialization calculations for output tracking will encounter division by zero (or by a number that is nearly zero).
- If left on automatic, the feedback trim controller is exposed to windup.

Liquid Flow Measurement Issues. To avoid a liquid flow of zero (or nearly zero), one approach is to use the two-stage cutoff block presented in Table 1.6. In Figure 6.12 a cutoff block is inserted between the liquid flow measurement and the wild flow input to the ratio computation. To force small flows to zero, the two-stage cutoff block is configured as follows (attributes are defined in Table 1.6):

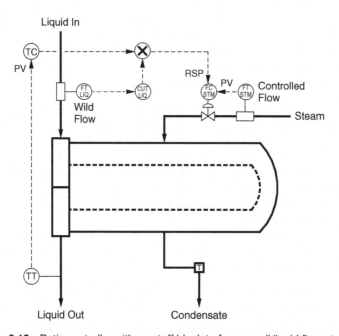

Figure 6.12 Ratio controller with a cutoff block to force small liquid flows to zero.

Input X: measured value of liquid flow

Input Y0: zero

Input Y1: measured value of liquid flow

Input XC: a flow less than this value is considered to be zero

For the steam-heated exchanger, the steam flow should be blocked when the liquid flow is zero. That is, if the liquid flow is zero, the control valve on the steam should be fully closed. Figure 6.12 attempts to achieve this by assuring a steam flow set point of zero. Another way to assure that the steam valve is closed is to configure output tracking for the flow controller. When the liquid flow is stopped, output Q0 of the cutoff block is true. To activate output tracking in the flow controller when the liquid flow is stopped, inputs TRKMN and MNI are configured as follows:

```
FCSTM.TRKMN = CUTLIQ.Q0
FCSTM.MNI = -2.0
```

The appropriate value for input MNI is the lower output limit for the steam flow controller, the value being -2% for most examples in this book.

However, configuring output tracking for the flow controller has one side effect. Regardless of the operational mode of the controller, the output of the controller will be set to the value specified by input MNI when input TRKMN is true. Even if the operator switches the flow controller to automatic and specifies a nonzero value for its set point, the steam valve remains closed. For the steam-heated exchanger, this behavior is probably desirable. With the liquid flow stopped, opening the steam valve quickly pressurizes the steam chest to the steam supply pressure. Thereafter, the steam flow must equal the steam condensation rate, which is determined by the heat loss (hopefully, small). For the exchanger, operating the steam flow controller in automatic with the liquid flow stopped makes no sense. However, this is not true for all ratio control applications.

Liquid Outlet Temperature Measurement. For the exchanger (and in many but not all other applications of ratio control), a wild flow of zero raises issues regarding the measured variable for the feedback trim controller. The measurement device for the liquid outlet temperature is usually installed in the piping immediately downstream of the exchanger. But when the liquid flow is zero, this measurement does not truly reflect the liquid outlet temperature. In fact, with no liquid flowing through the exchanger, it is not clear what "liquid outlet temperature" really means.

If the flow controller remains on cascade and the temperature controller remains on automatic, the consequences depend on the value of the liquid outlet temperature relative to its set point:

Liquid outlet temperature is above its set point. The temperature controller reduces the target for the steam-to-liquid ratio to its lower output limit.

Liquid outlet temperature is below its set point. The temperature controller increases the target for the steam-to-liquid ratio to its upper output limit.

Either case is windup (changes in the controller output have no effect on the process variable).

When the liquid flow is zero, the measured value of the liquid outlet temperature must not be used in any calculations. Considerations include the following:

- Enabling PV tracking in the liquid outlet temperature controller is not advisable.
- The computations for integral tracking and external reset require a value for the control error, which is computed from the liquid outlet temperature and its set point. This can be avoided by activating output tracking when the liquid flow is zero. Neither the normal control calculations nor windup protection calculations are performed when output tracking is active.

Output Tracking Issues. For the steam-heated exchanger, output tracking should be invoked in the liquid outlet temperature controller under two circumstances:

Steam flow controller is not on remote. The appropriate value for the temperature controller output is the current value of the ratio R_{SP}, which is computed from the current steam flow set point and the current liquid flow. Input TRKMN should be the inverse (logical NOT) of output RMT of the steam flow controller; input MNI should be the value computed for R_{SP}.

When the liquid flow is stopped. ($W \cong 0$). The appropriate value for the temperature controller output is the steam-to-liquid ratio R_0 for either normal operating conditions or design conditions. Input TRKMN should be the output Q0 from the cutoff block that detects when the wild flow is zero; input MNI should be the value of R_0.

Output tracking is active on two conditions, so the logic for inputs TRKMN and MNI to the liquid outlet temperature controller is as follows:

```
TC.TRKMN = (!FCSTM.RMT) | CUTLIQ.Q0
if (CUTLIQ.Q0)
   TC.MNI = R0
else
   TC.MNI = FCSTM.SP / CUTLIQ.Y
```

The objective of output tracking is to provide a smooth transition when the steam flow controller is switched from local to remote. In practice, a small bump is at most a nuisance; it is the large bumps that must be avoided. By always using the value of R_0 for input MNI, the output tracking logic for the liquid outlet temperature controller simplifies to the following:

```
TC.TRKMN = (!FCSTM.RMT) | CUTLIQ.Q0
TC.MNI = R0
```

With the simplified logic, a small bump will probably occur when the steam flow controller is switched from automatic to remote. In most ratio control applications, the following observations apply:

- The feedback trim controller makes only small adjustments in the ratio coefficient. Using a fixed value of the ratio coefficient for output tracking will result in a small change or bump in the steam flow set point when the flow controller is switched from local to remote.
- Switching the steam flow controller to local should be infrequent. If this occurs frequently, the reasons for doing so should be understood and addressed.

The consequences should be a minor nuisance that occurs infrequently, so keeping it simple deserves serious consideration.

Windup in the Feedback Trim Controller. This discussion pertains to the control configuration in Figure 6.12, which includes a cutoff block to force small values of the liquid flow to zero.

The purpose of the temperature controller is to provide feedback trim, so this controller surely contains reset action. As stated in Chapter 1, windup will occur in such controllers anytime the following statement is true:

Reset windup occurs in a controller when changes in the controller output have no effect on the process variable.

There are two conditions where this could occur:

Liquid flow (wild flow) is zero. Regardless of the value of the liquid outlet temperature controller output, multiplying by zero gives zero for the steam flow set point. Changes in the temperature controller output have no effect on the steam flow or the liquid outlet temperature. Activating output tracking when the liquid flow is zero will avoid windup. This justification is valid for all ratio applications. The previous justification for activating output tracking on a liquid flow of zero was because the liquid outlet temperature measurement is meaningless when the liquid flow is zero. This is not the case for all ratio control applications.

Steam flow has attained its maximum value. Three possible reasons are:
- The set point for the steam flow is at its upper limit.
- The control valve on the steam supply is fully open.
- Heat transfer limited conditions arise in the exchanger.

If any one of these has occurred, further increases in the liquid outlet temperature controller output have little or no effect on either the steam flow or the liquid outlet temperature. Consequently, windup protection is required.

In some applications of ratio control, windup is possible when the controlled flow attains a minimum value. For the exchanger, the minimum value is zero, which could be attained for the following reasons:

Liquid flow is zero. Output tracking is activated, making windup protection unnecessary.

Temperature controller output is at its lower output limit. (-2%). Windup protection is activated within a PID block should this occur.

In those applications where the lower limit is not zero, windup protection would be required at the minimum value. For example, in combustion processes a minimum firing rate must be enforced.

Windup Protection at Maximum Steam Flow. The maximum possible value for the steam flow is determined by the following:

Upper set point limit for steam flow controller. Although lower values are sometimes required by an application, the upper set point limit is usually the upper range value for the steam flow measurement. For the PID block in Table 1.1, attribute SPH is true should the set point attain its upper limit.

Steam control valve fully open. Before the steam flow set point has been increased to its upper limit, it is possible that the steam control valve is driven fully open (or actually, the steam flow controller output is driven to its upper output limit). For the PID block in Table 1.1, attribute QH is true should the controller output attain the upper output limit.

Heat transfer limited conditions. Before the control valve has been driven fully open, the temperature of the condensing steam may approach the steam supply temperature, which means that the exchanger is heat transfer limited. Opening the steam valve further only slightly increases the condensing steam temperature, the heat transfer rate, the steam flow, and the liquid outlet temperature. The effect is so small that the controller cannot function and windup ensues.

The "line in the sand" for heat transfer limited conditions is fuzzy, but usually is about 90% of the maximum possible heat transfer rate. Since the line of demarcation is fuzzy, the maximum heat transfer rate can be computed using the arithmetic average instead of the logarithmic mean:

$$\Delta T_{max} = \frac{(T_S - T_{in}) + (T_S - T_{out})}{2} = T_S - \frac{T_{in} + T_{out}}{2} = T_S - T_{avg}$$

where
ΔT_{max} = maximum possible temperature difference for heat transfer
T_S = saturation temperature at steam supply pressure
T_{in} = liquid inlet temperature

T_{out} = liquid outlet temperature

$T_{avg} = (T_{in} + T_{out})/2$ = average liquid temperature

The temperature difference for heat transfer limited conditions ΔT_{HTL} is approximately 90% of the maximum possible temperature difference ΔT_{max}:

$$\Delta T_{HTL} \cong 0.9 \Delta T_{max}$$

The corresponding condensing steam temperature $T_{C,HTL}$ is computed as follows:

$$T_{C,HTL} - T_{avg} \cong 0.9 \Delta T_{max} = 0.9(T_S - T_{avg})$$

$$T_{C,HTL} \cong T_{avg} + 0.9(T_S - T_{avg})$$

$$= T_S - 0.1(T_S - T_{avg}) = 0.9 T_S + 0.1 T_{avg}$$

To determine the onset of heat transfer limited conditions, the condensing steam temperature T_C must be measured and compared to $T_{C,HTL}$:

$$\text{heat transfer limited if } T_C \geq T_{C,HTL}$$

In most applications, a value for $T_{C,HTL}$ can be computed from normal operating conditions and treated as a coefficient in the calculations. However, occasionally, $T_{C,HTL}$ must be computed from measured inputs for T_{in}, T_{out}, and T_S.

Windup Prevention Using Integral Tracking. In simple cascades, input MRI to the outer loop is configured as the PV for the inner loop of the cascade. But for ratio control, the appropriate value of the ratio must be computed from the current value of the PV and the current value of the wild flow. This is the ratio R_{PV} defined previously.

Integral tracking must be active under any of the following conditions:

Steam flow set point is at its upper limit. Output SPH of the steam flow controller is true.

Steam flow controller output is at the upper output limit. Output QH of the steam flow controller is true.

Heat transfer limiting conditions have been attained. The condensing steam temperature (TTCOND.PV) equals or exceeds the condensing steam temperature for the onset of heat transfer limiting conditions (TCHTL).

The logic to configure both output tracking and integral tracking for the liquid outlet temperature controller is as follows:

```
TC.TRKMN = (!FCSTM.RMT) | CUTLIQ.QO
if (CUTLIQ.QO)
    TC.MNI = RO
```

```
else
   TC.MNI = FCSTM.SP / CUTLIQ.Y
TC.TRKMR = FCSTM.SPH | FCSTM.QH | (TTCOND.PV >= TCHTL)
if (!CUTLIQ.QO)
   TC.MRI = FTSTM.PV / CUTLIQ.Y
```

For output tracking, the value of input MNI depends on whether or not the liquid flow of zero (or nearly zero). Since output tracking takes priority over the control calculations and integral tracking, the value of input MRI is not used when the flow is zero or nearly zero.

Windup Prevention Using External Reset. The control configuration must be consistent with the following requirements:

- The value of input XRS for external reset input must be the ratio R_{PV}, which is the steam flow set point divided by the liquid flow.
- Output tracking must be activated if the steam flow controller is not on remote or if the liquid flow is zero.

The logic to configure both output tracking and external reset for the liquid outlet temperature controller is as follows:

```
TC.TRKMN = (!FCSTM.RMT) | CUTLIQ.QO
if (CUTLIQ.QO)
   TC.MNI = RO
else
   TC.MNI = FCSTM.SP / CUTLIQ.Y
if (!CUTLIQ.QO)
   TC.XRS = FTSTM.PV / CUTLIQ.Y
```

It is crucial that output tracking be configured. Input XRS is the value computed for the ratio R_{PV}, but this ratio is meaningless when the liquid flow is zero.

Windup Prevention Using Inhibit Increase/Inhibit Decrease. For the steam-heated exchanger, windup protection is required only when the maximum steam flow has been attained. Input NOINC must be true on any combination of the following:

Steam flow set point is at its upper limit. Output SPH of the steam flow controller is true.

Steam flow controller output is at the upper output limit. Output QH of the steam flow controller is true.

Heat transfer limiting conditions have been attained. The condensing steam temperature (TTCOND.PV) equals or exceeds the condensing steam temperature for the onset of heat transfer limiting conditions (TCHTL).

The logic to configure both output tracking and inhibit increase/inhibit decrease for the liquid outlet temperature controller is as follows:

```
TC.TRKMN = (!FCSTM.RMT) | CUTLIQ.QO
if (CUTLIQ.QO)
   TC.MNI = RO
else
   TC.MNI = FCSTM.SP / CUTLIQ.Y
TC.NOINC = FCSTM.SPH | FCSTM.QH | (TTCOND.PV >= TCHTL)
```

For ratio control applications that also require windup protection when the minimum steam flow has been attained, input NODEC must also be configured to prevent further decreases in the controller output should a low limit be encountered.

6.4. DYNAMIC COMPENSATION

For many applications, maintaining the controlled flow in algebraic proportion to the wild flow is all that is required. Although rarely perfect, the transient changes in the controlled variable are small and short-lived. However, there are exceptions.

The response in the controlled variable is a combination of its response to the change in the controlled flow and its response to the change in the wild flow. Two aspects of each response must be considered:

Magnitude. The value of the ratio coefficient reflects the difference in this aspect of the two responses.

Dynamics. This reflects how rapidly the change in each flow affects the controlled variable.

With regard to the dynamics, there are three possibilities:

Dynamics are approximately the same. In this case, changing the controlled flow in algebraic proportion to changes in the wild flow is appropriate.

Dynamics of the controlled flow are faster than the dynamics of the wild flow. If the controlled flow is changed in algebraic proportion to changes in the wild flow, the effect of the controlled flow appears in the controlled variable ahead of the effect of the wild flow. For this case, a lag should be incorporated into the ratio control logic.

Dynamics of the wild flow are faster than the dynamics of the controlled flow. If the controlled flow is changed in algebraic proportion to changes in the wild flow, the effect of the wild flow appears in the controlled variable ahead of the effect of the controlled flow. For this case, a lead should be incorporated into the ratio control logic.

Two options for providing the dynamic compensation are as follows:

Lead-lag block. (described in Chapter 1). Incorporating the block into the control configuration is the easy part; the difficulty is that the lead and lag times must be "tuned" to the process characteristics.

Ramp all changes. Some processes permit changes to be implemented at a rate that is at the discretion of the engineers. By slowing the rate at which changes are implemented, the steady-state relationships for feedforward control suffice.

In this section, only the former is examined.

Tuning the Lead-Lag Compensator. In the context of ratio control, the lead-lag block is providing dynamic compensation and is often referred to as the lead-lag compensator. The lead time τ_{LD} and the lag time τ_{LG} are coefficients whose values must reflect the characteristics of the process.

How does one tune a lead-lag compensator? Usually, with great difficulty. However, it is helpful to approach this endeavor using slightly different coefficients:

- τ_{LD}/τ_{LG}. The ratio basically determines the *shape factor* for the output of the compensator. If this ratio is 1.0, the output of the compensator is the same as the input. If this ratio is less than 1.0, the lead-lag compensator retards the correction. If this ratio is greater than 1.0, the lead-lad compensator advances the correction. Figure 1.9 illustrates this effect clearly.

- τ_{LG}. The lag time determines the time frame for adjusting the corrective action. To maintain a constant value for the ratio τ_{LD}/τ_{LG}, doubling the lag time τ_{LG} would require that the lead time τ_{LD} also be doubled. The result would be to double the time that the corrective action is applied.

In most applications, whether the lead-lag needs to advance ($\tau_{LD} > \tau_{LG}$) or retard ($\tau_{LD} < \tau_{LG}$) the corrective action can usually be ascertained by analyzing the process. However, it is not easy to determine by how much to advance or retard the corrective action. Nor is it easy to determine the time frame for applying the corrective action.

Steam-Heated Exchanger. For a steam-heated exchanger, maintaining a constant steam-to-liquid flow ratio effectively compensates for the long-term or steady-state effects of liquid flow changes on the liquid outlet temperature. But as Figures 6.6 and 6.10 illustrate, the dynamics of the process result in a transient period and a significant departure of the liquid outlet temperature from its desired value.

The magnitude of this transient effect depends on the difference between:

- The dynamics of the relationship between liquid flow (the wild flow) and liquid outlet temperature

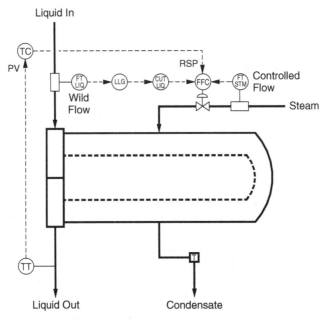

Liquid In

PV

TC

RSP

FT
LIQ — LLG — CUT
LIQ — FFC — FT
STM Controlled
Flow

Wild
Flow

Steam

TT

T

Liquid Out

Condensate

Figure 6.13 Dynamic compensation applied to the wild flow, flow-to-flow controller configuration.

- The dynamics of the relationship between steam flow (the controlled flow) and liquid outlet temperature

It is the difference that is important. For the exchanger, the dynamics are quite different, and the transient is very noticeable.

Ratio Configurations with Dynamic Compensation. For the flow-to-flow controller implementation of ratio control, the dynamic compensation must be applied to the measured value of the wild flow. As illustrated in Figure 6.13, the input to the lead-lag block (LLG) is the measured value of the liquid flow. The output of the lead-lag block is the input to the cutoff block that forces small flows to zero. Figure 6.14 is the corresponding configuration when ratio control is implemented by computing the set point to a flow controller. The input to the lead-lag block (LLG) is the measured value of the liquid flow. The output of the lead-lag block is the input to the cutoff block that forces small flows to zero.

An alternative approach is to insert the lead-lag block between the multiplier that computes the steam flow set point and the steam flow controller. Figure 6.15 presents this approach. As multiplication is not a linear operation, the configuration in Figure 6.15 is not exactly equivalent to the configuration in Figure 6.14. However, the difference in performance would only be significant when there are large changes in the ratio coefficient (the output of the temperature controller).

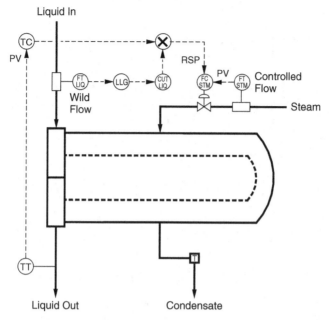

Figure 6.14 Dynamic compensation applied to wild flow, flow controller with computed set point.

But since the temperature controller is providing feedback trim, large changes in its output would not be expected.

Linear Approximations. The representation of the exchanger in Figure 6.16 is based on the principle of superposition:

- For a change in the liquid flow W, let the response in the exchanger outlet temperature be $T_{\text{out},W}$.
- For a change in the steam flow F, let the response in the exchanger outlet temperature be $T_{\text{out},F}$.

The response T_{out} in the liquid outlet temperature to a change in W and a change in F is

$$T_{\text{out}} = T_{\text{out},W} + T_{\text{out},F}$$

The principle of superposition is a property of linear systems. Heat transfer processes are not linear, so the principle of superposition would at best approximate the behavior of the exchanger. Generally, the larger the changes in the liquid and steam flows, the greater the error in linear approximations to nonlinear systems.

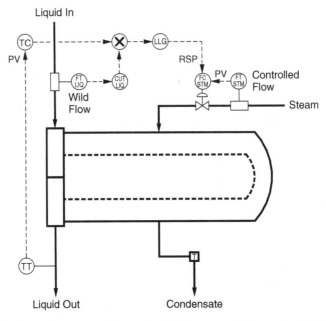

Figure 6.15 Dynamic compensation applied to the computed value for the flow controller set point.

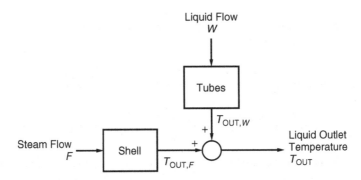

Figure 6.16 Linear approximation for the steam-heated exchanger.

Response to an Increase in the Liquid Flow. The response in Figure 6.17 is to a change in liquid flow from 1000 lb/min to 1200 lb/min with the steam flow held constant at 46.2 lb/min. There are two aspects of the response:

Steady-state. The liquid outlet temperature decreases from 280.0°F to 268.8°F for a change of 11.2°F. Ignoring signs, the gain K_W is approximately

$$K_W = \frac{11.2°F}{200 \text{ lb/min}} = 0.056°F/(\text{lb/min})$$

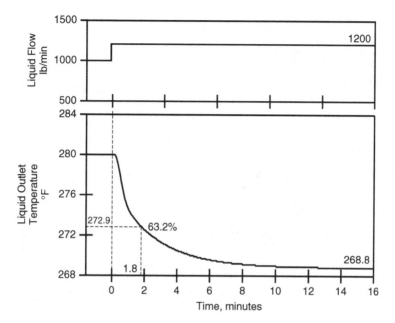

Figure 6.17   Response of liquid outlet temperature to a change in liquid flow.

Dynamics. A first-order lag is the simplest approximation for the dynamics. The value of the time constant is obtained most easily from the time
required for the response to change by 63.2% of the total change. The total
change is 11.2°F; 63.2% of this change is 7.1°F. A decrease of 7.1°F from
the initial value of 280.0°F gives a temperature of 272.9°F. As indicated on
the response in Figure 6.17, the 63.2% change is attained at approximately
1.8 min after the liquid flow is changed.

Response to a Decrease in the Steam Flow.   The response in Figure 6.18
is to a decrease in steam flow from 46.2 to 41.2 lb/min with a constant liquid
flow of 1000 lb/min. There are again two aspects of the response:

Steady-state. The liquid outlet temperature decreases from 280.0°F to 273.0°F
for a change of 7.0°F. The gain K_F is approximately

$$K_F = \frac{7.0°F}{5 \text{ lb/min}} = 1.4°F/(\text{lb/min})$$

Dynamics. A first-order lag will again be used to approximate the dynamics
and the time constant determined from the 63.2% point on the response.
The total change is 7.0°F; 63.2% of this change is 4.4°F. A decrease of
4.4°F from the initial value of 280.0°F gives a temperature of 275.6°F. As

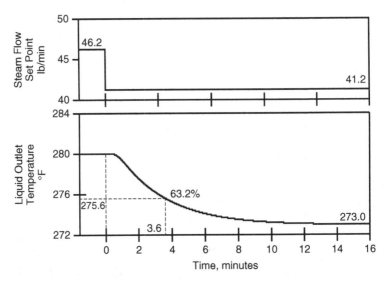

Figure 6.18 Response of liquid outlet temperature to a change in steam flow set point.

indicated on the response in Figure 6.18, the 63.2% change is attained at approximately 3.6 min after the liquid flow is changed.

Linear Model of the Exchanger. These test results provide the basis for a simple linear model for the exchanger. The dynamics of both tubes and shell are approximated by a first-order lag. The linear model is represented by the block diagram in Figure 6.19, the components being as follows:

Liquid flow. The steady-state gain K_W is 0.056°F/(lb/min); the time constant τ_W is 1.8 min. Increasing the liquid flow decreases the liquid outlet temperature, so the sign on the summer for the input from the tubes is negative.

Steam flow. The steady-state gain K_F is 1.4°F/(lb/min); the time constant τ_F is 3.6 min. Increasing the steam flow increases the liquid outlet temperature, so the sign on the summer for the input from the shell is positive.

A number of deficiencies can be cited for this simple model:

- The dynamics for neither the liquid flow nor the steam flow are approximated accurately by a first-order lag. However, the accuracy of a model must reflect the intended use for the model. In ratio control applications, the dynamic compensation need not be perfect. The simpler the model, the simpler the relationships for the dynamic compensation.

- Most processes, and the exchanger is no exception, are nonlinear. The values for both the gains and the time constants depend on the throughput. The effect of this on the ratio coefficient is addressed by the feedback trim

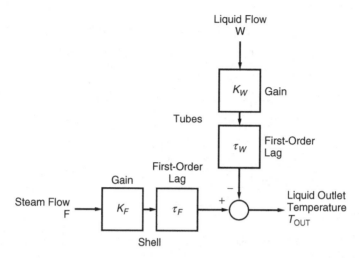

Figure 6.19 Simple linear model for the steam-heated exchanger.

controller. Changes in the dynamics affect the required dynamic compensation, but as discussed previously, the dynamic compensation need not be perfect.

As always, the best philosophy is to "keep it simple": Start with the simplest configuration that has some possibility of working, and then enhance as necessary to attain the performance required.

Ratio Control Formulation. The relationships for ratio control can be derived from the simple linear model in Figure 6.19. Since both linear models contain steady-state and dynamic components, the ratio controller must also contain these components:

Steady-state. If the liquid flow changes by 1 unit, that is, by 1 lb/min, what change in the steam flow is required to compensate for this change? A change in liquid flow of 1 lb/min will cause the liquid outlet temperature to change by K_W °F. To change the liquid outlet temperature by 1°F, the steam flow must change by $1/K_F$ lb/min. Consequently, for a change of 1 lb/min in the liquid flow, the ratio controller should change the steam flow by K_W/K_F lb/min.

Dynamics. The dynamics of a change in the liquid flow on the liquid outlet temperature is a lag whose time constant is τ_W. The corrective action must contain this same lag. The dynamics of a change in the steam flow on the liquid outlet temperature is a lag whose time constant is τ_F. The corrective must contain a lead to offset the effect of this lag. Consequently, the dynamic compensator should be a lead-lag element, the lead time being τ_F and the lag time being τ_W.

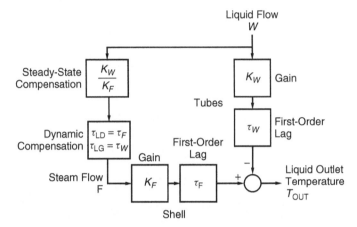

Figure 6.20   Ratio controller based on simple linear model for the steam-heated exchanger.

The block diagram in Figure 6.20 includes the ratio controller formulated on this basis. For every relationship in Figure 6.20, the steady-state component precedes the dynamic component. However, for linear models, the order is irrelevant.

Ratio Coefficient. When ratio control is implemented using a simple ratio, the ratio coefficient is the ratio of the controlled flow (the steam flow) to the wild flow (the liquid flow). For the tests in Figures 6.17 and 6.18, the starting conditions for each are as follows:

- *Liquid flow W*: 1000 lb/min
- *Steam flow F*: 46.2 lb/min
- *Liquid outlet temperature* T_{out} : 280.0°F

These data suggest that the appropriate value for the ratio coefficient R is as follows:

$$R = \frac{F}{W} = \frac{46.2 \text{ lb/min}}{1000 \text{ lb/min}} = 0.0462 \text{ lb/lb (actually, lb steam/lb liquid)}$$

However, one has to be a careful with this approach. To attain a liquid outlet temperature of 280°F, Figure 6.21 presents the graph of the required steam flow as a function of the liquid flow. When using a single point to compute the ratio coefficient, the assumption is made that the graph is essentially a straight line passing through the origin (that is, if the liquid flow is zero, the steam flow should also be zero). In Figure 6.21, the relationship for the simple ratio is represented by a dashed line that passes through the origin and the point ($W = 1000$, $F = 46.2$). This dashed line is a very good approximation to the graph of the required steam

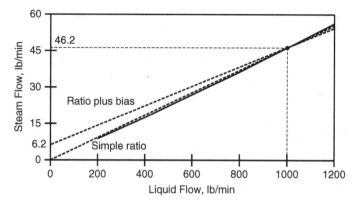

Figure 6.21 Required steam flow as a function of liquid flow (liquid outlet temperature of 280°F).

flow as a function of the liquid flow. Consequently, the simple ratio is appropriate. However, this is not always the case, and one must never just casually assume so.

The ratio coefficient can also be computed from the gains K_T and K_S of the linear model:

$$R = \frac{K_W}{K_F} = \frac{0.056°F/(\text{lb/min})}{1.4°F/(\text{lb/min})} = 0.04 \text{ lb/lb}$$

This value is significantly in error. If applied as a simple ratio, this relationship predicts that 40.0 lb/min of steam would be required to heat 1000 lb/min of liquid to 280°F.

However, applying this relationship as a simple ratio is not the correct interpretation of the linear approximation. The linear model does not relate the required steam flow directly to the liquid flow. Instead, from some reference point it relates the change in the required steam flow to a change in the liquid flow. Normally, the reference point is the process conditions for the data from which the model coefficients were derived. For Figures 6.17 and 6.18, the reference point is a liquid flow of 1000 lb/min and a steam flow of 46.2 lb/min. Consequently, the relationship for the linear model is as follows:

$$R = \frac{F - 46.2}{W - 1000} = \frac{K_W}{K_F} = 0.04 \text{ lb/lb}$$

Solving this relationship for F gives a ratio-plus-bias equation:

$$F = 0.04W + 6.2$$

In Figure 6.21, this relationship is represented by the dashed line that passes through the point ($W = 1000$, $F = 46.2$) and has a y-intercept of 6.2 (the bias). This relationship is not nearly as good as the simple ratio.

This example illustrates the usual pitfalls of applying linear approximations to nonlinear processes. Heat transfer processes are quite nonlinear, so it should be no surprise that the linear approximations do not yield especially good relationships. But as is always the case, the linear approximation is exact at the reference point, with the error increasing as process operating conditions move away from the reference point.

Dynamic Compensation. Based on the linear approximation, the formulation of the dynamic compensator is based on the following logic:

- The lag in the path from the wild flow to the controlled variable must be matched by a lag in the dynamic compensator:

$$\text{lag time} = \tau_{\text{LG}} = \tau_W = 1.8 \text{ min} = \text{lag on changes in liquid flow}$$

- The lag in the path from the controlled flow to the controlled variable must be offset by a lead in the dynamic compensator:

$$\text{lead time} = \tau_{\text{LD}} = \tau_F = 3.6 \text{ min} = \text{lag on changes in steam flow}$$

Figure 6.22 presents the performance of the ratio control configuration using this dynamic compensation. The change is an increase in the liquid flow from 1000 lb/min to 1200 lb/min, followed by a decrease in the liquid flow from 1200 lb/min to 1000 lb/min. The liquid outlet temperature controller is on manual, so the steam-to-liquid ratio is constant (0.0462 lb/lb).

In Figure 6.22, the measured value of the liquid flow exhibits a small lag. This value is the input to the lead-lag compensator. The output of the lead-lag compensator is multiplied by the ratio coefficient to obtain a value for the steam flow set point. Figure 6.22 presents the trend for the set point of the steam flow controller. Since the lead-lag compensator has a lead time of 3.6 min and a lag time of 1.8 min, lead is imparted to the steam flow set point. Without dynamic compensation, the increase in the liquid flow causes the steam flow set point to increase rapidly from 46.2 lb/min to 55.5 lb/min. But with dynamic compensation, the steam flow set point rapidly increases to 62.8 lb/min, and then slowly reduces to 55.5 lb/min.

For the liquid flow increase from 1000 lb/min to 1200 lb/min, the difference in performance with and without the lead-lag compensator is as follows:

Without dynamic compensation. The liquid outlet temperature drops to approximately 275.6°F, for a dip of 4.4°F.

With dynamic compensation. The liquid outlet temperature first decreases to approximately 276.4°F, for a dip of 3.6°F. However, the liquid outlet temperature then increases to 281.1°F, for an overshoot of just over 1°F.

Dynamic compensation is providing some improvement, but it is not impressive:

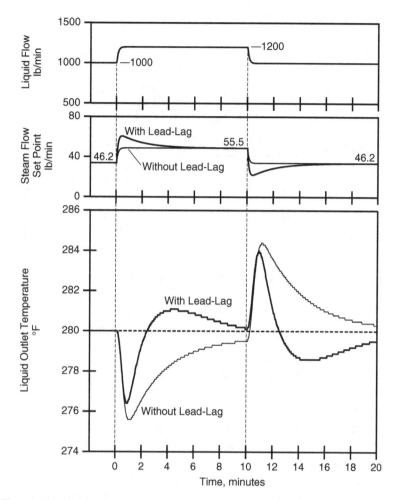

Figure 6.22 Performance of dynamic compensation, $\tau_{LD} = 3.6$ min, $\tau_{LG} = 1.8$ min.

- For the increase in liquid flow, dynamic compensation has reduced the peak error from 4.4°F to 3.6°F, a reduction of 18%. To reduce the peak error further, the ratio of lead to lag (τ_{LD}/τ_{LG}) must be increased.
- With dynamic compensation, the response is both below and above the original value. This suggests that the time interval over which the corrective action is being applied is too long. Both τ_{LG} and τ_{LD} must be reduced, maintaining a constant τ_{LD}/τ_{LG} ratio.

Tuning the Lead-Lag. As noted previously, tuning the lead-lag compensator is best approached using the lag time τ_{LG} and the lead-to-lag ratio τ_{LD}/τ_{LG}. The starting point is the response in Figure 6.22 for a lag time τ_{LG} of 1.8 min and

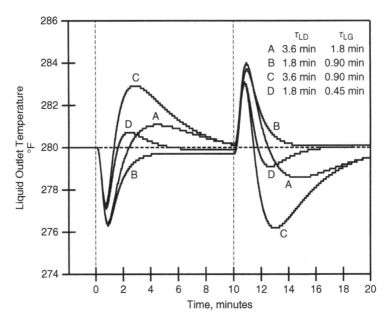

Figure 6.23 Tuning the lead-lag compensator.

a ratio τ_{LD}/τ_{LG} of 2.0, which is response A in Figure 6.23. The tuning effort proceeds as follows:

- As observed previously, the dynamic compensation is being applied for too long. The lag time τ_{LG} will be reduced to 0.9 min and the lead time τ_{LD} reduced to 1.8 min to maintain a lead-to-lag ratio τ_{LD}/τ_{LG} of 2.0. The result is response B in Figure 6.23.
- On the increase in the liquid flow, this response dips below the target but does not subsequently overshoot the target. The magnitude of the peak error is basically the same as before. To reduce the peak error, the lead-to-lag ratio will be increased to 4.0, giving a lead time τ_{LD} of 3.6 min and a lag time τ_{LG} of 0.9 min. The result is response C in Figure 6.23.
- The peak errors are smaller. But on the increase in the liquid flow, the dip below the target is followed by a significant overshoot, again suggesting that the dynamic compensation is being applied for too long. The lag time τ_{LG} will be reduced to 0.45 min and the lead time τ_{LD} reduced to 1.8 min to maintain a lead-to-lag ratio τ_{LD}/τ_{LG} of 4.0. The result is response D in Figure 6.23.

Clearly, dynamic compensation is now improving the performance. To obtain further improvement, the magnitude of the initial departure of the liquid outlet temperature from its initial value must be reduced. This can only be achieved by increasing the ratio τ_{LD}/τ_{LG}. However, a value of 4.0 for this ratio imparts significant lead to the dynamic compensation. Adding more lead to the lead-lag compensator is comparable to increasing the derivative time in a PID controller. One must be careful with large values for the derivative time, and one must also be careful with large leads in a lead-lag compensator.

The lead-lag compensator generally performs better when it must apply more lag than lead, that is, when $\tau_{LG} > \tau_{LD}$. When the lead-lag must apply more lead than lag, there are limits on what can be achieved. For the increase in the liquid flow presented in Figure 6.17, the outlet temperature drops very rapidly. The control system must attempt to offset this by making adjustments in the steam flow. But as Figure 6.18 indicates, the liquid outlet temperature responds more slowly to changes in the steam flow. The dynamic compensator attempts to offset the difference in speed initially by making larger than necessary changes in the steam flow set point (an overcorrection) and then backing out the overcorrection. However, only so much can be achieved with this approach.

Fortunately, the tuning for the lead-lag compensator does not have to be perfect—it only has to be "good enough," whatever that means. Before embarking on tuning the lead-lag compensator, the first step should be to quantify what "good enough" means. That is, for a given change in the wild flow, what deviation from the target is acceptable? In many cases, the performance of steady-state compensation alone will meet the criteria for "good enough." If one has to include dynamic compensation, the objective of the tuning effort should be to meet the requirements for "good enough"; there is no reward in optimizing the performance of the dynamic compensator.

Beyond Lead-Lag Compensation. The most common addition to the dynamic compensation is a dead time. This is very likely to be required in applications in the sheet processing industries, where the processes are dominated by dead time. Occasionally, dead time alone provides adequate dynamic compensation.

With digital systems, dynamic compensation beyond the lead-lag can be implemented. Configuring two lead-lag elements in series gives a second-order lead and a second-order lag. But before proceeding in this direction, ascertain how the values for the coefficients will be obtained. The responses must be approximated by second-order lags instead of first-order lags. However, these are still linear approximations. If the major source of error is the consequence of nonlinear behavior on the principle of superposition, more accurate approximations for the individual components may not yield any benefit.

Initialization. The lead-lag element is described by a differential equation along with an initial condition for the output of the lead-lag element. Consequently, most lead-lag elements provide an input to cause the lead-lag element to be initialized, which basically forces the output of the lead-lag element to a specified value. For the block in Table 1.3, input TRKMN serves this purpose. The value for the output can be either of the following:

- The current value of the input to the lead-lag element. In most applications (and for this feedforward example), this is the desired behavior.
- A value specified by another input to the function block. For the block in Table 1.3, input MNI provides this value. This provides a little more flexibility, but is rarely required.

First consider the control configuration in Figure 6.14, where the lead-lag compensator is on the liquid flow. The output of the liquid outlet temperature controller is to be initialized to the steam-to-liquid ratio computed from the current liquid flow (assuming that it is not zero). To achieve bumpless transfer of the steam flow controller to remote, the output of the lead-lag element must also be the current value of the liquid flow. Therefore, whenever output tracking is active in the liquid outlet temperature controller (input TRKMN is true), tracking must also be active in the lead-lag compensator. When tracking is active, the output of the lead-lag compensator must be the same as its input, which is the current liquid flow.

Next, consider the control configuration in Figure 6.15, where the lead-lag compensator is on the output of the ratio calculation. The output of the liquid outlet temperature controller is to be initialized to the steam-to-liquid ratio computed from the current liquid flow (assuming it is not zero). To achieve bumpless transfer of the steam flow controller to remote, the output of the lead-lag element must also be the value computed for the steam-to-liquid ratio. Therefore, whenever output tracking is active in the liquid outlet temperature controller (input TRKMN is true), tracking must also be active in the lead-lag compensator. When tracking is active, the output of the lead-lag compensator must be the same as its input, which is the value computed for the steam-to-liquid ratio.

Basically, dynamic compensation is required when the normal control actions are being performed. Whenever tracking or initialization is active, dynamic compensation must be "turned off," which usually means initializing the dynamic elements so that the output equals the input.

6.5. RATIO PLUS BIAS

Most applications of ratio control require only a simple ratio: The controlled flow is the wild flow multiplied by the appropriate ratio. But some applications require one of the following:

- Ratio plus bias
- Characterization function

These approaches have one aspect in common with simple ratio control. The logic is driven by only one input: specifically, the measured value of the wild flow.

Ratio-Plus-Bias Relationship. What applications would require a bias in the ratio equation? The answer is: where there is some aspect that requires a constant value for the controlled flow. In the heat exchanger, any significant heat loss requires steam to make up for the loss. Although not exactly constant, the heat loss depends primarily on the temperatures within the equipment. In practice, this is not a very good example of where ratio plus bias should be applied. If the heat loss is significant, additional insulation should be installed. However, the notation from this example is used in the following discussion.

The ratio-plus-bias computation is as follows:

$$F_{SP} = RW + B$$

where

F_{SP} = target for the controlled flow
W = wild flow
R = ratio coefficient of controlled flow to wild flow
B = bias for controlled flow

How does one obtain values for R and B? Although either or both could be computed from the process design equations, the customary approach is to rely on data from one or more steady-state operating points:

One point. Either R or B must be computed from the design equations. The unknown coefficient is computed from the operating point.

Two points. Both R and B are computed from the two operating points.

Three or more points. Such data can be used to construct a plot such as one in Figure 6.21 for the exchanger. If the plot is reasonably linear, values for R and B can be computed using linear regression. If there is a significant departure from linearity, the steady-state operating points provide the data for constructing a characterization function, as discussed in the next section.

Feedback Trim. When using the ratio-plus-bias equation, there are three options for the feedback trim:

• Adjust the value of the ratio coefficient R in the ratio-plus-bias calculation.
• Adjust the value of the bias B in the ratio-plus-bias calculation.
• Provide a multiplier configured as follows:

Input X1: Result of the ratio-plus-bias calculation.
Input X2: Output of the feedback trim controller. The controller must be configured such that its output is a value that is close to 1.0.

The normal control calculations must provide the following computation:

$$F_{SP} = RW + B$$

The ratio coefficient R may be either a variable or a coefficient, depending on how the feedback trim is applied.

Ratio coefficient R is the output of feedback trim controller. The ratio coefficient R is a variable, which necessitates multiplication to compute the product RW. A multiplier block, a general arithmetic equation block, or similar is required to compute the product RW. If the block cannot add the coefficient B to the result, a summer block must also be configured.

Other approaches for feedback trim. The ratio coefficient R is a coefficient. The ratio-plus-bias relationship can be implemented using a summer block that is described by the following equation:

$$Y = k_0 + k_1 X_1 + k_2 X_2$$

where

$Y =$ output of summer (the controlled flow set point F_{SP})
$k_0 =$ coefficient
$X_1 =$ input 1 (the wild flow W)
$k_1 =$ coefficient for input 1 (the ratio coefficient R)
$X_2 =$ input 2
$k_2 =$ coefficient for input 2

If the output of the feedback trim is the bias B, the summer must be configured as follows:

- Input X_2 is the bias B.
- Coefficient k_2 is 1.0.
- Coefficient k_0 is 0.0.

Otherwise, the summer can be configured as follows:

- Input X_2 is not configured.
- Coefficient k_2 is not used when input X_2 is not configured.
- Coefficient k_0 is the bias B.

Output Tracking. Output tracking must be active in the feedback trim controller under at least one condition and possibly two:

1. *Controller for controlled flow is not on remote.* This is always required.
2. *Wild flow is stopped.* This is an issue in many but not all applications for ratio plus bias.

These two considerations determine the configuration for input TRKMN to the feedback trim controller. The configuration for input MNI depends on how feedback trim is configured:

- *Adjust the value of the ratio coefficient R in the ratio-plus-bias calculation.* The value of input MNI must be the value for the ratio coefficient R that is computed as follows:

$$R = \frac{F_{SP} - B}{W}$$

Obviously, issues arise if W is zero, but these will be examined shortly.

- *Adjust the value of the bias B in the ratio-plus-bias caalculation.* The value of input MNI must be the value for the bias B that is computed as follows:

$$B = F_{SP} - RW$$

No issues arise if W is zero.

- *Multiply the output of the ratio-plus-bias calculation by a coefficient k that is the output of the feedback trim controller.* The value of input MNI must be the value for the coefficient k that is computed as follows:

$$k = \frac{F_{SP}}{RW + B}$$

If the bias B is positive, no issues arise if W is zero.

Wild Flow Is Zero. The issues that arise for the ratio-plus-bias configuration are similar to those that arise for simple ratios:

- When the wild flow is zero, the value computed for the controlled flow set point F_{SP} is the value of the bias B. In some applications, this value is appropriate, but in some applications, the controlled flow must be stopped when the wild flow is stopped. As discussed previously for the simple ratio, a cutoff block can be used to detect when the wild flow is small. Output Q0 from the cutoff block can be used to activate output tracking to force the flow controller output to its lower output limit.

- The measured value of the process variable for the feedback trim controller may be meaningless when the wild flow is stopped. If this is the case, the normal control calculations must not continue in the feedback trim controller. Either output tracking must be active or the controller must be switched to manual.

- When the output of the feedback trim controller is the ratio coefficient R, division by zero (or by a value close to zero) can occur in the computation for input MNI. The issues are the same as discussed previously for simple ratios.

Windup Protection. Windup is possible in the feedback trim controller. The considerations depend on the limit that is imposed:

Output of the feedback trim controller. For example, if the output of the feedback trim controller is the ratio coefficient R, the output of the feedback trim controller may be restricted to the design value for the ratio coefficient plus or minus 10%. Such limits should be imposed through the controller output limits so that windup protection is activated should either limit be attained.

Set point for controlled flow. The result of the ratio-plus-bias computation must be limited to the measurement range of the controlled flow. Even more restrictive limits may be imposed. Windup protection must be activated should either limit be attained.

Output of controller for the controlled flow. In most applications it is possible for the control valve to become fully open (actually controller output attains its upper output limit) before the set point is attained. If so, windup protection must be activated. For some applications, analogous issues arise at the lower output limit.

Since the logic for the windup protection for ratio-plus-bias applications is similar to the logic for simple ratio applications, it is not repeated.

6.6. CHARACTERIZATION FUNCTION

A common application of characterization functions within a control configuration is for steam boilers. The objective herein is to illustrate various issues pertaining to the application of feedforward and ratio control. A steam boiler will be used only as the example. In no way is this discussion intended to be a complete treatment of boiler control.

Steam Boiler. For the purposes herein, the simplified flowsheet in Figure 6.24 is sufficient. Water from the steam drum and the feedwater enter the exchanger, where the liquid is partially vaporized. The drum provides for vapor–liquid disengagement, the vapor being the steam product. Although not included in Figure 6.24, another stream exiting the drum is the *blowdown*, a very small flow used to control the hardness of the water in the drum.

Air and fuel (natural gas for this example) are supplied to the burners, which supply the heat to partially vaporize the water entering the exchanger. The boiler illustrated in Figure 6.24 has both an induced-draft fan and a forced-draft fan, both of which are equipped with a variable-speed drive and a speed controller. In older boilers, dampers were installed in lieu of variable-speed drives. Large boilers are normally equipped with both fans; smaller boilers may have only one (usually, an induced-draft fan).

Where multiple boilers supply steam to a common header, an individual boiler may be operated in either of the following modes:

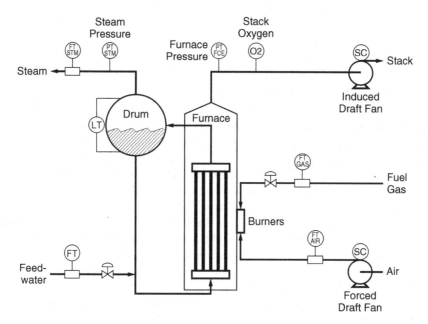

Figure 6.24 Steam boiler.

Base-loaded. The steam flow (or steam demand) is controlled directly.

Swing. The steam pressure is controlled, with the steam demand measured by the steam flow transmitter.

One boiler is the swing boiler; the others are base-loaded.

The following variables are controlled to specified targets:

- Steam pressure (swing boiler) or steam flow (base-loaded boiler).
- Drum level.
- Stack oxygen (for combustion efficiency).
- Furnace pressure. Most boilers with both forced-draft and induced-draft fans are operated just below atmospheric pressure (outside air leaks into the boiler instead of the hot combustion gases leaking out).

The manipulated variables are the following:

- Feedwater control valve opening
- Fuel control valve opening
- Forced-draft fan speed (or forced-draft damper opening)
- Induced-draft fan speed (or induced-draft damper opening)

In large boilers, simple feedback loops are rarely installed. For example, the speeds of the forced-draft and induced-draft fans determine the following two variables:

- Airflow
- Furnace pressure

This is a multivariable control problem—both fans affect the airflow, and both fans affect the furnace pressure.

Characterization Function. Maintaining the proper ratio of air and fuel is crucial in every combustion process. In older boilers the fuel valve and the dampers were linked mechanically through a mechanism known as a *jackshaft* that was driven by a single master actuator. In newer boilers, individual actuators are provided and their positions linked electronically. With variable-speed drives the fan speeds can only be linked electronically.

One of the methods suggested previously for determining values for the ratio and bias coefficients involved obtaining data from several equilibrium states for the process. A steam boiler is one example where such data are often readily available. If the customer specifies that the boiler is to operate from 30% of rated capacity to full rated capacity, part of the acceptance tests involve operating the boiler at the extremes and at several intermediate points. If the tests are conducted at 10% intervals, the following data could be recorded during the acceptance tests:

- Throughput, % of max (30%, 40%, 50%, ..., 100%)
- Fuel flow (mscfh)
- Forced-draft fan speed (rpm)
- Induced-draft fan speed (rpm)

These data relate the speed of each fan to either the throughput or the fuel flow. Assuming that each relationship exhibits a significant departure from linearity, characterization functions are required.

In the control configuration in Figure 6.25, the fuel flow is measured and controlled. If the fuel flow changes, the speed of each fan must change in an appropriate manner, the objective being:

- To maintain the proper ratio of airflow to fuel flow
- To maintain the desired furnace pressure

The functional equivalent to the traditional jackshaft mechanical linkage between the fuel valve opening and the damper openings is provided by the two characterization functions in Figure 6.25 that electronically link the fan speeds to the fuel flow. The fuel flow set point is the input to each characterization function:

PYID: Induced-draft fan speed as a function of fuel flow. This fan is used to control furnace pressure.

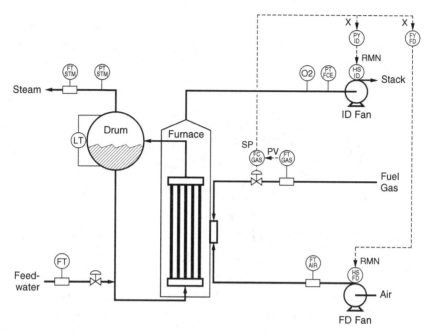

Figure 6.25 Induced-draft fan speed and forced-draft fan speed as a functions of fuel flow set point.

> *FYFD:* Forced-draft fan speed as a function of fuel flow. This fan is used to control the airflow. But if a constant furnace pressure is maintained, the induced-draft fan speed must change along with the forced-draft fan speed.

The jackshaft mechanical system linked the fuel valve opening and the damper openings. Often, neither fuel flow nor airflow was measured. A fuel flow measurement is now installed on most boilers, so the damper openings or fan speeds are preferably electronically linked to the fuel flow (or the fuel flow set point, as in Figure 6.25).

For each fan, Figure 6.25 indicates a hand station instead of a speed controller. The speed controller is normally part of the variable-speed drive control logic, not part of the process controls. To provide the capability for the process operator to specify the speed of each fan manually, a hand station for each fan is included in Figure 6.25:

> *Input to hand station:* The output of the respective characterization function.
> *Output from hand station:* The set point for the speed controller for the variable-speed drive.

Feedback Trim for Induced-Draft Fan. The configuration in Figure 6.26 provides feedback trim using a summer. The output of the furnace pressure controller (PCFCE) is a small positive or negative value that is added to the

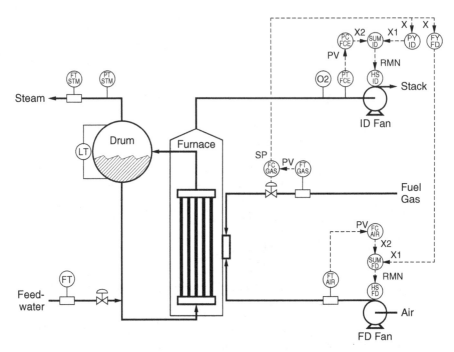

Figure 6.26   Feedback trims for both the induced-draft fan speed and the forced-draft fan speed.

output of the characterization function (PYID) to provide the set point for the induced-draft fan speed. The characterization function responds rapidly to any changes in the fuel flow, so the pressure controller needs to make only small trim adjustments to maintain the furnace pressure at its set point.

Most digital systems permit an output range in engineering units to be specified for a PID controller. Suppose that the feedback trim controller is to be capable of adjusting the output by ±400 rpm (approximately 10% based on the maximum fan speed of 3600 rpm). Specifying −400 to +400 rpm for the output range permits the feedback trim controller to increase or decrease the speed by 400 rpm. Specifying −400 to +200 rpm is also possible. However, it is traditional for an output of midrange (50% of span) to correspond to no adjustment. Consequently, this capability is more likely to be obtained by specifying an output range of −400 to +400 rpm and an upper output limit of +200 rpm.

An alternative approach to providing feedback trim is to use a multiplier. The output of the furnace pressure controller (PCFCE) is a number close to 1.0 that is multiplied by the output of the characterization function (PYID) to provide the set point for the induced-draft fan speed. Specifying 0.9 to 1.1 for the output range permits the feedback trim controller to increase or decrease the speed by 10% of its current value.

In conventional pneumatic and electronic analog controls, the simplicity of summers over multipliers led to their predominant use to provide feedback trim.

A multiplier is as easy to configure as a summer in digital systems, but the use of summers continues to be more common.

Feedback Trim for Forced-Draft Fan. Using a similar approach, Figure 6.26 provides feedback trim for the airflow. The output of the airflow controller (FCAIR) is a small positive or negative value that is added to the output of the characterization function (FYFD) to provide the set point for the forced-draft fan speed.

The characterization function responds rapidly to any changes in the fuel flow, the objective being to quickly change the airflow. However, the configuration in Figure 6.26 will function properly only provided that the airflow set point is also changed to a value consistent with the fan speed from the characterization function. For the relationship between the airflow and the fuel flow, a simple ratio usually suffices. This is much simpler than the characterization function.

Assuming that the airflow set point is adjusted to reflect the current fuel flow, there are two options for controlling the airflow:

Characterization function with feedback trim. (Figure 6.26). The characterization function quickly changes the forced-draft fan speed to reflect the fuel flow. As the airflow controller is providing feedback trim, it responds more slowly than the typical flow controller.

Simple feedback control. (Figure 6.27). When its output is a fan speed, the airflow controller can be tuned to respond rapidly. Being mechanical, some dampers move relatively slowly, and the flow controller must respond accordingly.

For flow controllers, simple feedback configurations usually provide adequate performance, making unnecessary the characterization function for the forced-draft damper fan speed, as in Figure 6.26. In subsequent P&I diagrams, the simple feedback configuration in Figure 6.27 will be used. If no airflow measurement is available, using the characterization function to link the forced-draft fan speed to the fuel flow is appropriate. A feedback trim adjustment similar to that subsequently proposed for the ratio of airflow to fuel flow can be used in conjunction with the characterization function.

Output Tracking. When a hand station is incorporated into the control configuration (as in Figure 6.27), output tracking must be configured, the objective being a bumpless transition from hand station on local to hand station on remote. For the furnace pressure controller that provides the feedback trim, output tracking must be configured as follows:

Input TRKMN. Output tracking must be active when the hand station is not on remote (HSID.RMT is false).

Input MNI. For the transfer from local to remote to be bumpless, the output of the pressure controller must be initialized to the current output of the hand station (HSID.MN) less the output of the characterization function (PYID.Y).

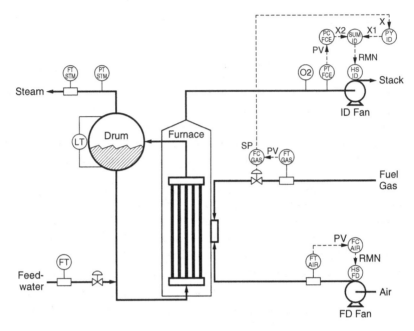

Figure 6.27   Simple feedback control for airflow.

The logic for output tracking for the furnace pressure controller in Figure 6.27 must be as follows:

```
PCFCE.TRKMN = !HSID.RMT
PCFCE.MNI = HSID.MN - PYID.Y
```

Output tracking is also required for the airflow controller should the hand station for the forced-draft fan be on local. If the characterization function for the forced draft fan speed is used as in Figure 6.26, equations analogous to the above are required for the airflow controller. However, if a simple feedback configuration as in Figure 6.27 is used for airflow, the logic for output tracking is as follows:

```
FCAIR.TRKMN = !HSFD.RMT
FCAIR.MNI = HSFD.MN
```

Windup Prevention for the Furnace Pressure Controller. At high steam demands, a high airflow is required, which could potentially result in a fan running at full speed (output QH of the respective hand station is true). On the lower end, a minimum speed is normally specified for both fans; there must always be some airflow, so both fans must run at all times. A lower output limit is required for each hand station, and a corresponding limit may be configured in the speed controller for each fan. At low steam demands, it is possible that one of the fans could be running at its minimum speed (output QL of the hand station is true).

Should the output of the induced-draft fan hand station be driven to an output limit, windup will occur in the feedback trim controller. This can be prevented by any of the following approaches:

Integral tracking. Preventing windup using integral tracking requires the following two components:

- *Input TRKMN.* Integral tracking must be activated if the output of the induced-draft fan hand station is at either its upper limit (output QH is true) or at its lower limit (output QL is true).
- *Input MRI.* The appropriate value for the controller output bias is the output of the induced-draft fan hand station (HSID.MN) less the output of the characterization function (PYID.Y).

The following logic implements output tracking and integral tracking in the furnace pressure controller in Figure 6.27:

```
PCFCE.TRKMN = !HSID.RMT
PCFCE.MNI = HSID.MN - PYID.Y
PCFCE.TRKMR = HSID.QH | HSID.QL
PCFCE.MRI = HSID.MN - PYID.Y
```

External reset. Preventing windup using external reset involves only one input:

- *Input XRS.* The appropriate value for the input to the reset mode is the output of the induced-draft fan hand station (HSID.MN) less the output of the characterization function (PYID.Y).

The following logic implements output tracking and external reset in the furnace pressure controller in Figure 6.27:

```
PCFCE.TRKMN = !HSID.RMT
PCFCE.MNI = HSID.MN - PYID.Y
PCFCE.XRS = HSID.MN - PYID.Y
```

Inhibit increase/inhibit decrease. Preventing windup using inhibit increase/ inhibit decrease requires the following two components:

- *Input NOINC.* If the output of the induced-draft fan hand station has been driven to its upper output limit (output QH is true), the feedback trim controller must not further increase its output.
- *Input NODEC.* If the output of the induced-draft fan hand station has been driven to its lower output limit (output QL is true), the feedback trim controller must not further decrease its output.

The following logic implements output tracking and inhibit increase/inhibit decrease in the furnace pressure controller in Figure 6.27:

```
PCFCE.TRKMN = !HSID.RMT
PCFCE.MNI = HSID.MN - PYID.Y
```

```
PCFCE.NOINC = HSID.QH
PCFCE.NODEC = HSID.QL
```

Windup Prevention for the Airflow Controller. For the configuration in Figure 6.27 that uses simple feedback control for airflow, the simplest approach to provide windup protection in the airflow controller is to set the output limits for the airflow controller at the same values as the output limits in the hand station. By doing this, the standard features of the PID block provide the windup protection.

When the values for certain configuration parameters for two blocks must be the same, concerns arise. Given long enough, someone will change one but not the other (some say "We never pass up the opportunity to make a mistake"). To avoid the need to also specify output limits for the airflow controller, windup protection must be invoked in the airflow controller when the output of the forced-draft fan hand station is at an output limit.

The appropriate windup protection can be provided by any of the following approaches:

Integral tracking. Preventing windup using integral tracking requires the following two components:

  • *Input TRKMN.* Integral tracking must be activated if the output of the forced-draft fan hand station is at either its upper limit (output QH is true) or at its lower limit (output QL is true).

  • *Input MRI.* The appropriate value for the controller output bias is the output of the forced-draft fan hand station (HSFD.MN).

The following logic implements output tracking and integral tracking in the airflow controller in Figure 6.27:

```
FCAIR.TRKMN = !HSFD.RMT
FCAIR.MNI = HSFD.MN
FCAIR.TRKMR = HSFD.QH | HSFD.QL
FCAIR.MRI = HSFD.MN
```

External reset. Preventing windup using external reset involves only one input:

  • *Input XRS.* The appropriate value for the input to the reset mode is the output of the forced-draft fan hand station (HSID.MN).

The following logic implements output tracking and external reset in the airflow controller in Figure 6.27:

```
FCAIR.TRKMN = !HSFD.RMT
FCAIR.MNI = HSFD.MN
FCAIR.XRS = HSFD.MN
```

Inhibit increase/inhibit decrease. Preventing windup using inhibit increase/ inhibit decrease requires the following two components:

- *Input NOINC.* If the output of the forced-draft fan hand station has been driven to its upper output limit (output QH is true), the airflow controller must not increase its output further.
- *Input NODEC.* If the output of the forced-draft fan hand station has been driven to its lower output limit (output QL is true), the airflow controller must not decrease its output further.

The following logic implements output tracking and inhibit increase/inhibit decrease in the airflow controller in Figure 6.27:

```
FCAIR.TRKMN = !HSFD.RMT
FCAIR.MNI = HSFD.MN
FCAIR.NOINC = HSFD.QH
FCAIR.NOINC = HSFD.QL
```

6.7. CROSS-LIMITING

In the ensuing discussion, combustion control for a steam boiler is used as an example where a variety of limits are often incorporated into the control logic, which in turn requires that consideration be given to bumpless transfer and windup protection. The focus is on how limits can be incorporated into the control logic and the consequences that occur when a limit is encountered. The expectation is that the reader can then apply such logic to other applications. The discussion that follows is neither a comprehensive nor a complete treatment of combustion control.

Combustion is a rapid chemical reaction involving two components: fuel and oxygen (from air). Both are necessary, and in the proper ratio. The fuel-to-air ratio is crucial in every combustion process, including the steam boiler illustrated in Figure 6.24. From the perspective of efficiency, there is an optimum value for the fuel-to-air ratio:

- Ratios off the optimum affect efficiency negatively. If the air is in excess, more heat is lost with the stack gases. If the fuel is in excess, the combustion is incomplete. Logic generally referred to as *combustion control* is responsible for maintaining good combustion efficiency.
- If the fuel is in excess, unburned fuel is present in the stack gases, creating the potential for a fire or explosion. Logic known as *burner management* is responsible for avoiding unsafe process operating conditions, usually by initiating a trip (a rapid process shutdown).

Herein only combustion control is discussed.

Combustion Control. Combustion control has two major objectives:

- Provide the heat required to meet the demands of the process. For example, providing the required steam flow from the boiler entails providing the appropriate heat input from the combustion of the fuel.

- Maintain the combustion efficiency at or near its peak.

The logic for combustion control should also adhere to the following statement:

The process controls should never take an action that would necessitate a response from the safety system.

For the boiler, the combustion control logic must never take an action that would cause the burner management logic to initiate a trip. In addition to the disruption to production operations, a sudden process shutdown places more stress on the equipment than an orderly shutdown, resulting in some risk of major damage to the equipment.

In this section we examine the following two components of the combustion control logic:

- For the current fuel flow, the airflow must exceed some minimum value: specifically, the airflow required to combust the fuel completely.
- For the current airflow, the fuel flow must not exceed some maximum value: specifically, the fuel flow that would completely consume all of the oxygen provided by the current airflow.

Should the combustion control logic be unable to achieve either of these objectives, the burner management system should initiate a trip, thereby avoiding unsafe operation of the process.

Airflow and Fuel Flow. For peak efficiency, most combustion processes are operated near the stoichiometric ratio of the two reactants: the fuel and the oxygen from air. In practice, the air is always in an excess that depends on the nature of the fuel. The ratio is nearest to stoichiometric for fuels such as natural gas, where only a small excess of air is required.

There are two ways to approach controlling a combustion process:

- Set the fuel flow and then adjust the airflow to provide the oxygen required to consume all of the fuel.
- Set the airflow and then adjust the fuel flow so that nearly all of the oxygen in the air is consumed.

Most have a better "warm, fuzzy feeling" with the former approach. However, combustion requires both fuel and air, making the latter equally viable.

Combustion Efficiency. The composition of the stack gases provides a good indication of combustion efficiency. There are two approaches:

Stack oxygen. Depending on the nature of the fuel, targets can be established for the stack oxygen. For example, efficient combustion of natural gas requires little excess air, so targets for the stack oxygen concentration will be low. The term *stack oxygen* is used routinely, but one must be especially careful when the boiler is operated below atmospheric pressure. Whatever air leaks from outside to inside contains 21% oxygen. In a sense, the stack gas is being "contaminated" by the leaks, resulting in a higher oxygen concentration in the stack gas than in the combustion zone. In such processes, the oxygen concentration must be sensed within or near the combustion zone.

Carbon monoxide (CO). This is a direct measure of the incomplete products of combustion. If measured as the ratio of CO to CO_2, the value is not affected by leaks. Where multiple fuels are involved, the appropriate CO concentration is less affected by the type of fuel than is the appropriate O_2 concentration.

Most early efforts to assess combustion efficiency relied on oxygen concentration, but the use of CO concentration has merits in some applications.

Ratio Airflow to Fuel Flow. For a swing boiler Figure 6.28 presents the basic combustion control configuration to ratio airflow to fuel flow. The key aspects are as follows:

- A flow controller (FCGAS) is provided for the fuel flow.
- The fuel flow set point is provided by the steam pressure controller (PCSTM).
- The airflow set point is the fuel flow set point multiplied by the desired air-to-fuel ratio.
- The composition controller (CCO2) for the stack oxygen provides feedback trim by adjusting the desired air-to-fuel ratio.

A couple of additional features are often incorporated into the logic:

- Cross-limits on the fuel flow and the airflow. In this context, the following air-to-fuel ratios arise:

    Actual air-to-fuel ratio R. This is the airflow divided by the fuel flow.

    Desired air-to-fuel ratio R_{SP}. In the configuration Figure 6.28, this ratio is the output of the stack oxygen composition controller that provides the feedback trim.

    Minimum permitted air-to-fuel ratio R_{min}. As long as the actual ratio is greater than the minimum ratio, the burner management logic should not initiate a trip. The lower output limit for the stack oxygen concentration controller should be R_{min}.

    Design ratio R_0. This air-to-fuel ratio is computed from either design data or from normal operating conditions. This ratio will be used when the ratio computed from the current airflow and fuel flow is unrealistic.

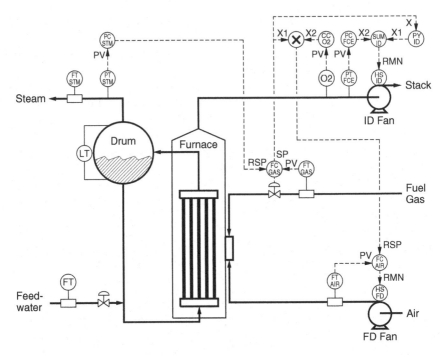

Figure 6.28   Combustion control configuration that ratios the airflow to the fuel flow.

- Directional lags on the fuel flow and the airflow.

Cross-limits are examined below; directional lags are examined in the next section.

Heat of Combustion.   The heat of combustion can be based on any reactant; the basis can be either a unit mass or a unit volume. For combustion applications, the flow meters are usually volumetric meters, and if so, volume is the most logical basis. Therefore, the options for stating the heat of combustion are as follows:

Btu per unit volume of fuel. The value depends on the nature of the fuel. For gases, the unit of fuel is either ft^3 or m^3.

Btu per unit volume of oxygen. This value also depends on the nature of the fuel. The unit of oxygen may be ft^3 or m^3. As air is 21% oxygen by volume, multiplying the volumetric value for oxygen by 0.21 gives "Btu per unit volume of air."

Both values are influenced by the composition of the fuel, but not to the same degree.

To maintain constant throughput, the ratio should be applied to the flow with the least variability in the value of the heat of combustion. The possibilities are:

Btu per unit volume of fuel is constant. Determine the throughput by adjusting the fuel flow and then the ratio of airflow to fuel flow.

Btu per unit volume of oxygen (or air) is constant. Determine the throughput by adjusting the airflow and then ratio fuel flow to airflow.

For fuels such as natural gas, the composition is constant, so the heat of combustion will be constant regardless of the basis. But not all fuel gas streams have a constant composition. The values for the heat of combustion for the usual components of fuel gas are as follows:

Fuel	Btu/ft^3 Fuel	Btu/ft^3 O$_2$	Btu/ft^3 Air
Methane (CH$_4$)	961	481	101
Ethane (C$_2$H$_6$)	1710	489	103
Propane (C$_3$H$_8$)	2450	489	103
n-Butane (C$_4$H$_{10}$)	3180	490	103
n-Pentane (C$_5$H$_{12}$)	3890	486	102
Hydrogen (H$_2$)	289	579	122

When expressed on the basis of a unit volume of O$_2$ or air, the heat of combustion is nearly the same for all light hydrocarbons, which means that it is not affected significantly by changes in the composition of these components. However, the value is significantly different for hydrogen, which is a variable component of some fuel gas streams.

Ratio Fuel Flow to Airflow. For a swing boiler Figure 6.29 presents the basic combustion control configuration to ratio fuel flow to airflow. The key aspects are as follows:

- A flow controller (FCGAS) is provided for the fuel flow.
- The airflow set point is provided by the boiler pressure controller (PCSTM).
- The fuel flow set point is the airflow set point multiplied by the desired fuel-to-air ratio (the reciprocal of the desired air-to-fuel ratio R_{SP}).
- The composition controller (CCO2) for the stack oxygen provides feedback trim by adjusting the fuel-to-air ratio.

Cross-limits and directional lags are often incorporated into the control configuration. Herein these are discussed only in the context of the control configuration in Figure 6.28 that ratios the airflow to the fuel flow. However, they can be incorporated into the control configuration in Figure 6.29 in an analogous manner.

Minimum Fuel Flow. For all combustion processes, a minimum firing rate is stipulated by the burner designers. If the fuel flow is too low, instabilities appear in the flame from the burners. The burner management system is responsible for

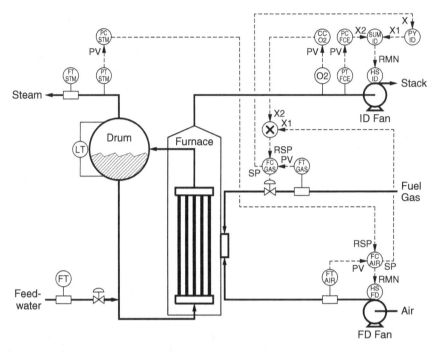

Figure 6.29 Combustion control configuration that ratios the fuel flow to the airflow.

initiating a trip should the firing rate drop too low. To avoid trips, logic must be incorporated into the combustion controls to prevent unreasonably low fuel flow set points.

For the configuration in Figure 6.28, imposing a lower limit on the fuel flow set point can be implemented by two alternatives:

Lower output limit for the steam pressure controller. The PID block activates windup protection when the output attains the lower output limit.

Set point lower limit for the fuel flow controller. Windup protection must be activated in the steam pressure controller when output SPL from the fuel flow controller is true.

In Figure 6.28 the input to the multiplier that computes the airflow set point is output SP from the fuel flow controller, not output MN of the steam pressure controller. When the lower limit on the fuel flow set point is imposed by a lower output limit for the steam pressure controller, either approach can be used. But when the limit is imposed by a lower set point limit for the fuel flow controller, output SP must be used, as in Figure 6.28.

Minimum Airflow. The minimum airflow A_{min} may be higher than the airflow corresponding to the fuel flow for the minimum firing rate. The airflow set point is the output of the multiplier block whose inputs are:

- Fuel flow set point (output SP of fuel flow controller)
- Desired air-to-fuel ratio (output MN of the stack oxygen composition controller)

In the configuration in Figure 6.28, this limit can be imposed only by specifying a lower set point limit on the airflow controller. Output SPL is required by the windup protection logic.

Should the PID block provided by the control system not provide set point limits (or not provide outputs that indicate that the set point is at a limit), a high select can be inserted between the output of the multiplier block and the RSP input to the airflow controller. The inputs to the high select are:

- Airflow set point computed by the multiplier block
- Value of A_{min}

The high select block provides an indication that A_{min} is being selected.

Maximum Fuel Flow for Current Airflow. To increase the throughput (the steam demand for a boiler), the configuration in Figure 6.28 would increase both the fuel flow and the airflow. The following is possible:

- The fuel flow increases as the combustion control logic specifies.
- The airflow does not increase. Perhaps the forced-draft fan speed is at its maximum, or for some other reason, the drive does not respond to changes in the input signal for the set point. If dampers are installed instead of variable-speed drives, dampers occasionally become "stuck" and do not move.

If the result is more fuel than can be burned by the current airflow, there are two options:

1. *Initiate a trip.* This is the responsibility of the burner management logic.
2. *Do not increase the fuel flow.* The combustion controls must maintain the actual air-to-fuel ratio above that at which the trip is initiated by the burner management logic.

The combustion control logic must not increase the fuel flow if it would cause the burner management logic to initiate a trip.

The set point for the fuel flow controller must be the smaller of the following two values:

- The fuel flow set point specified by the steam pressure controller
- The maximum allowable fuel flow computed by dividing the actual airflow by the minimum air-to-fuel ratio R_{min}

Figure 6.30 is Figure 6.28 with the addition of a low select (SELGAS) whose inputs are as follows:

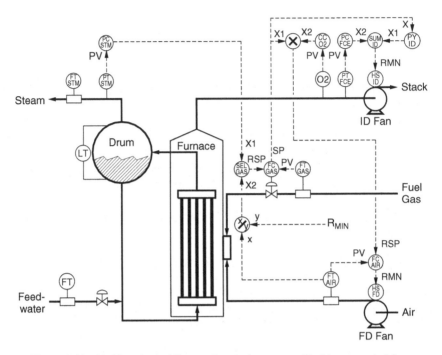

Figure 6.30   Limiting the fuel flow to the maximum permitted for current airflow.

Input X1: The steam pressure controller output PCSTM.MN. The output of the selector should normally be this input.

Input X2: The actual airflow FTAIR.PV divided by the minimum air-to-fuel ratio $R_{\min}$. This input is selected only when the fuel flow specified by input X1 would exceed what could be combusted by the current airflow.

Input RSP to the fuel flow controller is the output of this select block.

Minimum Airflow for Current Fuel Flow. To decrease the throughput (the steam demand for a boiler), the configuration in Figure 6.28 (and Figure 6.30) would decrease both the fuel flow and the airflow. This creates the following possibility:

- The airflow decreases as the combustion control logic specifies.
- The fuel flow does not decrease. Valves occasionally "stick" and do not respond to the control signal to the valve.

If the result is insufficient air to consume the fuel, there are two options:

1. *Initiate a trip.* This is the responsibility of the burner management logic.

2. *Do not decrease the airflow.* The combustion controls must maintain the actual fuel-to-air ratio below that at which the trip is initiated by the burner management logic.

The combustion control logic must not reduce the airflow if it would cause the burner management logic to initiate a trip.

The set point for the airflow controller must be the larger of the following two values:

- The airflow set point obtained by multiplying the fuel flow set point by the desired air-to-fuel ratio R_{SP}
- The airflow computed by multiplying the actual fuel flow by the minimum air-to-fuel ratio R_{min}

Figure 6.31 is Figure 6.30 with the addition of a high select (SELAIR) whose inputs are as follows:

Input X1: The product of the current fuel flow set point FCGAS.SP and the stack oxygen composition controller output CCO2.MN, which is the desired air-to-fuel ratio R_{SP}. The output of the selector should normally be this input.

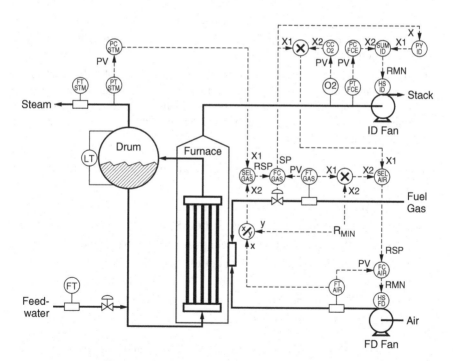

Figure 6.31 Combustion control with cross-limits.

Input X2: The product of the actual fuel flow FTGAS.PV and the minimum air-to-fuel ratio R_{min}. This input is selected only when the airflow provided by input X1 would be inadequate for the current fuel flow.

Input RSP to the airflow controller is the output of this select block.

Cross-limiting occurs when the combustion control logic provides for both of the following, as in Figure 6.31:

- The minimum airflow is what is required to combust the current fuel flow.
- The maximum fuel flow is what can be combusted by the current airflow.

Preventing Windup in the Steam Pressure Controller. Windup occurs in the steam pressure controller in Figure 6.31 under the following conditions (the output of the steam pressure controller has no effect on the steam pressure):

- The low select is imposing the maximum fuel flow consistent with the current airflow. For the low selector configured as in Figure 6.31, output SELGAS.Q1 is false (steam pressure controller output is not selected).
- The set point for the fuel flow controller is at its upper set point limit, which usually corresponds to the upper range value of the fuel flow transmitter. Output FCGAS.SPH is true.
- The set point for the fuel flow controller is at the flow corresponding to the minimum firing rate. Output FCGAS.SPL is true.
- The fuel flow controller has driven the fuel flow control valve fully open (actually, the controller output has attained its upper output limit). Output FCGAS.QH is true.
- Often, a mechanical stop is fitted to the fuel flow control valve so that the valve cannot fully close. The lower output limit for the fuel flow controller should correspond approximately to the opening of the mechanical stop. When the fuel flow controller has driven its output to the lower output limit, output FCGAS.QL is true.

To provide windup protection by integral tracking, two inputs to the steam pressure controller must be configured:

Input TRKMR: True if any limit is encountered:
- Output Q1 from the low select (SELGAS) is false.
- Output SPH from the fuel flow controller (FCGAS) is true.
- Output SPL from the fuel flow controller (FCGAS) is true.
- Output QH from the fuel flow controller (FCGAS) is true.
- Output QL from the fuel flow controller (FCGAS) is true.

Input MRI: Current value of the fuel flow.

The following logic provides both output tracking and integral tracking for the steam pressure controller:

```
PCSTM.TRKMN = !FCGAS.RMT
PCSTM.MNI = FCGAS.SP
PCSTM.TRKMR = (!SELGAS.Q1) | FCGAS.SPH | FCGAS.SPL | FCGAS.QH |
    FCGAS.QL
PCSTM.MRI = FTGAS.PV
```

To provide windup protection by external reset, only one input to the steam pressure controller must be configured:

Input XRS: Current value of the fuel flow.

The following logic provides both output tracking and external reset for the steam pressure controller:

```
PCSTM.TRKMN = !FCGAS.RMT
PCSTM.MNI = FCGAS.SP
PCSTM.XRS = FTGAS.PV
```

To provide windup protection by inhibit increase/inhibit decrease, two inputs to the steam pressure controller must be configured:

Input NOINC: True if any high limit is encountered:
- Output Q1 from the low select (SELGAS) is false.
- Output SPH from the fuel flow controller (FCGAS) is true.
- Output QH from the fuel flow controller (FCGAS) is true.

Input NODEC: True if any low limit is encountered:
- Output SPL from the fuel flow controller (FCGAS) is true.
- Output QL from the fuel flow controller (FCGAS) is true.

The following logic provides both output tracking and inhibit increase/inhibit decrease for the steam pressure controller:

```
PCSTM.TRKMN = !FCGAS.RMT
PCSTM.MNI = FCGAS.SP
PCSTM.NOINC = (!SELGAS.Q1) | FCGAS.SPH | FCGAS.QH
PCSTM.NODEC = FCGAS.SPL | FCGAS.QL
```

Preventing Windup in the Stack Oxygen Composition Controller. Should the airflow be increased to its maximum value or decreased to its minimum value, windup occurs in the stack oxygen composition controller in Figure 6.31. If any limiting condition is encountered, the output of the composition controller ceases to affect the airflow and stack oxygen composition, so windup results.

Should any of the following be true, the maximum airflow has been attained:

- The set point for the airflow controller is at the upper set point limit (usually, corresponding to the upper range value of the airflow transmitter). Output FCAIR.SPH is true.
- The airflow controller has driven the induced-draft fan speed to its upper output limit. Normally, this corresponds to the forced-draft fan running at maximum speed. Output HSFD.QH is true.
- The induced-draft fan speed has been driven to its upper output limit. Normally, this corresponds to the induced-draft fan running at maximum speed. Output HSID.QH is true.

Should any of the following be true, the minimum airflow has been attained:

- The high select is imposing the minimum airflow consistent with the current fuel flow. For the high selector configured as in Figure 6.31, output SELAIR.Q1 is false (composition controller output is not selected).
- The set point for the airflow controller is at the lower set point limit, which corresponds to the minimum airflow A_{min}. Output FCAIR.SPL is true.
- The airflow controller has driven the induced-draft fan speed to its lower output limit. The corresponding airflow should be approximately A_{min}. Output HSFD.QL is true.
- The induced-draft fan speed has been driven to its lower output limit. The corresponding airflow should be approximately A_{min}. Output HSID.QL is true.

In a typical application, windup protection would be based on the actual air-to-fuel ratio R computed from the current airflow and fuel flow. But when a limit is being imposed on the airflow, the value computed for the ratio R might not be realistic. The consequences depend on which limit is being imposed:

Low limit. An example is when the airflow is at the minimum A_{min}. Further reductions in the fuel flow are permitted (down to the fuel flow for the minimum firing rate). Such reductions increase the current air-to-fuel ratio R. Basing windup protection on $\min(R_0, R)$ avoids unreasonably large values of R.

High limit. An example is the forced-draft fan running at full speed. Further increases in the fuel flow are permitted, provided that the actual air-to-fuel ratio exceeds the minimum ratio R_{min} imposed by the cross-limiters. Such increases reduce the current air-to-fuel ratio R. Basing windup protection on $\max(R_0, R)$ avoids unreasonably small values of R.

To provide windup protection by integral tracking, two inputs to the stack oxygen composition controller must be configured:

Input TRKMR: True if any limit is imposed:

- Output Q1 from the high select (SELAIR) is false.

- Output SPH or output SPL from the airflow controller (FCAIR) is true.
- Output QH or output QL from the forced-draft fan hand station (HSFD) is true.
- Output QH or output QL from the induced-draft fan hand station (HSID) is true.

Input MRI: Value for the air-to-fuel ratio:

- If a high limit is imposed, use $\max(R_0, R)$ for the air-to-fuel ratio.
- If a low limit is imposed, use $\min(R_0, R)$ for the air-to-fuel ratio.

The following logic provides both output tracking and integral tracking for the stack oxygen composition controller:

```
CCO2.TRKMN = !FCAIR.RMT
if (FCAIR.SPH | HSFD.QH | HSID.QH)
   CCO2.MNI = max(FTAIR.SP / FTGAS.PV, RO)
else if ((!SELAIR.Q1) | FCAIR.SPL | HSFD.QL | HSID.QL)
   CCO2.MNI = min(FTAIR.SP / FTGAS.PV, RO)
else
   CCO2.MNI = FCAIR.SP / FTGAS.PV
CCO2.TRKMR = (!SELAIR.Q1) | FCAIR.SPH | FCAIR.SPL | HSFD.QH |
      HSFD.QL |HSID.QH | HSID.QL
if (FCAIR.SPH | HSFD.QH | HSID.QH)
   CCO2.MRI = max(FTAIR.PV / FTGAS.PV, RO)
else if ((!SELAIR.Q1) | FCAIR.SPL | HSFD.QL | HSID.QL)
   CCO2.MRI = min(FTAIR.PV / FTGAS.PV, RO)
```

To provide windup protection by external reset, only one input to the stack oxygen composition controller must be configured:

Input XRS: Value for the air-to-fuel ratio:

- If no limit is being imposed, use the actual fuel-to-air ratio R.
- If a high limit is imposed, use $\max(R_0, R)$ for the air-to-fuel ratio.
- If a low limit is imposed, use $\min(R_0, R)$ for the air-to-fuel ratio.

The following logic provides both output tracking and integral tracking for the stack oxygen composition controller:

```
CCO2.TRKMN = !FCAIR.RMT
if (FCAIR.SPH | HSFD.QH | HSID.QH)
   CCO2.MNI = max(FTAIR.SP / FTGAS.PV, RO)
else if ((!SELAIR.Q1) | FCAIR.SPL | HSFD.QL | HSID.QL)
   CCO2.MNI = min(FTAIR.SP / FTGAS.PV, RO)
else
   CCO2.MNI = FCAIR.SP / FTGAS.PV
```

```
if (FCAIR.SPH | HSFD.QH | HSID.QH)
   CCO2.MRI = max(FTAIR.PV / FTGAS.PV, RO)
else if ((!SELAIR.Q1) | FCAIR.SPL | HSFD.QL | HSID.QL)
   CCO2.MRI = min(FTAIR.PV / FTGAS.PV, RO)
else
   CCO2.MRI = FTAIR.PV / FTGAS.PV
```

To provide windup protection by inhibit increase/inhibit decrease, two inputs to the stack oxygen composition controller must be configured:

Input NODEC: True if a low limit is being imposed:
  - Output Q1 from the high select (SELAIR) is false.
  - Output SPL from the airflow controller (FCAIR) is true.
  - Output QL from the forced-draft hand station (HSFD) is true.
  - Output QL from the induced-draft hand station (HSID) is true.

Input NOINC: True if a high limit is being imposed:
  - Output SPH from the airflow controller (FCAIR) is true.
  - Output QH from the forced-draft hand station (HSFD) is true.
  - Output QH from the induced-draft hand station (HSID) is true.

The following logic provides both output tracking and inhibit increase/inhibit decrease for the stack oxygen composition controller:

```
CCO2.TRKMN = !FCAIR.RMT
if (FCAIR.SPH | HSFD.QH | HSID.QH)
    CCO2.MNI = max(FTAIR.SP / FTGAS.PV, RO)
else if ((!SELAIR.Q1) | FCAIR.SPL | HSFD.QL | HSID.QL)
   CCO2.MNI = min(FTAIR.SP / FTGAS.PV, RO)
else
  CCO2.MNI = FCAIR.SP / FTGAS.PV
CCO2.NOINC = FCAIR.SPH | HSFD.QH | HSID.QH
CCO2.NODEC = (!SELAIR.Q1) | FCAIR.SPL | HSFD.QL | HSID.QL
```

6.8. DIRECTIONAL LAGS

In most feedforward control applications, dynamic compensation is provided by the lead-lag compensator possibly coupled with a dead time. Both elements apply the same dynamic compensation to increases as well as decreases. Some applications require a lag in one direction but not in the other. Herein, a combustion process is used as an example of a process that requires directional lags, but the same or similar requirement occasionally arises in other ratio control applications.

Combustion Processes. In combustion processes, mixtures containing an excess of fuel must be avoided under all situations. Mixtures containing an excess of air affect the efficiency negatively but do not pose risks. There are two aspects of these considerations:

Steady-state. The issues are addressed by the air-to-fuel ratio or the fuel-to-air ratio supplemented with the logic for cross-limiting.

Dynamic. During any transient, a higher excess of air is appropriate. For most processes, the economics are determined by the steady-state conditions. Short-duration transients have little impact.

To obtain an excess of air on any change in throughput, the behavior must be as follows:

Increasing firing rate (increasing fuel flow and airflow). The airflow must increase ahead of the fuel flow. A lag must be applied to the fuel flow but not the airflow.

Decreasing firing rate (decreasing fuel flow and airflow). The fuel flow must decrease ahead of the airflow. A lag must be applied to the airflow but not the fuel flow.

Suppose that the usual lead-lag compensator is used to lag the fuel flow. On an increase in firing rate, the behavior is as desired. But on a decrease in firing rate, the airflow decreases faster than the fuel flow, resulting in a mixture that is low on air. This is not acceptable.

Each lag must be applied in one direction only. The lag on fuel flow must only be applied to increases. The lag on airflow must only be applied to decreases.

Directional Lag. One way to obtain a directional lag is to use a selector block in conjunction with a lead-lag compensator that is tuned to provide a lag. Figure 6.32 presents the logic for two types of lags:

Lag on increases only. In a combustion process, this would be applied to the fuel flow. A low selector is configured with the following inputs:
- The fuel flow.
- The lagged fuel flow. This is the output of a lag function block (a lead-lag compensator tuned to provide a pure lag) whose input is the fuel flow.

On increases, the lagged fuel flow will be less than the fuel flow, so the output of the selector will be the lagged fuel flow. On decreases, the fuel flow will be less than the lagged fuel flow, so the output of the selector will be the fuel flow. Consequently, the lag is applied only to increases in the fuel flow.

Lag on decreases only. In a combustion process, this would be applied to the airflow. A high selector is configured with the following inputs:
- The airflow.

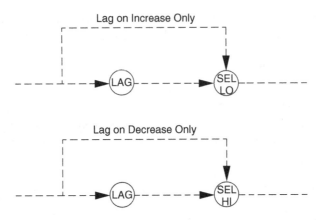

Figure 6.32 Directional lags.

- The lagged airflow. This is the output of a lag function block (a lead-lag compensator tuned to provide a pure lag) whose input is the airflow.

On increases, the airflow will be greater than the lagged airflow, so the output of the selector will be the airflow. On decreases, the lagged airflow will be greater than the airflow, so the output of the selector will be the lagged airflow. Consequently, the lag is applied only to decreases in the airflow.

The configuration in Figure 6.33 includes directional lags on both the fuel flow and the airflow. The directional lags are applied before applying the cross-limiters. But since the objective of the directional lags is temporarily to provide an excess of air relative to the fuel, this should not cause either limit to be imposed.

6.9. FEEDFORWARD CONTROL

Like ratio control, feedforward control inserts a computation into the control configuration. This computation is more complex than a ratio calculation:

- The feedforward relationship is not merely multiplying one input by the value of the desired ratio.
- The feedforward relationship requires more than one input. Often, one of the inputs is a flow, but the other inputs commonly include temperatures and pressures.

Like ratio control, the output of the feedforward calculation is preferably the set point for a flow controller. The output of the feedforward calculation can be the opening of a control valve, but there are two drawbacks to this approach:

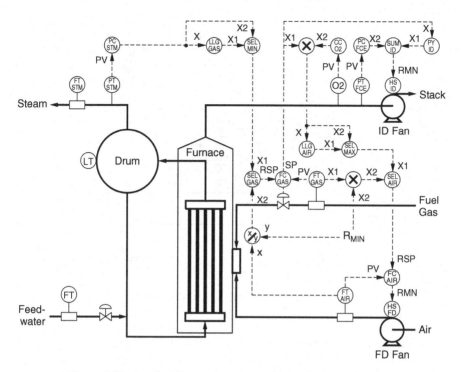

Figure 6.33   Lag fuel flow on increase and lag airflow on decrease.

- A relationship that reflects the installed characteristics of the control valve is required.
- Any maintenance performed on the control valve can alter the relationship.

Consequently, outputting a control valve opening should be considered only when measuring the flow through the control valve is impractical. The cost of the engineering effort coupled with the cost of degraded performance easily offsets the cost of a flow meter.

Supercritical Boiler. Although in common use, this term is a contradiction. At supercritical conditions, there is no distinction between the liquid and gas phases, so there is no phase change and no boiling. However, these units were designed to replace the customary steam boiler in a power plant. The output of a conventional boiler is steam at subcritical conditions; the output of a supercritical "boiler" can be thought of as steam under supercritical conditions.

Figure 6.34 presents a simplified version of a supercritical boiler that provides the steam to drive the turbine for generating electricity. Feedwater is pumped into the unit at 3500 psig, which is above the critical pressure of water. There is no phase change within the unit. Water enters at or near ambient temperature

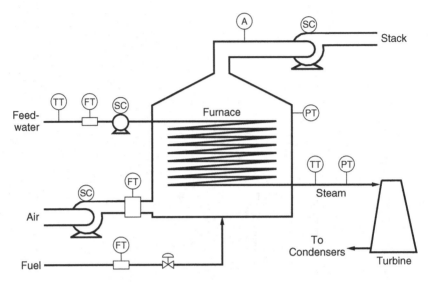

Figure 6.34 Simplified diagram of a supercritical "boiler."

and leaves as a supercritical fluid at a high temperature. There is no steam drum; the water flows through tubes, making *furnace* a more appropriate description for the unit.

The major simplifications for Figure 6.34 are the following:

Only one type of fuel is used. Most power plants are capable of burning at least two different types of fuel, such as fuel oil and fuel gas.

No reheat loop is provided. The exhaust from one of the turbine stages is returned to the furnace to be heated so that it enters the next turbine stage as superheated steam.

In an integrated power grid, the dispatcher monitors the demands for power on the grid and attempts to adjust operations so that the power required is generated in the most efficient manner. Being the larger and more efficient units on the grid, most supercritical units are operated in the base-loaded mode. Each is given a target for power production, usually referred to as the *megawatt demand* (MW_{SP}). The control system has to adjust the conditions within the supercritical unit so that this power is delivered by the turbine. For the furnace part of the electric generating plant, the crucial variables to maintain at proper values are:

Fuel. The firing rate must be consistent with the demand for power.

Air. The air rate must be appropriate to the fuel rate.

Feedwater. The feedwater rate must be consistent with the demand for power.

Feedforward strategies are presented for all three.

Feedwater. The turbine manufacturer provides turbine performance data in a graphical form generally referred to as the *turbine curves*. Based on these data, the power delivered by the turbine can be determined from the flow through the turbine, the turbine inlet pressure, the inlet steam enthalpy, and the turbine outlet pressure (depends on condenser temperatures). Assuming that all process conditions (notably the furnace exit temperature and pressure) are constant, a relationship can be derived for the megawatts generated (MW) as a function of flow (W) through the turbine:

$$MW = f(W)$$

Within the furnace there is no phase change. Therefore, the flow through the turbine is the same as the feedwater flow. Inverting the relationship for megawatts as a function of flow enables one to determine the feedwater rate W_{SP} required to generate the specified megawatts MW_{SP}:

$$W_{SP} = f^{-1}(MW_{SP})$$

The feedwater pumps provide this flow of water through both the furnace and the turbine.

Fuel. The fuel flow is calculated from an energy balance. The energy required to heat W_{SP} lb/min of feedwater from its inlet enthalpy of H_W to the target exit enthalpy of $H_{F,SP}$ is

$$W_{SP}(H_{F,SP} - H_W)$$

The heat required from the fuel is the heat to be added to the feedwater divided by the furnace efficiency ε:

$$\frac{W_{SP}(H_{F,SP} - H_W)}{\varepsilon}$$

Dividing by the fuel heating value H_V gives the following expression for the target F_{SP} for the fuel flow rate:

$$F_{SP} = \frac{W_{SP}(H_{F,SP} - H_W)}{\varepsilon H_V}$$

This expression is a good example of a situation in which a dilemma arises for feedforward control.

There are two options for obtaining a value for the feedwater enthalpy H_W:

- Measure the feedwater temperature and compute the feedwater enthalpy H_W.
- Provide a constant value for the feedwater enthalpy H_W.

The issues are as follows:

- The measurement device for the feedwater temperature is relatively inexpensive.
- Significant changes in the feedwater temperature do not occur during normal operations.
- Feedforward control responds aggressively to a change in any input. Unfortunately, the change may be the result of errors and/or failures of a measurement device.

The latter issue is often the major concern. Applying reasonableness tests to the inputs to a feedforward controller definitely has merit. In this regard, smart transmitters offer distinct advantages in that they perform a variety of checks to detect when the measured value is suspect. When interfaced via a current loop, the transmitter can be configured to fail upscale, fail downscale, or hold the last value. But with a network interface (commonly referred to as a *fieldbus*), the value can be reported as being suspect, and the decision regarding the action to take can be implemented in the controls.

Air. Where the fuel is a high-quality fuel such as fuel oil or fuel gas, the airflow set point $F_{A,\text{SP}}$ can be computed from the fuel flow set point F_{SP} and the air-to-fuel ratio R_A:

$$F_{A,\text{SP}} = R_A F_{\text{SP}}$$

Where the fuel is of variable quality, the ratio must be adjusted for variations in fuel heat content. Where mixed fuels are used (such as fuel oil and natural gas), separate ratios are required for air-to-fuel oil and air-to-fuel gas.

Feedback Trims. No feedforward relationship is perfect. Therefore, feedback control must be incorporated to provide the small adjustments required to obtain the desired process conditions. Three feedback controllers are required:

Megawatt demand. The feedwater rate is adjusted until the power output equals the current megawatt demand.

Furnace exit temperature. The fuel rate is adjusted until the temperature of the exiting supercritical fluid is the desired value.

Stack oxygen. The air rate is adjusted until the concentration of oxygen in the flue gas equals the desired value.

Dynamic Compensation. Although the dispatcher changes the megawatt demand from one value to another, implementing the change in a ramp fashion is acceptable. The fuel, air, and feedwater flows required at the new megawatt demand are calculated, and ramps are initiated to move the unit to the new operating level. The rate is determined by the characteristics of the furnace. With the high temperatures and pressures within the unit, thermal stresses are a concern. To protect against excess stresses, maximum allowable temperature differences are specified for certain pieces of equipment. The ramp rate must be sufficiently slow that no maximum allowable temperature difference is exceeded. If one is exceeded, the ramp must be stopped immediately.

Another consideration pertains to the fuel and air. Additional excess air is desirable during any transition. Therefore, on an increase in firing rate, the air ramp is started before the fuel ramp. On a decrease in firing rate, the fuel ramp is started before the air ramp.

6.10. FEEDFORWARD CONTROL EXAMPLE

In a previous example we discussed the application of ratio control to a steam-heated exchanger. In this section the example is extended to a feedforward control application. Basically, one or more variables that affect the ratio coefficient significantly will be incorporated into the equation used to compute the target for the steam flow.

Energy Balance. Assuming no heat loss, the energy balance states that the heat added to the liquid stream equals the heat released by the steam. The heat added to the liquid stream is

$$W c_P (T_{out} - T_{in})$$

The heat released by the steam is

$$F \Delta H_F$$

The energy balance around the heat exchanger is

$$W c_P (T_{out} - T_{in}) = F \Delta H_F$$

where

c_P = liquid heat capacity (Btu/lb–°F)
F = steam flow (lb/min)
T_{in} = liquid inlet temperature (°F)
T_{out} = liquid outlet temperature (°F)
W = liquid flow (lb/min)
ΔH_F = enthalpy of steam less enthalpy of condensate (Btu/lb)

Feedforward Control Equation. To derive the feedforward control equation, the steps are as follows:

• Solve the energy balance for the manipulated variable:

$$F = \frac{W c_P (T_{out} - T_{in})}{\Delta H_F}$$

- Note that the calculated value for the manipulated variable is a set point. The measured value of the steam flow is F; the target for the steam flow is F_{SP}:

$$F_{SP} = \frac{W c_P (T_{out} - T_{in})}{\Delta H_F}$$

- Replace the controlled variable by its set point. The measured value of the liquid outlet temperature is T_{out}; the set point for the liquid outlet temperature is $T_{out,SP}$. The objective is to calculate the steam flow required to heat the liquid to the desired liquid outlet temperature:

$$F_{SP} = \frac{W c_P (T_{out,SP} - T_{in})}{\Delta H_F}$$

To be usable as a control equation, values are required for the remaining four quantities in the equation: W, c_P, T_{in}, and ΔH_F. These variables are broadly classified as either measured disturbances or coefficients.

Measured Disturbances. For feedforward control to perform effectively, measurements must be provided for the major disturbances. For the steam-heated exchanger, these are:

Liquid flow W. For processes subject to throughput changes, the flow through the process is often the major disturbance. Where the flow is the only major disturbance, the feedforward configuration usually reduces to a ratio control configuration.

Liquid inlet temperature T_{in}. The need to measure the liquid inlet temperature depends on the frequency and magnitude of the changes in the liquid inlet temperature. If the liquid inlet temperature changes very slowly, these changes can be addressed by feedback trim.

This example assumes that significant and rapid changes occur in the liquid inlet temperature.

Coefficients. In the context of feedforward control, coefficients include those quantities that change so slowly that the feedback trim can respond effectively to their changes. For the steam-heated exchanger, the following quantities are considered to be coefficients:

Liquid heat capacity c_P. In most applications, liquid heat capacity variations are small. The exception is when the liquid is a mixture whose composition varies considerably. As no direct measurement of heat capacity is available, these situations are usually difficult to address.

Enthalpy change from steam to condensate ΔH_F. This value depends on the steam supply pressure and on the condensate temperature (which is a function of the steam pressure in the shell of the exchanger). Although it is possible to instrument the exchanger sufficiently to determine the steam enthalpy (from steam supply pressure) and condensate enthalpy (from condensate temperature or steam pressure in the shell), this is rarely necessary.

Control Equation. The steady-state feedforward controller is based on the equation derived previously from the energy balance:

$$F_{SP} = \frac{W c_P (T_{out,SP} - T_{in})}{\Delta H_F}$$

In the schematic in Figure 6.35, this equation is implemented within the block labeled "Feedforward Controller." Inputs are provided for the liquid flow W and the liquid inlet temperature T_{in}. A value must also be provided for the set point $T_{out,SP}$ for the liquid outlet temperature. The output from the feedforward controller is the set point F_{SP} to the steam flow controller.

The term *steady-state* is applied to this type of feedforward controller because it is based on the steady-state energy balance. Any change in an input is translated immediately to a change in the feedforward controller output. No consideration is provided for process dynamics.

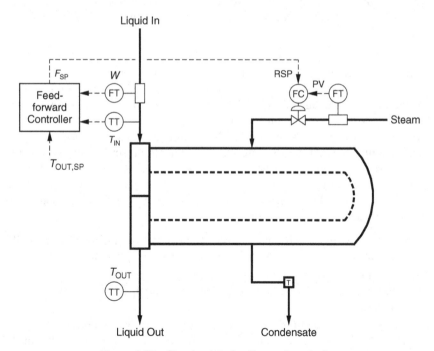

Figure 6.35 Steady-state feedforward controller.

There is no generic feedforward controller. The implementation of each must be based on a relationship that is specific to the application. This requires considerable flexibility on the part of the control system. Digital controls have this flexibility; their pneumatic and electronic predecessors were more restrictive.

In the schematic in Figure 6.36, the feedforward controller is implemented using two function blocks:

Summer. This block computes $T_{\text{out,SP}} - T_{\text{in}}$.

Multiplier. The block multiplies the output of the summer by the liquid flow W. The multiplier block also provides a coefficient, the calculation being

$$Y = kX_1X_2$$

where
Y = output of multiplier block
X_1 = input 1, which is the liquid flow W
X_2 = input 2, which the output of the summer $(T_{\text{out,SP}} - T_{\text{in}})$
k = coefficient, which must be $c_P/\Delta H_F$

If the multiplier block does not provide a coefficient, a second multiplier block would be required for the coefficient.

As the complexity of the feedforward relationship increases, the function block approach becomes less attractive than an implementation using programmed logic.

Feedback Trim. For a change in the liquid flow, the performance of the feedforward controller would be exactly the same as illustrated previously for the ratio controller. The advantage of the feedforward controller is that it also responds to changes in the liquid inlet temperature.

Feedforward controllers require feedback trim for the same reasons as those for ratio controllers. At steady-state, the adjustments proscribed by the feedforward controller will maintain the controlled variable close to the desired value, but rarely exactly at the desired value. Even in a simple application such as in Figure 6.36 for the steam-heated exchanger, numerous sources potentially contribute to the steady-state error:

- Errors in the measured inputs to the feedforward computation: specifically, errors in the liquid flow W and the liquid inlet temperature T_{in}.
- Errors in the coefficients in the feedforward computation: specifically, the liquid heat capacity c_P and the change in enthalpy ΔH_F from steam to condensate.
- Errors in the equations from which the feedforward computation was derived. For the steam-heated exchanger, no heat loss was included in the equation.

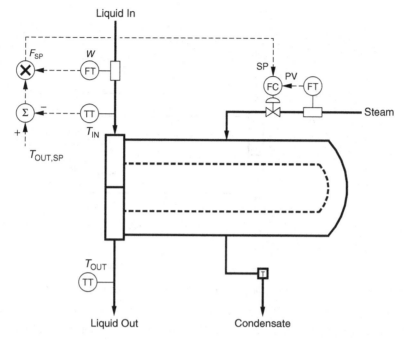

Figure 6.36 Steady-state feedforward controller implemented using function blocks.

- Error in the measurement of the steam flow F. As a consequence, the actual steam flow is slightly different from that which the feedforward computation suggests.
- Error in the measurement of the controlled variable T_{out}. The results of the feedforward computation must be trimmed to compensate for this error.

To provide the feedback trim, a temperature controller compares the measured value for the liquid outlet temperature to the target for the liquid outlet temperature. Different mechanisms are available for incorporating feedback trim into a feedforward control configuration. For the steam-heated exchanger, the following four approaches are possible:

1. Add or subtract a small value from the output of the feedforward computation to obtain the steam flow set point.
2. Multiply the output of the feedforward computation by a coefficient close to 1.0 to obtain the steam flow set point.
3. Adjust a coefficient in the feedforward computation.
4. Let the feedback trim controller adjust the set point to the feedforward computation.

Rarely is there is a distinct advantage of one over another. For most applications, it seems that if one of the above proves to be successful, all would be

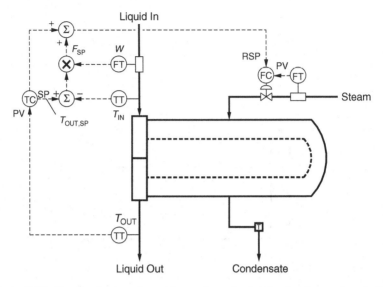

Figure 6.37   Feedback trim using a bias on the output of the feedforward controller.

successful. That is, if one approach does not work, trying another approach is unlikely to succeed. If one of the approaches is currently in use for other feedforward applications at a site, this creates an incentive to use that approach in future feedforward applications.

Summer. In the configuration in Figure 6.37a summer has been inserted between the output of the feedforward computation and the set point to the steam flow controller. This summer obtains the set point for the steam flow controller by adding a small positive or negative value to the result of the feedforward computation. The feedback trim controller provides the value to be added.

In configuring the feedback trim controller, the following considerations apply:

- An output of 50% of the output span (or midrange) is a trim adjustment of zero.
- The output of the feedforward calculation is a flow in engineering units (lb/min), so the trim input to the summer must also be in engineering units.

Both can be achieved by configuring the output range appropriately for the feedback trim controller. To permit the feedback trim controller to increase the steam flow set point by 10 lb/min and to decrease the steam flow set point by 10 lb/min, the appropriate output range is -10 to $+10$ lb/min. The measurement range for the steam flow transmitter is 0 to 80 lb/min, so the maximum feedback trim adjustment is a little over 10% of the upper range value of the measurement range.

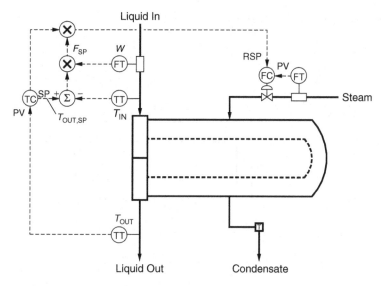

Figure 6.38   Feedback trim using a multiplier on the output of the feedforward controller.

Multiplier. In the configuration in Figure 6.38a multiplier has been inserted between the output of the feedforward computation and the set point to the steam flow controller. This multiplier obtains the set point for the steam flow controller by multiplying the result of the feedforward computation by a number that is close to 1.

Suppose that the feedback trim controller is to be able to increase the steam flow set point by 10% and to decrease the steam flow set point by 10% from the value computed by the feedforward controller. An appropriate output range for the feedback trim controller is 0.9 to 1.1. An output of midrange provides an input of 1.0 to the multiplier, which means that the set point for the steam flow controller will be the value computed by the feedforward controller.

Model Coefficient. Basically, the idea is for the feedback trim controller to recalibrate the model so that the results match the behavior of the process. That is, if the liquid outlet temperature is different from the set point for the liquid outlet temperature, the value computed by the feedforward controller for the steam flow set point must be in error. This error can be reduced and potentially eliminated by adjusting a coefficient in the model.

The choice of the coefficient to adjust takes the following into consideration:

- The degree to which changes in the value of the coefficient affect the output of the feedforward computation.
- The accuracy to which the coefficient is known.

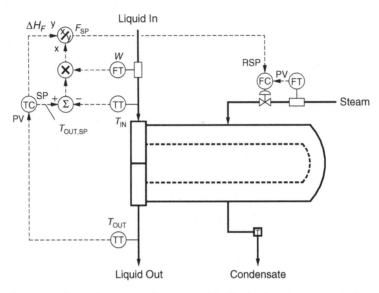

Figure 6.39   Feedback trim by adjusting a coefficient in the feedforward control equation.

For most exchanger applications, these considerations suggest adjusting the change in enthalpy ΔH_F from steam to condensate. However, if the liquid composition is highly variable, the error in the liquid heat capacity c_P is potentially larger.

In the control configuration in Figure 6.39, the value of ΔH_F is specified by the feedback trim controller. The coefficient for the multiplier is the liquid heat capacity c_P. Consequently, the output of the multiplier is $W c_P (T_{\text{out,SP}} - T_{\text{in}})$. The divider block divides the output of the multiplier by the output of the feedback trim controller to obtain the set point for the steam flow controller. For the configuration in Figure 6.39, the output range for the feedback trim controller must be in the engineering units of enthalpy: specifically, Btu/lb. If the computations suggested above give a value of 950 Btu/lb for ΔH_F, a logical output range for the feedback trim controller could be 850 to 1050 lb/min. This permits the feedback trim controller to increase or decrease the enthalpy change ΔH_F by approximately 10%.

Basically, the feedback trim controller is recalibrating the steady-state model by adjusting the value of the enthalpy change ΔH_F. When recalibrating a steady-state model, the usual approach is to group as many individual coefficients into a single parameter whose value will be adjusted. For the steam-heated exchanger, the logical choice for the parameter to be adjusted is $\Delta H_F / c_P$, or its inverse $c_P / \Delta H_F$. When choosing the latter, the divider block in Figure 6.39 must be replaced by a multiplier block. The resulting configuration is equivalent to the configuration in Figure 6.38, where the output of the feedforward calculation is multiplied by a coefficient close to 1.0. That two approaches are basically equivalent is common for simple applications such as the steam-heated exchanger.

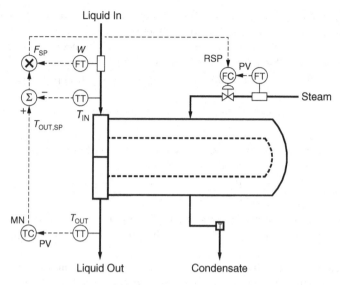

Figure 6.40 Feedback trim by adjusting the set point for the feedforward controller.

Adjusting the Set Point of the Feedforward Controller. The configurations in Figures 6.37 to 6.39 have one characteristic in common: The set point for the feedback trim controller must also be used as the set point for the feedforward controller. In current digital systems this is easily configured, but this was not the case for conventional pneumatic and electronic controls (and some early digital systems).

The control configuration in Figure 6.40 is an easy way around this limitation. The output of the feedback trim controller is the set point for the feedforward controller. For this configuration, the engineering units for the output range of the feedback trim controller must be temperature units: specifically, °F. The output range in engineering units for the feedback trim controller could be the same as the measurement range for the liquid outlet temperature, but a narrower range could be specified if desired.

The concept behind this approach is conceptually very simple. Suppose that the set point for the feedforward controller is 280°F but the liquid outlet temperature is 282°F when the process lines-out. How can a liquid outlet temperature of 280°F be obtained? Try a set point of 278°F for the feedforward controller. Most processes are reasonably linear over a narrow range, so the result is very likely a liquid outlet temperature of 280°F plus or minus a few tenths of a degree.

The configuration in Figure 6.40 applies the feedback trim by offsetting the set point of the feedforward controller from the desired value of the liquid outlet temperature. This configuration was used almost universally with conventional pneumatic and electronic controls. It gets the job done. As it is difficult to argue with success, this approach continues to be used.

Transient Error. When the liquid flow changes from 1000 lb/min to 500 lb/min, the output of the feedforward controller changes immediately. If the liquid flow decreases by a factor of 2, the feedforward controller immediately decreases the steam flow set point by a factor of 2. In a similar fashion, any change in the liquid inlet temperature would be translated immediately to an appropriate change in the steam flow.

The transient error depends on the following dynamic characteristics of the process:

- Effect of a change in the steam flow on the liquid outlet temperature. This behavior was illustrated in Figure 6.18.
- Effect of a change in the liquid flow on the liquid outlet temperature. This behavior was illustrated in Figure 6.17.
- Effect of a change in the liquid inlet temperature on the liquid outlet temperature. A test could be conducted on the process to determine this behavior.

The steady-state feedforward controller makes no provision for process dynamics. To reduce the transient errors, dynamic compensation must be incorporated into the control configuration. In a manner similar to that for ratio control, the usual approach is to provide dynamic compensation by a lead-lag element, occasionally accompanied by a dead-time element. There are two options for applying dynamic compensation:

- *To the output of the feedforward controller.* This approach requires only one lead-lag element whose output is the set point for the flow controller. For the steam-heated exchanger, are the dynamics of a change in the liquid inlet temperature on the liquid outlet temperature similar to the dynamics of a change in the liquid flow on the liquid outlet temperature? Probably not. Consequently, the tuning of the lead-lag compensator reflects primarily the change that has the largest impact during process operations (probably the change in the liquid flow), and the results used for the other input. Usually, this is satisfactory, but not always.
- *To each input to the feedforward controller.* This approach requires a lead-lag element for each input, and each lead-lag element must be tuned. Given the difficulty of tuning the lead-lag compensator, this approach is pursued only when the single lead-lag compensator on the output of the feedforward controller proves unsatisfactory.

Tuning the lead-lag compensator for a feedforward controller is approached in the same manner as tuning the lead-lag compensator for a ratio controller. This was discussed in detail earlier and will not be repeated.

7

LOOP INTERACTION

Most processes are multivariable in nature. For example, a simple two-product distillation column has five variables that could potentially be controlled:

- Accumulator level
- Bottoms level
- Column pressure
- Overhead composition (or the temperature of an upper stage)
- Bottoms composition (or the temperature of a lower stage)

The customary approach is to provide a single-loop PID controller for each of the variables to be controlled. Usually, this works, but occasionally the result is one or more controllers that cannot be tuned. Applying automatic tuning to such controllers is merely throwing technology at the problem, and is rarely successful. Tuning difficulties are symptoms of other problems, in this case interaction between the variables to be controlled.

Applying single-loop controllers to a multivariable process can be successful only if the degree of interaction is no more than modest. In most applications the initial effort is to understand the qualitative nature of the interaction between the process variables. But in processes such as distillation, quantitative measures are required.

Advanced Process Control: Beyond Single-Loop Control   By Cecil L. Smith
Copyright © 2010 John Wiley & Sons, Inc.

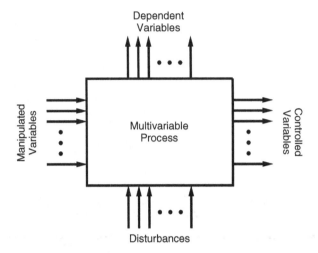

Figure 7.1 Generic representation of a multivariable process.

7.1. MULTIVARIABLE PROCESSES

Figure 7.1 provides a generic representation of a multivariable process. The inputs can be divided into two categories:

Manipulated variables. The values of these variables are at the discretion of the control system.

Disturbances. The values of these variables are determined by external factors.

The outputs can also be divided into two categories:

Controlled variables. The values of these variables are to be maintained at or near their targets. These variables must be independent; that is, the process must not establish a relationship between two or more of the controlled variables (for example, the temperature and pressure of a boiling liquid are not independent).

Dependent variables. The values of these variables are influenced by the manipulated variables and the disturbances. They are not controlled to specified targets, but in some applications, constraints may apply to the dependent variables.

Controlled Variables and Manipulated Variables. Figure 7.2 presents a schematic of a chlorine vaporizer with an internal steam coil. The purpose of the vaporizer is to supply chlorine to the process at a specified flow rate, so a flow measurement is provided on the chlorine vapor discharge stream. Being a pressure vessel, the vaporizer is equipped with a pressure-relief device. Potentially, the

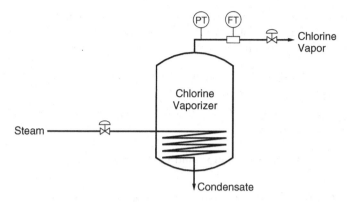

Figure 7.2 Chlorine vaporizer.

chlorine within the vaporizer could be heated to the steam supply temperature. The corresponding vaporizer pressure is the vapor pressure of chlorine at this temperature. If this pressure exceeds the setting on the pressure-relief device, releasing chlorine through the relief device becomes a possibility. The vaporizer pressure is measured and is to be controlled at a set point that is above the process pressure and below the pressure setting on the relief device.

Control valves are installed on the chlorine vapor line and on the steam supply. This gives the following set of controlled and manipulated variables:

Manipulated Variable	Controlled Variable
Chlorine vapor valve opening	Chlorine vapor flow
Steam supply valve opening	Vaporizer pressure

Dimensionality and Notation. Let the number of controlled variables be n. Let the number of manipulated variables be m. The dimensionality of a multivariable process is stated as $n \times m$. The process in Figure 7.2 would be a 2×2 multivariable process. The generic notation for the controlled and manipulated variables is as follows:

$$C_i = \text{controlled variable } i, \qquad \text{where } i = 1, 2, \ldots, n$$
$$M_j = \text{manipulated variable } j, \qquad \text{where } i = 1, 2, \ldots, m$$

The numeric index for the respective variables is at the user's discretion. For the process in Figure 7.2, the subscripts are assigned arbitrarily as follows:

Manipulated Variable	Controlled Variable
M_1: chlorine vapor valve opening	C_1: chlorine vapor flow
M_2: steam supply valve opening	C_2: vaporizer pressure

Often, the subscripts for the controlled variables and manipulated variables are assigned in a manner consistent with the configuration of the individual loops, as least as contemplated originally. As assigned above, the subscripts suggest that:

- The chlorine vapor flow C_1 will be controlled by manipulating the chlorine vapor valve opening M_1.
- The vaporizer pressure C_2 will be controlled by manipulating the steam supply valve opening M_2.

However, this is not assured and should never be assumed.

Square Configurations. A multivariable process is said to be "square" if the number of controlled variables n equals the number of manipulated variables m. Provided that no constraints are encountered (for example, control valve fully open or fully closed), it is possible for the control system to drive each controlled variable to its respective target. Single-loop controllers can only be applied to square configurations. If there are four manipulated variables and four controlled variables, four single-loop controllers are required.

Pairing refers to how the controllers are arranged. A manipulated variable must be selected for each controlled variable. The controlled variable is the measured variable for that controller; the controller output drives the selected manipulated variable. In some applications, the selection is obvious. In others, the choice is not clear at all. In yet others, subtle characteristics of the process cause the "obvious" selection to be not as good as thought initially.

Pairings. For a 2×2 process, there are two possible pairings:

- Control C_1 by manipulating M_1; control C_2 by manipulating M_2.
- Control C_1 by manipulating M_2; control C_2 by manipulating M_1.

For a 3×3 process there are six possible pairings; for an $n \times n$ process, the number of possible pairings is

$$n! = n \times (n-1) \times (n-2) \times \cdots \times 1$$

From a multivariable control perspective, all pairings are potential candidates. However, other considerations may eliminate one or more of the possible pairings.

For a 2×2 process only one of the possible pairings can potentially perform properly. However, it is not assured that either will perform properly. Unless one is willing to sacrifice performance by tuning one loop to respond much more slowly than the other, the possibilities for a 2×2 process are as follows:

- Controlling C_1 by manipulating M_1 (and C_2 by manipulating M_2) performs properly, but controlling C_1 by manipulating M_2 (and C_2 by manipulating M_1) does not.

- Controlling C_1 by manipulating M_2 (and C_2 by manipulating M_1) performs properly, but controlling C_1 by manipulating M_1 (and C_2 by manipulating M_2) does not.
- Neither configuration performs properly.

Skinny Configurations. A multivariable process is said to be *skinny* if the number of controlled variables exceeds the number of manipulated variables. In such configurations, it is not possible for the control system to drive all of the controlled variables to their targets. Skinny configurations are not commonly encountered in process applications.

It is only possible to control such a process in some "best" sense. This raises the question of what is meant by "best." A generic definition might be to control so as to minimize the sum of squares of the deviations from target. That is, "best" means

$$\min\{\Sigma E_k^2\}$$

where $E_k = \text{SP} - \text{PV}$ for controlled variable k. This weighs all errors equally, which is probably not appropriate. A weighting factor for each error could be incorporated. Alternatively, an objective function specific to the application could be developed.

Fat Configurations. A multivariable process is said to be *fat* if the number of manipulated variables exceeds the number of controlled variables. Not only is it possible for the control system to drive all the controlled variables to their targets, but there are many combinations of the manipulated variables that would do so. Many process applications prove to be fat.

An objective function is required to determine which combination of the manipulated variables is "best" (minimum energy, maximum throughput, or whatever). One approach to accomplish this is as follows:

- Obtain a square configuration by "fixing" (that is, set to a constant value) the necessary number of the manipulated variables.
- Adjust the remaining manipulated variables so as to drive the controlled variables to their targets. This can be done with single-loop controllers, provided that the degree of interaction is no more than modest.
- Use steady-state optimization routines to adjust the "fixed" manipulated variables so as to attain the "best" process performance.

When approached in this manner, the control issues for the fat configuration are the same as for the square configuration.

Purified Water Supply Process. An example of a *fat* configuration is the purified water supply process illustrated in Figure 7.3. The purified water from the feed tank is first cooled to a specified temperature (such as 5°C) for delivery

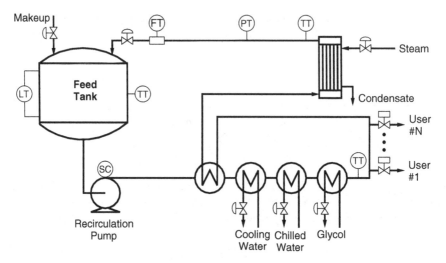

Figure 7.3 Purified water supply process.

to multiple users, who take as much or as little water as they like. A minimum flow of water back to the feed tank is required, and this water is heated so as to maintain the water in the feed tank at a desired temperature (high enough to prevent bacterial growth). The water from the feed tank is cooled in four stages:

- Recirculated water (to provide a measure of energy conservation)
- Cooling water
- Chilled water
- Glycol

The controlled and manipulated variables are as follows:

Manipulated Variable	Controlled Variable
Purified water makeup valve opening	Feed tank level
Recirculation valve opening	Feed tank temperature
Recirculation pump speed	Recirculation flow
Steam valve opening	Recirculation pressure
Cooling water valve opening	Purified water temperature
Chilled water valve opening	
Glycol valve opening	

This process has five controlled variables and seven manipulated variables, making this a *fat* configuration. The objective function should be to cool the purified water in the least expensive manner (glycol is more expensive than chilled water, which is more expensive than cooling water).

The purified water supply process is also typical of industrial control applications in another respect: specifically, constraints are present. For any water-based process at atmospheric pressure, water temperatures are limited by 0°C and 100°C:

Freezing. The set point for the temperature of the purified water delivered to the users must be above 0°C. However, rapid decreases in the water withdrawn by the users cause the purified water temperature to drop below its set point. Excessive dips could lead to problems.

Boiling. The purified water feed tank is at atmospheric pressure. Consequently, the temperature of the recirculated water leaving the steam-heated exchanger must not exceed 100°C. With the control valve on the recirculated water located as in Figure 7.3, an upper limit must be imposed on the recirculated water temperature to avoid flashing in the control valve.

7.2. ISSUES WITH THE P&I DIAGRAM

Often, the history behind a P&I diagram is obscure in the sense that little or no documentation is available on the rationale for the arrangement of the loops. On repeat projects, the arrangement of the controls is usually copied from a previous installation, and it is even possible that those controls never functioned properly. For new process designs, P&I diagrams are being developed by less experienced people in shorter times, neither of which is comforting. This industry also has a history of not operating processes under the conditions for which they were designed. As modifications to the process are made, evolution of the P&I diagram does not always properly reflect the current behavior of the process. In short, never casually assume that the current P&I diagram is appropriate to the process as it is now constructed and operated.

Tuning Problems. The usual approach is first to attempt to tune the loops as they appear on the P&I diagram. Most loops can be tuned successfully, which means that the designers get the P&I diagram right most of the time. But there are exceptions. If the P&I diagram does not properly reflect the characteristics of the process, tuning difficulties will arise in one or more loops. In a sense, the P&I diagram must be "tuned" to the process just as the PID controller must be tuned to reflect the process characteristics.

Loop tuning difficulties are the symptoms of some problem within the controls, one possibility being loop interaction. A control loop that functions properly when used alone is not assured to function when used in conjunction with other loops. A common situation in a control room is to have a loop that performs in an acceptable manner when another loop is on manual, but when this other loop is switched to automatic, neither loop functions properly. The probable explanation for this is interaction between the two loops.

Unless prior experience is available, the initial control configuration usually relies on single-loop controllers, mainly because plant personnel are comfortable with them. Multivariable control technology of various forms is readily available today, but these technologies are normally applied only in the following situations:

- No single-loop controller configuration proves successful.
- A significant incentive exists to operate the process in the most efficient manner.

For the latter, the multivariable control technology is justified in support of a process optimization endeavor with significant economic benefits. The reason is simple: You have to be able to control a process in order to operate the process at the optimum conditions.

Off-Gas Process. Figure 7.4 presents a multivariable process and a proposed P&I diagram for this process. The input to this process is a gas stream containing the following three materials:

- Small amount of particulate matter, that is, dust.
- Sufficient water vapor that the dew point is above ambient temperature.
- Sulfur dioxide (SO_2) or similar component that would cause any condensate to be acidic.

The processing of the off-gas involves the following equipment:

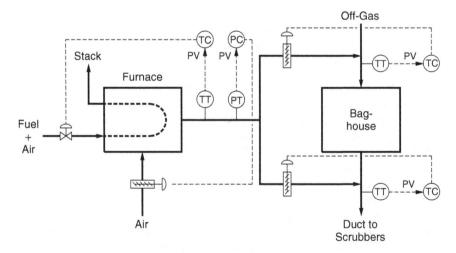

Figure 7.4 Proposed single-loop control configuration for the off-gas process.

Baghouse (included in Figure 7.4). This removes the particulate matter.
Scrubber (not included in Figure 7.4). This removes the sulfur dioxide.

The presence of significant water vapor complicates process operations. Basically, condensation must be avoided by keeping the off-gas temperature above its dew point until it arrives at the scrubbers. How can a gas stream containing dust and sulfur dioxide be heated? The answer is: by adding hot gas. The process in Figure 7.4 includes a furnace that heats outside air to provide a hot gas stream. So that the off-gas will remain above the dew point through the baghouse, some of the hot air is added upstream of the baghouse. The remaining hot air is added downstream of the baghouse so that the off-gas remains above its dew point until it arrives at the scrubbers.

Loop Pairing. The controlled and manipulated variables for the off-gas process in Figure 7.4 are as follows:

Manipulated Variable	Controlled Variable
Opening of hot gas damper on baghouse inlet	Baghouse inlet temperature
Opening of hot gas damper on baghouse outlet	Scrubber duct inlet temperature
Fresh air damper opening	Hot gas pressure
Furnace fuel control valve opening	Hot gas temperature

The control configuration in Figure 7.4 is one possible pairing of controlled and manipulated variables, but as for most multivariable processes, there are other possibilities.

The logic followed in selecting the pairing of manipulated variables with controlled variables is not easily expressed. The pairing has traditionally been developed by experienced control engineers, who undoubtedly draw upon their past experiences as much as following any rigorous logical rules. But if one examines enough P&I diagrams, the common denominator can be summarized as follows:

Control each variable with the nearest final control element that significantly affects that variable.

Most developers of P&I diagrams do not rely on such a rule. However, it usually turns out that way, probably because dynamic behavior receives most of the attention. And from a dynamic perspective, the rule is probably appropriate. But controls must reflect both the steady-state and the dynamic behavior of the process. This is the problem: The steady-state effect of a final control element on the variable to be controlled can be much less than it appears initially. This is reflected in the advice of the "old hands" in this business:

You have to understand the process.

Deficiency. The control configuration in Figure 7.4 has a serious deficiency. One would tune the controllers in the following order:

- Hot gas pressure controller. This controller can be tuned with only the air blowers in operation.
- Hot gas temperature controller. The other temperature loops require hot gas in order to function.
- Baghouse inlet temperature controller.
- Scrubber duct inlet temperature controller.

This works well until tuning the last controller. With this controller on automatic, cycles are induced into all the other loops.

The configuration in Figure 7.4 proposes to control the temperature of the gas entering the ducts to the scrubber using the damper on the hot gas added downstream of the baghouse. Superficially, one would think that this damper has an effect on that temperature. However, its long-term (steady-state) effect is nil, as explained in a companion book [1]. But when this controller moves its damper, the primary effect is to upset the other three loops. This can be mitigated by tuning the controller to respond very slowly, but the loop performance degrades to the point that the loop might as well remain on manual.

One Loop Depending on Another Loop. The following logic suggests that the controller for the temperature of the gas entering the duct to the scrubbers will perform properly:

- Opening the hot gas damper to baghouse outlet drops the hot gas pressure.
- The hot gas pressure controller will open the fresh air damper, which decreases the hot gas temperature.
- The hot gas temperature controller increases the fuel to the furnace.
- The additional hot gas will increase the temperature of the gas entering the duct to the scrubbers.

In effect, this suggests that the temperature controller can achieve its objective by working through two other controllers. This will perform properly only under one condition: Any controller that depends on another controller must be slower than that other controller, preferably by a factor of 5 (same as for cascade controls). Actually, the reasoning above has a two-level dependency: The controller for the temperature of the gas entering the duct to the scrubbers depends on the hot gas pressure controller, which in turn depends on the hot gas temperature controller. The loops in Figure 7.4 all have about the same dynamics. The dynamic separation can be achieved only by tuning controllers very conservatively, which also degrades performance.

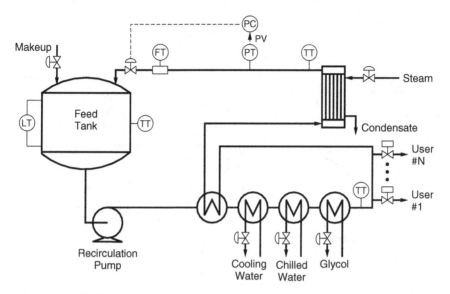

Figure 7.5 Purified water supply process with a constant-speed drive for the pump.

Purified Water Supply Process. The purified water supply process was originally installed with a constant-speed drive for the recirculation pump, as illustrated in Figure 7.5. At that time, energy was cheap and variable-speed drive technology was much more expensive than today.

The following requirements pertain to the recirculation flow and pressure:

- The recirculation flow must be turbulent at all times.
- The purified water must be delivered to users at a specified pressure.

To deliver the water to users at the required pressure, the configuration in Figure 7.5 controls the recirculation pressure by adjusting the recirculation valve opening. When the users are withdrawing no purified water, the recirculation valve opening is large and the recirculation flow is also large. To maintain the recirculation pressure at its set point, the recirculation valve opening must decrease as the users withdraw more water. Provided that the recirculation flow is in the turbulent regime when the users are withdrawing the maximum amount of water, the configuration in Figure 7.5 meets all process requirements.

Especially when users are withdrawing little purified water, the pump in Figure 7.5 is pumping far more water than required to keep the recirculation flow in the turbulent regime. With the combination of increased energy costs and lower costs for variable-speed drive technology, the economics justified replacing the constant-speed drive with a variable-speed drive. In keeping with the "if it ain't broke, don't fix it" philosophy, the pressure loop in Figure 7.5 was retained and a flow loop added, giving the configuration in Figure 7.6.

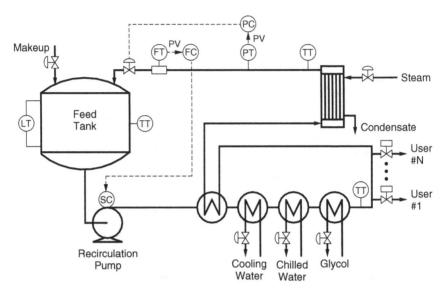

Figure 7.6   Original control configuration for purified water supply process with a variable-speed drive for the pump.

With the addition of the variable-speed drive, the process becomes a 2×2 multivariable process. The following notation reflects the configuration in Figure 7.6:

Manipulated Variable	Controlled Variable
M_1: recirculation valve opening	C_1: recirculation pressure
M_2: recirculation pump speed	C_2: recirculation flow

Unfortunately, the rationale for retaining the pressure loop as in Figure 7.5 is flawed. A loop that performs properly when used alone is not assured of performing properly when used in conjunction with another loop. Even two loops that perform properly when used individually are not assured of performing satisfactorily when used together.

Loops in a Multivariable Process. The purpose of the three diagrams in Figure 7.7 is to show that three loops are present when controlling a 2×2 multivariable process. But first, a word about the representation of the process. The effect of each final control element in Figure 7.6 is as follows:

- Increasing the recirculation valve opening M_1 decreases the recirculation pressure C_1 but increases the recirculation flow C_2.
- Increasing the recirculation pump speed M_2 increases both the recirculation pressure C_1 and the recirculation flow C_2.

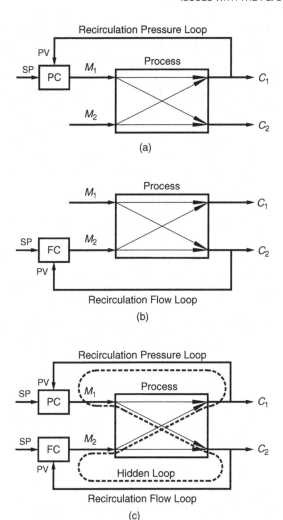

Figure 7.7 Loops in a 2 × 2 multivariable process: (a) recirculation pressure loop; (b) recirculation flow loop; (c) hidden loop.

To convey that both manipulated variables affect both controlled variables, the simplest approach is to represent the process by a rectangle that contains four arrows, one connecting each manipulated variable to each controlled variable. Figure 7.7 uses such a representation.

For Figure 7.7(a) the recirculation pressure controller is in automatic and the recirculation flow controller is in manual. Only one loop exists: the recirculation pressure loop. The pressure controller changes the recirculation valve opening M_1 so as to maintain the recirculation pressure C_1 near the target. Changes in M_1 also affect the recirculation flow C_2. However, there is no automatic response to these changes.

For Figure 7.7(b) the recirculation pressure controller is in manual and the recirculation flow controller is in automatic. Only one loop exists: the recirculation flow loop. The flow controller changes the recirculation pump speed M_2 so as to maintain the recirculation flow C_2 near the target. Changes in M_2 also affect the recirculation pressure C_1. However, there is no automatic response to these changes.

For Figure 7.7(c) the recirculation pressure controller and the recirculation flow controller are in automatic. Both the recirculation pressure loop and the recirculation flow loop exist, but there is a third loop, sometimes referred to as the *hidden loop*. Starting at the recirculation valve opening M_1, this loop is constructed as follows:

- Changes in the recirculation valve opening M_1 affect the recirculation flow C_2.
- The recirculation flow controller responds to changes in the recirculation flow C_2 by changing the recirculation pump speed M_2.
- Changes in the recirculation pump speed M_2 affect the recirculation pressure C_1.
- The recirculation pressure controller responds to changes in the recirculation pressure C_1 by changing the recirculation valve opening M_1.

A few observations pertaining to the hidden loop:

- The hidden loop exists only when both controllers are in automatic. If the pressure loop and the flow loop function properly when used individually but not when both are in automatic, the hidden loop is the source of the problems.
- When both controllers contain the integral mode (as they usually do), the hidden loop contains two integrators. Stability issues are a potential problem in any loop containing double integration.

For a 2×2 multivariable process, there is only one hidden loop. However, the number of hidden loops increases with the dimensionality of the process.

Analysis of Interaction. When a loop interaction problem arises, the customary first effort is to qualitatively analyze the interaction within the process. Ultimately, the success of this depends on how well those doing the analysis really understand the nature of the process. As the complexity of the process increases, this task becomes more daunting, which increases the probability of an incorrect conclusion. This approach does not always produce correct results even for a simple process, or at least a process that is thought to be simple.

The analysis of interaction must focus on the process. The analysis should be based on the process flowsheet with only measurement devices and final control elements indicated. For the purified water supply process, Figure 7.3 presents such a flowsheet. At this point, the focus is only on controlling the recirculation pressure and recirculation flow by adjusting the recirculation valve opening and the recirculation pump speed. The analysis begins by understanding how each final control element affects each controlled variable, or conversely, how each controlled variable is affected by each final control element.

An analysis of the flow loop reveals an interesting behavior. In the relationships for the recirculation flow, there are only two variables:

- The recirculation pressure
- The recirculation valve opening

If the recirculation pressure can be maintained at its target, a constant recirculation valve opening will give a constant recirculation flow. Purified water is clean, so buildups do not occur (or if they do, their effect on the recirculation flow is minor relative to other consequences).

This analysis suggests that perhaps the recirculation flow controller is not necessary. Just set the recirculation valve opening to the appropriate value and maintain the proper recirculation pressure. A more realistic approach is to configure the recirculation flow controller as follows:

- The output of the recirculation flow controller should be the recirculation valve opening, not the recirculation pump speed.
- Tune the recirculation flow controller to respond more slowly than the recirculation pressure controller.

This suggests the loop pairing presented in Figure 7.8, which is as follows:

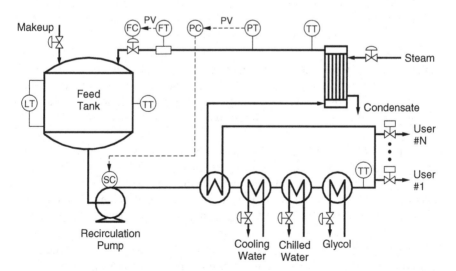

Figure 7.8 Revised control configuration for purified water supply process.

- Control the recirculation pressure C_1 with the recirculation pump speed M_2.
- Control the recirculation flow C_2 with the recirculation control valve opening M_1.

With modern digital controls, changing the loop pairing from that in Figure 7.6 to that in Figure 7.8 is relatively easy. The ability to make such changes easily was cited as an incentive to migrate from conventional electronic analog controls to digital controls. But never assume that such changes will be made readily. In fact, such changes should never just be implemented casually. Reviews are always appropriate, and are mandatory if management of change is in effect. But such reviews can go on and on, to the extent that the configuration is never changed.

7.3. STEADY-STATE SENSITIVITIES OR GAINS

Quantitative measures of interaction are computed from the steady-state gains for the process. For a process with only a single input and a single output, the steady-state gain K is the sensitivity of the output C (the controlled variable) to the input M (the manipulated variable). This sensitivity is expressed by a partial derivative but is often computed using a finite-difference approximation:

$$K = \frac{\partial C}{\partial M} \cong \frac{\Delta C}{\Delta M}$$

For a multivariable process, two aspects greatly complicate this subject:

- There is a sensitivity or gain from each input to each output. For a 2×2 multivariable process, there are four gains; for a 3×3 multivariable process, there are nine gains; for an $n \times m$ multivariable process, there are $n \times m$ gains.
- Most initially assume that the gain is the sensitivity of an output to one of the inputs, with all other inputs maintained constant. However, this is not the only possibility. Specifically, the gain could be the sensitivity of an output to one of the inputs with all other outputs maintained constant.

For measures of interaction, the first issue adds complexity in numbers. But the latter distinction proves to be fundamental to quantifying the degree of interaction.

The objective of this section is to examine the various aspects of steady-state gains or sensitivities, including how to obtain numerical values for the various gains. This provides the basis for developing a quantitative measure of interaction.

Block Diagram. Figure 7.9 presents a block diagram for a 2×2 multivariable process. Only the steady-state gains or sensitivities are of interest, so no dynamics are included in the block diagram. Such a block diagram applies to the

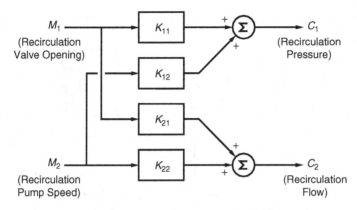

Figure 7.9   Block diagram for 2 × 2 multivariable process.

recirculation pressure and recirculation flow loops for the purified water supply process. The block diagram in Figure 7.9 involves four gains:

Gain	Explanation
K_{11}	Sensitivity of C_1 (recirculation pressure) to M_1 (recirculation valve opening)
K_{12}	Sensitivity of C_1 (recirculation pressure) to M_2 (recirculation pump speed)
K_{21}	Sensitivity of C_2 (recirculation flow) to M_1 (recirculation valve opening)
K_{22}	Sensitivity of C_2 (recirculation flow) to M_2 (recirculation pump speed)

Constructing a block diagram is only manageable for a 2 × 2 multivariable process. For a 3 × 3 multivariable process, the block diagram becomes very complex (there are nine gains). For higher-order multivariable processes, drawing a block diagram is out of the question.

Steady-State Sensitivities. For a 2 × 2 multivariable process, the statement "sensitivity of C_1 to M_1" is ambiguous. This gain or sensitivity always relates a change in C_1 (recirculation pressure) to a change in M_1 (recirculation valve opening). But for a 2 × 2 multivariable process, sensitivity of C_1 to M_1 can be obtained under two different conditions. One way to understand the difference in the two sensitivities is to examine the test procedures for obtaining values for the two sensitivities.

Figure 7.10 illustrates a test configuration for which no control loops are in automatic. The recirculation control valve opening M_1 is increased by ΔM_1, which leads to a decrease in ΔC_1 in the recirculation pressure C_1 and an increase in the recirculation flow C_2. As for simple loops, the value of the sensitivity of

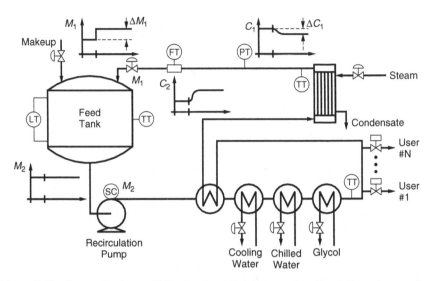

Figure 7.10   Test to obtain sensitivity of recirculation pressure to recirculation valve opening, recirculation flow controller in manual.

C_1 to M_1 is the ratio of ΔC_1 to ΔM_1. But a multivariable process requires more specificity. In Figure 7.10, the value of the recirculation pump speed M_2 is held constant throughout the test. When the sensitivity is evaluated in this manner, the result is the sensitivity of C_1 to M_1 at constant M_2.

Figure 7.11 illustrates an alternative test configuration for obtaining a value of the sensitivity of C_1 to M_1. In Figure 7.11, the recirculation flow controller is in automatic. The recirculation valve opening M_1 is again increased by ΔM_1. Initially, the recirculation pressure C_1 decreases and the recirculation flow C_2 increases. But in this test configuration, the recirculation flow controller responds to the increase in the recirculation flow C_2 by increasing the recirculation pump speed M_2. Once the transients have passed, the flow controller has returned the recirculation flow C_2 to its set point, but to do so, the recirculation pump speed M_2 has increased.

Will the change ΔC_1 in the recirculation flow be the same as for the test configuration in Figure 7.10 (flow controller in manual)? Depending on the nature of the process, there are three possibilities:

1. ΔC_1 *is the same.* This occurs if there is no interaction within the process.
2. ΔC_1 *is less.* Although not the case for the purified water supply process, some processes behave in this manner.
3. ΔC_1 *is greater.* This is the case for the purified water supply process. Opening the recirculation valve decreases the recirculation pressure and increases the recirculation flow. To return the flow to its original value, the pump speed must be decreased, with further decreases the recirculation pressure.

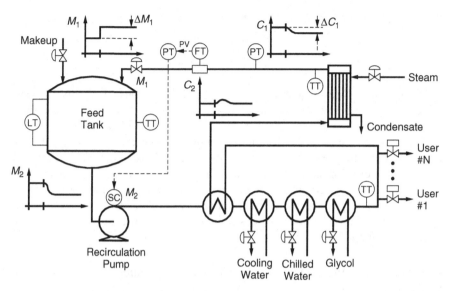

Figure 7.11 Test to obtain sensitivity of recirculation pressure to recirculation valve opening, recirculation flow controller in automatic.

The test configuration in Figure 7.11 obtains a value for the sensitivity of C_1 to M_1 at constant C_2. There is a transient in the response for the recirculation flow C_2. But when evaluating the steady-state sensitivity, only long-term changes are relevant. With the recirculation flow controller in automatic, the final value for the recirculation flow is the same as its initial value. From a steady-state perspective, the value of the recirculation flow C_2 is constant.

Notation. In a multivariable process, sensitivities must be expressed by partial derivatives, which can in turn be approximated by finite differences. The sensitivity of C_1 to M_1 is expressed by the following partial derivative:

$$K_{11} = \frac{\partial C_1}{\partial M_1} \cong \frac{\Delta C_1}{\Delta M_1}$$

But as noted previously, the "sensitivity of C_1 to M_1" is ambiguous. The sensitivity obtained using the test configuration in Figure 7.10 with the recirculation pressure controller in manual is properly referred to as the "sensitivity of C_1 to M_1 at constant M_2." This sensitivity is expressed as follows:

$$K_{11} = \left.\frac{\partial C_1}{\partial M_1}\right|_{M_2} \cong \left.\frac{\Delta C_1}{\Delta M_1}\right|_{M_2}$$

The sensitivity obtained using the test configuration in Figure 7.11 with the recirculation pressure controller in automatic is properly referred to as the "sensitivity of C_1 to M_1 at constant C_2." This sensitivity is expressed as follows:

$$K'_{11} = \frac{\partial C_1}{\partial M_1}\bigg|_{C_2} \cong \frac{\Delta C_1}{\Delta M_1}\bigg|_{C_2}$$

This sensitivity is designated as K'_{11}, whereas the previous sensitivity is designated as K_{11}. These two sensitivities are not the same, and whenever interaction is present within the process, their numerical values will be different.

The notation gives the appearance of more complexity than is actually the case. The only difference between these two sensitivities is that one is evaluated with the recirculation flow controller on manual, and the other is evaluated with the recirculation flow controller on automatic. Despite the complex notation, this is a simple concept. Why is this of interest? As explained in the next section, the difference between these two sensitivities is a good measure of the degree of interaction within the process.

Process Gain Matrix. For the block diagram in Figure 7.9, the steady-state gains are as follows:

Gain	Notation	Explanation
K_{11}	$\dfrac{\partial C_1}{\partial M_1}\bigg|_{M_2}$	Sensitivity of C_1 (recirculation pressure) to M_1 (recirculation valve opening) at constant M_2 (recirculation pump speed)
K_{12}	$\dfrac{\partial C_1}{\partial M_2}\bigg|_{M_1}$	Sensitivity of C_1 (recirculation pressure) to M_2 (recirculation pump speed) at constant M_1 (recirculation valve opening)
K_{21}	$\dfrac{\partial C_2}{\partial M_1}\bigg|_{M_2}$	Sensitivity of C_2 (recirculation flow) to M_1 (recirculation valve opening) at constant M_2 (recirculation pump speed)
K_{22}	$\dfrac{\partial C_2}{\partial M_2}\bigg|_{M_1}$	Sensitivity of C_2 (recirculation flow) to M_2 (recirculation pump speed) at constant M_1 (recirculation valve opening)

These gains relate changes in the controlled variables to changes in the manipulated variables. In equation form, the block diagram in Figure 7.9 is expressed as follows:

$$\Delta C_1 = K_{11}\Delta M_1 + K_{12}\Delta M_2$$

$$\Delta C_2 = K_{21}\Delta M_1 + K_{22}\Delta M_2$$

This set of equations can be expressed using vectors and matrices as follows:

$$\begin{bmatrix} \Delta C_1 \\ \Delta C_2 \end{bmatrix} = \begin{bmatrix} K_{11} & K_{12} \\ K_{21} & K_{22} \end{bmatrix} \begin{bmatrix} \Delta M_1 \\ \Delta M_2 \end{bmatrix}$$

$$\mathbf{c} \quad = \quad \mathbf{K} \quad\quad \mathbf{m}$$

where

$$\mathbf{c} = \begin{bmatrix} \Delta C_1 \\ \Delta C_2 \end{bmatrix} = \text{vector of changes in the controlled variables}$$

$$\mathbf{m} = \begin{bmatrix} \Delta M_1 \\ \Delta M_2 \end{bmatrix} = \text{vector of changes in the manipulated variables}$$

$$\mathbf{K} = \begin{bmatrix} K_{11} & K_{12} \\ K_{21} & K_{22} \end{bmatrix} = \text{process gain matrix}$$

Note that K_{ij} is the sensitivity of C_i to M_j with all controllers on manual, which means that all manipulated variables other than M_j are held constant.

As for notation, one approach is to use uppercase letters for the actual values of the controlled and manipulated variables and lowercase letters for changes. This is the basis for the use of $\mathbf{c}$ for the vector of changes in the controlled variables and $\mathbf{m}$ for the vector of changes in the manipulated variables.

Evaluating the Sensitivities. There are two options for evaluating the sensitivities:

Process testing. In practice, this can only be considered for processes that respond rapidly. The recirculation pressure and recirculation flow would do so. But even so, a time must be found where the tests can be conducted with an acceptable degree of interference with production operations.

Process models. For these sensitivities, a steady-state model is sufficient. For applications such as distillation, most modern designs are based on a steady-state column simulation, which means that some prior work should be available. But for other applications, the availability of a model is not assured, especially for processes that have been in operation for several years.

Although testing is a viable approach, process testing usually proves more difficult that anticipated initially. Evaluating the sensitivities from a process model usually proves the most viable approach. For assessing the degree of interaction, extreme accuracy is not required of the model. The absolute value of the sensitivities is actually not of interest; the degree of interaction is assessed from the change in the sensitivity that results from switching the other controller(s) between manual and automatic.

In deciding which approach to use, one other factor must enter the decision process. For the purified water supply process, the demand from the users varies from low to high. Some users require purified water 24 hours a day, 7 days a week. Some do not require purified water on the weekends; a few require purified water only during normal working hours. The extremes are as follows:

• No user is taking purified water (all is returned to the feed tank).
• The users are consuming 380 gpm of purified water.

Is the degree of interaction the same for these two extremes? Not necessarily. To answer this question using process testing requires significant additional effort. To answer this question using a process model requires only a nominal additional effort.

Model for a Purified Water Supply Process. The objective is to model only the recirculation pressure and flow. Consequently, the following simplifications will be made:

- No thermal aspects will be modeled. Temperature changes have only a nominal effect on the recirculation flow and pressure.
- The level in the feed tank is assumed to be constant. The hydrostatic head is only a nominal component of the pump head, and will be neglected.

The model will be based on the simplified flowsheet in Figure 7.12 that is an approximation to the flowsheet in Figure 7.3. The components in the approximation are as follows:

- A pump curve is available for the recirculation pump. The pump head H will be related to the pump flow F_P by computing the coefficients in a quadratic equation from the following three points on the pump curve:

Flow (gpm)	Head (ft)
0	250
400	203
800	118

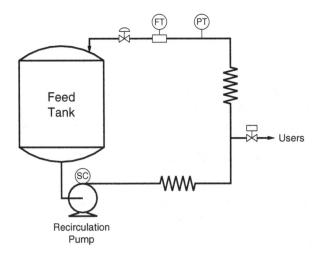

Figure 7.12 Simplified purified water supply process.

These points are for a pump speed of 3500 rpm. The affinity laws will be used to model the effect of pump speed N on head and flow:

$$\frac{N}{N_C} = \frac{F}{F_C} = \left(\frac{H}{H_c}\right)^2$$

where

N = pump speed (rpm)
F = flow at speed N (gpm)
H = head at speed N (ft)
N_C = pump speed for pump curve (rpm)
F_C = flow at speed N_C (gpm)
H_C = head at speed N_C (ft)

- The four exchanger stages on the discharge of the recirculation pump will be approximated by an orifice coefficient. At a flow of 280 gpm, the pressure drop is 5.0 psi. The orifice coefficient can be computed from these data.
- The users will be approximated by a solenoid valve. When closed, the users are taking no purified water (user demand is zero). When open, the users are taking 380 gpm of purified water.
- The steam-heated exchanger and the exchanger stage in the recirculation to the feed tank will also be approximated by an orifice coefficient. For a flow of 30 gpm, the pressure drop is 6.0 psi. The orifice coefficient will be computed from these data.
- The C_V of the control valve is 7.7, with the flow in gpm and the pressure drop in psi. The valve has equal-percentage characteristics with a proportionality constant of approximately 4.

For a specified user demand (either 0 or 380 gpm), the model must relate the controlled variables (recirculation pressure C_1 and recirculation flow C_2) and the manipulated variables (recirculation valve opening M_1 and recirculation pump speed M_2). If values are specified for any two of these, the remaining two can be computed from the model equations.

Consider how the process itself behaves. Suppose that one specifies the recirculation valve opening and the recirculation pump speed (the manipulated variables). Assuming that the user demand for purified water is constant, the process will seek an equilibrium state. The values of the recirculation pressure and the recirculation flow (the controlled variables) at this equilibrium state are the solution of the model equations. To emulate this behavior of the process, the model equations will be arranged so that values of the manipulated variables (recirculation valve opening and recirculation pump speed) are specified and corresponding values for the controlled variables (recirculation pressure and recirculation flow) are computed.

Occasionally, the model equations can be simplified sufficiently that an analytic solution can be formulated. However, this is not possible for the purified water supply process. The equations are nonlinear, which requires an iterative approach to obtain the solution. The following procedure can be used for the model:

- Assume a flow F_P through the pump.
- Compute the pump head H_P from pump flow F_P and pump speed M_2.
- Compute the pressure drop ΔP_1 through the four exchanger stages.
- Subtract the purified water flow F_U to the users from the pump flow F_P to obtain the recirculation flow F.
- Compute the pressure drop ΔP_2 through the steam-heated exchanger and the exchanger stage.
- Calculate $C_{V,M}$, the C_V for the control valve at its current opening, from the fully open C_V, the valve opening M_1, and the inherent characteristics of the valve.
- Compute the pressure drop ΔP_V across the control valve from $C_{V,M}$ and recirculation flow F.
- The purified water feed tank is at atmospheric pressure, so the recirculation pressure is ΔP_V.
- The pump head H_P less ΔP_1, ΔP_2, and ΔP_V should sum to zero. If the result is positive, the assumed value for F_P is too low; if the result is negative, the assumed value for F_P is too high.

The model can be implemented in a variety of ways, ranging from a spreadsheet model to a custom program written in Fortran or C++.

Normal Operating Conditions. For the purified water supply process, the normal operating conditions are a recirculation pressure of 60 psig and a recirculation flow of 30 gpm. The sensitivities will be calculated from two model solutions using finite differences. One of the model solutions will correspond to the normal operating conditions, and this solution will be referred to as the *base case*. The other model solution depends on which sensitivity is being calculated.

The equilibrium conditions for the base case depend on the consumption of purified water by users. The results are as follows:

Variable	Base Case	Base Case
Purified water flow to users (gpm)	0	380
Recirculation valve opening M_1 (%)	82.82	82.82
Recirculation pump speed M_2 (rpm)	2749.5	3315.2
Recirculation pressure C_1 (psig)	60.00	60.00
Recirculation flow C_2 (gpm)	30.00	30.00

The demand for purified water affects the pump speed but not the recirculation valve opening. In both cases, the recirculation flow is 30 gpm. For this flow, a valve opening of 82.82% gives a pressure drop of 60 psi across the valve, which is also the recirculation pressure.

The precision (digits after the decimal point, as in C++) of the values in the solutions above deserves attention:

- Such precision is not available from commercial control systems.
- The model is relatively crude, so why solve the equations so precisely?

This issue is examined at the end of the section.

Sensitivities K_{11} and K_{21}. Process sensitivity K_{11} is the sensitivity of the recirculation pressure C_1 to the recirculation valve opening M_1 at constant recirculation pump speed M_2. Using finite differences, the sensitivity will be computed from the model solution for the base case and a second model solution computed for the following values:

Recirculation valve opening M_1. The value for the valve opening must be different than for the base case. The valve opening will be increased by 5%.

Recirculation pump speed M_2. As the recirculation pump speed must be the same for the two solutions, the recirculation pump speed will be the same as for the base case.

From the base case and this solution (designated as case 1), the sensitivities K_{11} and K_{21} can be computed using finite differences to approximate the partial derivative. The following values are obtained from the solution of the model equations for case 1:

Variable	Base Case	Case 1	Base Case	Case 1
Water flow to users (gpm)	0	0	380	380
Valve opening M_1 (%)	82.82	87.82	82.82	87.82
Pump speed M_2 (rpm)	2749.5	2749.5	3315.2	3315.2
Recirculation pressure C_1 (psig)	60.00	57.27	60.00	56.82
Recirculation flow C_2 (gpm)	30.00	35.80	30.00	35.66
Sensitivity K_{11} (psig/%)	$\dfrac{57.27 - 60.00}{87.82 - 82.82} = -0.546$		$\dfrac{56.82 - 60.00}{87.82 - 82.82} = -0.636$	
Sensitivity K_{21} (gpm/%)	$\dfrac{35.80 - 30.00}{87.82 - 82.82} = 1.160$		$\dfrac{35.66 - 30.00}{87.82 - 82.82} = 1.132$	

The signs on the sensitivities are consistent with how the process should behave. Increasing the recirculation valve opening decreases the recirculation pressure and increases the recirculation flow. The values of these sensitivities are not affected significantly by the demand for purified water.

Sensitivities K_{12} and K_{22}. Process sensitivity K_{12} is the sensitivity of the recirculation pressure C_1 to the recirculation pump speed M_2 at constant recirculation valve opening M_1. Using finite differences, the sensitivity will be computed from the model solution for the base case and a second model solution computed for the following values:

Recirculation valve opening M_1. As the recirculation valve opening must be the same for the two solutions, the recirculation valve opening will be the same as for the base case.

Recirculation pump speed M_2. The value for the recirculation pump speed must be different than for the base case. The recirculation pump speed will be increased by 100 rpm.

From the base case and this solution (designated as case 2), the sensitivities K_{12} and K_{22} can be computed using finite differences. The following values are obtained from the solution of the model equations for case 2:

Variable	Base Case	Case 2	Base Case	Case 2
Water flow to users (gpm)	0	0	380	380
Valve opening M_1 (%)	82.82	82.82	82.82	82.82
Pump speed M_2 (rpm)	2749.5	2849.5	3315.2	3415.2
Recirculation pressure C_1 (psig)	60.00	64.44	60.00	64.95
Recirculation flow C_2 (gpm)	30.00	31.09	30.00	31.21
Sensitivity K_{12} (psig/rpm)	$\dfrac{64.44 - 60.00}{2849.5 - 2749.5} = 0.0444$		$\dfrac{64.95 - 60.00}{3415.2 - 3315.2} = 0.0495$	
Sensitivity K_{22} (gpm/rpm)	$\dfrac{31.09 - 30.00}{2849.5 - 2749.5} = 0.0109$		$\dfrac{31.21 - 30.00}{3415.2 - 3315.2} = 0.0121$	

Increasing the recirculation pump speed increases both the recirculation pressure and the recirculation flow, which is consistent with the signs for the sensitivities. The values of these sensitivities are not affected significantly by the demand for purified water either.

Sensitivity K'_{11}. Process sensitivity K'_{11} is the sensitivity of the recirculation pressure C_1 to the recirculation valve opening M_1 at constant recirculation flow C_2. This sensitivity will be computed from the values for the base case and the values from a solution computed from the following values:

> *Recirculation valve opening M_1.* The value for the recirculation valve opening must be different than the value for the base case. The recirculation valve opening will be increased by 5%.
>
> *Recirculation flow C_2.* The value at normal operating conditions is 30 gpm. As the recirculation flow must be the same for the two solutions, this value will be retained.

As formulated, the model computes the outputs (the controlled variables) from the inputs (the manipulated variables). Consequently, the solutions for cases 1 and 2 could easily be obtained. But for sensitivity K'_{11}, there are two options:

1. *Reformulate the model.* Separate formulations would be required for K'_{11}, K'_{12}, K'_{21}, and K'_{22}. This makes this option unrealistic.
2. *Iteratively solve the model as currently formulated.* To determine sensitivity K'_{11}, assume a value for the recirculation pump speed M_2, compute the value of the recirculation flow C_2, and then adjust M_2 until the desired value is computed for C_2. Despite requiring an iteration within an iteration, this approach will be used.

From the base case and this solution (designated as case 3), the sensitivity K'_{11} can be computed using finite differences. The following values are obtained from the solution of the model equations for case 3:

Variable	Base Case	Case 3	Base Case	Case 3
Water flow to users (gpm)	0	0	380	380
Valve opening M_1 (%)	82.82	87.82	82.82	87.82
Pump speed M_2 (rpm)	2749.5	2304.0	3315.2	2932.3
Recirculation pressure C_1 (psig)	60.00	40.22	60.00	40.22
Recirculation flow C_2 (gpm)	30.00	30.00	30.00	30.00
Sensitivity K'_{11} (psig/%)	$\dfrac{40.22 - 60.00}{87.82 - 82.82} = -3.956$		$\dfrac{40.22 - 60.00}{87.82 - 82.82} = -3.956$	

For no water flow to the users and a constant recirculation pump speed, increasing the recirculation valve opening from 82.82% to 87.82% increases the recirculation flow from 30 gpm to 35.80 gpm but reduces the recirculation pressure from 60 psig to 57.27 psig (solution for case 1). To reduce the recirculation flow back to 30 gpm, the recirculation pump speed must decrease from 2749.5 rpm to 2304.0 rpm. This decreases the recirculation pressure further, the equilibrium value being 40.22 psig.

The user demand has no effect on the value of sensitivity K'_{22}. If the recirculation flow is 30 GPM and the recirculation valve opening is 87.82%, the pressure drop across the recirculation valve is 40.22 psi and the recirculation pressure is 40.22 psig. None of these values are affected by the user demand for purified water.

The sensitivities are summarized as follows:

Water Flow to Users (gpm):	0	380
Sensitivity K_{11} (psig/%)	−0.546	−0.636
Sensitivity K'_{11} (psig/%)	−3.956	−3.956

The values for sensitivity K'_{11} are significantly different from the values for sensitivity K_{11}. As explained in the next section, such difference suggests significant interaction.

Sensitivity K'_{11} is the only sensitivity that can be computed from the solution for case 3. Additional solutions are required to compute values for sensitivities K'_{12}, K'_{21}, and K'_{22}. The calculations will not be repeated, but the values are as follows:

Water Flow to Users (gpm):	0	380
Sensitivity K'_{12} (psig/rpm)	0.0492	0.0559
Sensitivity K'_{21} (gpm/%)	1.328	1.328
Sensitivity K'_{22} (gpm/rpm)	0.0987	0.0950

None of these is significantly affected by the user demand for purified water.

Relationship of the Gains. With the model formulated to compute values for the controlled variables from specified values of the manipulated variables, the following observations apply to the sensitivities:

K_{11}, K_{12}, K_{21}, and K_{22}. These sensitivities can be computed from two model solutions plus a solution for the base case. If plant testing is applied, two plant tests are required. For an $n \times n$ multivariable process, $n + 1$ solutions or n tests are required.

K'_{11}, K'_{12}, K'_{21}, and K'_{22}. Four model solutions plus a solution for the base case are required to compute these sensitivities. If plant testing is applied, four plant tests are required. For an $n \times n$ multivariable process, $n^2 + 1$ solutions or n^2 tests are required.

The difficulty of obtaining values for the sensitivities K'_{11}, K'_{12}, K'_{21}, and K'_{22} provides an incentive to compute these sensitivities from the sensitivities K_{11}, K_{12}, K_{21}, and K_{22}. To do so, the starting point is the equations that correspond to the block diagram in Figure 7.9:

$$\Delta C_1 = K_{11} \, \Delta M_1 + K_{12} \, \Delta M_2$$

$$\Delta C_2 = K_{21} \, \Delta M_1 + K_{22} \, \Delta M_2$$

The sensitivity K'_{11} is the change in C_1 (that is, ΔC_1) divided by the change in M_1 (that is, ΔM_1) for a constant value of C_2 (that is, $\Delta C_2 = 0$). Substituting 0 for ΔC_2 in the equations above gives the following:

$$\Delta C_1 = K_{11} \, \Delta M_1 + K_{12} \, \Delta M_2$$

$$0 = K_{21} \, \Delta M_1 + K_{22} \, \Delta M_2$$

Eliminating ΔM_2 gives the following equation that relates ΔC_1 to ΔM_1:

$$\Delta C_1 = \frac{K_{11} K_{22} - K_{12} K_{21}}{K_{22}} \, \Delta M_1 = K_{11} \left(1 - \frac{K_{12} K_{21}}{K_{11} K_{22}} \right) \Delta M_1$$

The sensitivity K'_{11} is related to the sensitivities K_{11}, K_{12}, K_{21}, and K_{22} by the following equation:

$$K'_{11} = \left. \frac{\partial C_1}{\partial M_1} \right|_{C_2} \cong \left. \frac{\Delta C_1}{\Delta M_1} \right|_{C_2} = \frac{K_{11} K_{22} - K_{12} K_{21}}{K_{22}} = K_{11} \left(1 - \frac{K_{12} K_{21}}{K_{11} K_{22}} \right)$$

The degree of interaction depends on the difference between K_{11} and K'_{11}, which depends on the ratio $K_{12} K_{21} / K_{11} K_{22}$. This ratio could be used as a measure of the degree of interaction in a 2×2 process, but there is no counterpart for higher-order processes.

Matrix Equations. Alternatively, values for the sensitivities can be obtained by computing the inverse of the process gain matrix. As presented previously, the process is described by the following matrix equation:

$$\mathbf{c} = \mathbf{Km}$$

Alternatively, this can be expressed as follows:

$$\mathbf{K}^{-1} \mathbf{c} = \mathbf{m}$$

where $\mathbf{K}^{-1}$ is the inverse of the process gain matrix $\mathbf{K}$. For a 2×2 process, this is expressed by the following two equations:

$$(\mathbf{K}^{-1})_{11}\Delta C_1 + (\mathbf{K}^{-1})_{12}\Delta C_2 = \Delta M_1$$

$$(\mathbf{K}^{-1})_{21}\Delta C_1 + (\mathbf{K}^{-1})_{22}\Delta C_2 = \Delta M_2$$

where $(\mathbf{K}^{-1})_{ij}$ is the element on row i and column j of the inverse $\mathbf{K}^{-1}$ of the process gain matrix $\mathbf{K}$. In terms of sensitivities, these two equations would be expressed as follows:

$$\left.\frac{\partial M_1}{\partial C_1}\right|_{C_2}\Delta C_1 + \left.\frac{\partial M_1}{\partial C_2}\right|_{C_1}\Delta C_2 = \Delta M_1$$

$$\left.\frac{\partial M_2}{\partial C_1}\right|_{C_2}\Delta C_1 + \left.\frac{\partial M_2}{\partial C_2}\right|_{C_1}\Delta C_2 = \Delta M_2$$

The elements of $\mathbf{K}^{-1}$ are related to the sensitivities K'_{11}, K'_{12}, K'_{21}, and K'_{22} as follows:

$$(\mathbf{K}^{-1})_{11} = \left.\frac{\partial M_1}{\partial C_1}\right|_{C_2} = \frac{1}{K'_{11}}$$

$$(\mathbf{K}^{-1})_{12} = \left.\frac{\partial M_1}{\partial C_2}\right|_{C_1} = \frac{1}{K'_{21}}$$

$$(\mathbf{K}^{-1})_{21} = \left.\frac{\partial M_2}{\partial C_1}\right|_{C_2} = \frac{1}{K'_{12}}$$

$$(\mathbf{K}^{-1})_{22} = \left.\frac{\partial M_2}{\partial C_2}\right|_{C_1} = \frac{1}{K'_{22}}$$

The relationship can be summarized as follows:

$$K'_{ij} = \frac{1}{(\mathbf{K}^{-1})_{ji}}$$

In addition to the reciprocal, the subscripts are reversed. In matrix parlance, reversing the subscripts of the elements of a matrix is known as the *transpose* of a matrix. The notation is as follows:

$(\mathbf{K}^{-1})^{\mathbf{T}}$ is the transpose of the inverse of the process gain matrix $\mathbf{K}$.

$(\mathbf{K}^{-1})^{\mathbf{T}}_{ij}$ is the element on row i and column j of the transpose of the inverse of the process gain matrix $\mathbf{K}$.

Using this notation, the sensitivities K'_{11}, K'_{12}, K'_{21}, and K'_{22} can be computed as follows:

$$K'_{ij} = \frac{1}{(\mathbf{K}^{-1})^{\mathrm{T}}_{ji}}$$

Numerical Example. The sensitivities K_{11}, K_{12}, K_{21}, and K_{22} were previously computed for the purified water supply process. Using the values for a purified water demand of 380 gpm, the process gain matrix $\mathbf{K}$ is as follows:

$$\mathbf{K} = \begin{bmatrix} K_{11} & K_{12} \\ K_{21} & K_{22} \end{bmatrix} = \begin{bmatrix} -0.636 & 0.0495 \\ 1.132 & 0.0121 \end{bmatrix}$$

The matrix inverse $\mathbf{K}^{-1}$ of the process gain matrix $\mathbf{K}$ is

$$\mathbf{K}^{-1} = \begin{bmatrix} -0.1899 & 0.7767 \\ 17.763 & 9.980 \end{bmatrix}$$

The transpose of this matrix is

$$(\mathbf{K}^{-1})^{\mathrm{T}} = \begin{bmatrix} -0.1899 & 17.763 \\ 0.7767 & 9.980 \end{bmatrix}$$

Sensitivity K'_{ij} is the reciprocal of $(\mathbf{K}^{-1})^{\mathrm{T}}_{ij}$:

Sensitivity	Calculated from Process Gain Matrix	From Model Solution
K'_{11}	$(-0.1899)^{-1} = -5.27$	-3.956
K'_{12}	$(17.763)^{-1} = 0.0563$	0.0559
K'_{21}	$(0.7767)^{-1} = 1.29$	1.328
K'_{22}	$(9.980)^{-1} = 0.100$	0.0950

Except for K'_{11}, the values computed from the process gain matrix compare favorably to the values computed from the model. Even the values for K'_{11} differ only by about 25%. The relationships for computing the sensitivities K'_{11}, K'_{12}, K'_{21}, and K'_{22} from the sensitivities K_{11}, K_{12}, K_{21}, and K_{22} assume that the process is linear, which it certainly is not. This explains the differences between the values for K'_{11}, K'_{12}, K'_{21}, and K'_{22} obtained by the two approaches.

Linear Approximations to Nonlinear Processes. In obtaining the solutions for cases 1 through 4 presented previously, the manipulated variables were changed by an arbitrary amount, specifically 5% for the recirculation valve opening and 100 rpm for the recirculation pump speed. In selecting these values, there are two issues:

1. The larger the change in the manipulated variable, the larger the change in the controlled variable(s).
2. The departure from linearity increases with the magnitude of the change in the manipulated variable.

Discourses on numerical methods always advise against subtracting two large numbers to obtain a small number. Any error in either of the large numbers is magnified in the small number. Unfortunately, a finite-difference calculation does exactly this. The change in the manipulated variable leads to a change in the controlled variable. This change is computed as the difference in the controlled variable before and after the change. Any errors in the controlled variable are magnified in the difference. This consideration is especially important for data from plant testing, where the data values are always in error by some amount.

In model solutions, the values for the manipulated and controlled variables can be computed to a higher precision (again as digits after the decimal point, as in C++). This was the case for all solutions presented previously in this section. This permits smaller changes in the manipulated variable to be made, which will yield smaller differences. However, another issue arises with regard to any iterative solution. These solutions are never exact, the errors being generally referred to as *convergence errors*. When calculating values to higher and higher precisions, ever smaller convergence errors are required for accurate differences to be computed. The convergence of multicomponent distillation calculations is a good example of where these issues arise.

7.4. QUANTITATIVE MEASURES OF INTERACTION

Quantitative measures of interaction must take the following two aspects into consideration:

Steady-state. The simplest measure to understand and apply is the relative gain [2].

Dynamic. No effective measure of the dynamic aspects of interaction is available. It is possible to formulate a dynamic relative gain, but so far this has attracted only academic interest.

Basically, if one loop is faster than another loop, the degree of steady-state interaction between the two loops is irrelevant. The issues are as follows:

- One must first tune the fast loop and then tune the slow loop.
- If for any reason the fast loop is switched to manual, it is possible that the slow loop will not function properly, and will possibly be unstable.
- How much dynamic separation is required for this to be successful? A factor of 5 is usually adequate, but if the degree of steady-state interaction is moderate, a smaller separation will be sufficient.

The preferable situation is that the process provides the dynamic separation. However, the dynamic separation can be achieved by tuning one controller to respond as rapidly as the process allows, and then slowing down the response of the other controller until the dynamic separation is adequate. But this comes at a price: namely, degraded performance in the slow loop.

Sensitivity K_{11}. In a 2×2 multivariable process, how does one go about tuning the loops? The answer is: one at a time. Choosing which loop to tune first involves the following issues:

Loop response speed. If one loop is faster than the other loop, start by tuning the fast loop.

Loop importance. If both loops respond in approximately the same time frame, start by tuning the loop whose control variable has the most impact on process operations.

As the choice of the subscripts for the manipulated and controlled variables is arbitrary, the loop whose controlled variable is C_1 and manipulated variable is M_1 will be tuned first. While tuning this loop, the control configuration is as illustrated in Figure 7.7(a). The controller changes the manipulated variable M_1; the process responds with changes in C_1 and C_2. The loop being tuned responds to the changes in C_1. But with the second controller on manual, there is no response to the changes in C_2. While tuning the first loop, the values for controlled variable C_2 for the other loop must remain within the range deemed to be acceptable, but otherwise these changes are ignored.

What is the process sensitivity to which the first controller is being tuned? Changes are made in M_1, but M_2 is fixed. This sensitivity is K_{11}:

$$K_{11} = \left. \frac{\partial C_1}{\partial M_1} \right|_{M_2}$$

Can the first controller be tuned successfully? Problems can potentially arise. However, interaction with the second loop is not one of these problems. As long as the second loop remains on manual, interaction with that loop cannot occur.

Sensitivity K'_{11}. Once the first loop is tuned successfully, the next step is to switch the second loop to automatic and attempt to tune that loop. While tuning this loop, the control configuration is as illustrated in Figure 7.7(c). The controller changes manipulated variable M_2; the process responds with changes in C_1 and C_2. The loop being tuned responds to the changes in C_2 by making changes in the manipulated variable M_2. The objective of the controller being tuned is to maintain C_2 at its set point.

What is the process sensitivity that the first controller is now experiencing? Changes are made in M_1, but C_2 is fixed. This sensitivity is K'_{11}:

$$K'_{11} = \left. \frac{\partial C_1}{M_1} \right|_{C_2}$$

With the second controller in manual, the first controller is tuned to the process sensitivity K_{11}. But the instant the second controller is switched to automatic, the process sensitivity abruptly changes to K'_{11}.

The controller was initially tuned to a process sensitivity of K_{11}. Will this tuning be satisfactory when the process sensitivity is K'_{11}? Provided that the difference in the sensitivities is modest, the initial tuning will be satisfactory. But if these sensitivities differ significantly, the initial tuning will not be satisfactory.

What is meant by "modest" and "significant"? These are difficult to quantify, mainly because these sensitivities only quantify the steady-state aspect of interaction. A difference of less than 20% in the sensitivities rarely causes problems. In the field, controllers are at best tuned to within 20% of the most appropriate values for the tuning coefficients. Even differences in the sensitivities of up to 30% can usually be resolved with no more than minor adjustments in the tuning of the controller that was initially tuned. With some separation in the dynamics, a difference of 50% in the sensitivities can be tolerated, usually with a nominal relaxation of the tuning of the slower loop. The greater the difference in the sensitivities, the greater the separation in the dynamics required for acceptable performance. If this difference is attained by relaxing the tuning in the slower loop, at some point the performance degrades to the point that it is unacceptable.

Relative Gain. The relative gain is simply the ratio of the two sensitivities above. The relative gain is normally designated by λ with subscripts designating the manipulated and controlled variable in the same manner as for the gain or sensitivity. That is, λ_{11} is the relative gain for controlling C_1 using manipulated variable M_1. The relative gain λ_{11} is the ratio of sensitivity K_{11} to K'_{11}:

$$\lambda_{11} = \frac{K_{11}}{K'_{11}}$$

A value of 1.0 for the relative gain means no interaction. As the value departs from 1.0 in either direction (increases or decreases), the degree of interaction increases. That is, a relative gain of 0.5 denotes as much interaction as a relative gain of 2.0. It is even possible for the relative gain to be negative.

The relative gain can be extended to multivariable processes beyond 2×2. For controlling C_1 using manipulated variable M_1, the relative gain λ_{11} is defined as follows:

$$\lambda_{11} = \frac{\text{sensitivity of } C_1 \text{ to } M_1 \text{ with all other loops on manual}}{\text{sensitivity of } C_1 \text{ to } M_1 \text{ with all other loops on automatic}}$$

The relative gain is the most commonly applied measure of interaction. Its major limitation is that it only assesses the steady-state degree of interaction.

Because the relative gain only assesses the steady-state aspects of interaction, it is difficult to answer the following question: How close must the relative gain be to 1.0 for single-loop controllers to perform satisfactorily? A previous discussion considered what difference between sensitivities K_{11} and K'_{11} could be tolerated. The values for the relative gain are analogous. For a 2×2 process, rarely do tuning difficulties arise for relative gains between 0.8 and 1.2. In most cases, relative gains as low as 0.7 and as high as 1.4 prove to be acceptable, although some tuning adjustments may be required in the controller that was initially tuned. For relative gains down to 0.5 and up to 2.0 to be acceptable, some separation in dynamics between the two loops is required. If this is achieved with some relaxation in the controller tuning, the degradation in loop performance can become an issue.

With significant separation in the dynamics between the two loops, the degree of steady-state interaction is irrelevant. The loops can be tuned even if the relative gain is very large or even negative. One has to tune the fast loop first, and then tune the slow loop. But if the fast loop must be switched to manual for any reason, the slow loop is unlikely to perform satisfactorily and could even be unstable.

A negative value for the relative gain means that the sensitivities K_{11} and K'_{11} have opposite signs. The sign of the sensitivity reflects the directionality of the process. To have a negative value for the relative gain, the directionality of the process must reverse when the other loop is switched between manual and automatic.

Relative Gain Array. For a multivariable process, a relative gain can be defined for each controlled variable and each manipulated variable:

$$\lambda_{ij} = \frac{K_{ij}}{K'_{ij}} = \text{relative gain for controlled variable } i \text{ and manipulated variable } j$$

Furthermore, these relative gains can be arranged in an array that is known as the relative gain array Λ. This array will be defined as follows:

$$\Lambda = \begin{bmatrix} \lambda_{11} & \lambda_{12} \\ \lambda_{21} & \lambda_{22} \end{bmatrix} = \text{relative gain array}$$

Often, the relative gain array is presented as follows (controlled variables as rows and manipulated variables as columns):

	M_1	M_2
C_1	λ_{11}	λ_{12}
C_2	λ_{21}	λ_{22}

Occasionally, the array is presented in a transposed fashion, with the controlled variables as columns and the manipulated variables as rows. However, this will not be done in this book.

Two very useful properties of this array are the following:

- Each row must sum to unity:

$$\lambda_{11} + \lambda_{12} = 1$$

$$\lambda_{21} + \lambda_{22} = 1$$

- Each column must sum to unity:

$$\lambda_{11} + \lambda_{21} = 1$$

$$\lambda_{12} + \lambda_{22} = 1$$

For a 2×2 multivariable process, a numerical value must be obtained for only one of the relative gains. Values of the three remaining relative gains can be computed using the fact that each row and each column must sum to unity. For a 2×2 multivariable process, the relative gain array will be symmetrical. However, this does not extend to higher-order processes.

Evaluating the Relative Gains. There are three possibilities for obtaining values for the relative gains:

Analytical. The advantage of the analytical approach is that an expression is obtained for the relative gain. Often, this expression provides considerable insight into the behavior of the process. Unfortunately, this approach is restricted to simple processes and often requires assumptions that may be questionable.

Process models. The fact that the relative gain is based on steady-state relationships permits the relative gains to be computed from a steady-state model of the process, such as the stage-by-stage simulation model for a distillation column.

Process tests. Relative gains can be evaluated from process test data. However, this is only practical for processes that respond quickly. Even so, the potential disruption to the process means that such testing must be done only at carefully selected times and must be completed as quickly as possible.

For most applications, a steady-state model is the most practical approach for obtaining values for the relative gains.

Calculation of Relative Gains: Approach 1. The two approaches differ in the values that must be obtained either from tests conducted on the process or from solutions of a process model. The starting point is either with the process lined-out at its normal operating conditions or with a model solution corresponding to the normal operating conditions.

For a 2×2 process, one approach is as follows:

- Determine sensitivity K_{11} (other options are K_{12}, K_{21}, and K_{22}). This will require one process test or one model solution (such as case 1 presented earlier).
- Determine sensitivity K'_{11}. This will require one process test or one model solution (such as case 3 presented earlier).
- Compute λ_{11} as the ratio of these two sensitivities.
- Compute the remaining relative gains from the fact that each column and each row of the relative gain array must sum to unity.

In the preceding section, the following values were obtained from the solutions (cases 1 and 3) of the steady-state process model for the purified water supply process:

Water Flow to Users (gpm)	0	380
Sensitivity K_{11} (psig/%)	−0.546	−0.636
Sensitivity K'_{11} (psig/%)	−3.956	−3.956
Relative gain λ_{11}	0.14	0.16
Relative gain array Λ	$\begin{bmatrix} 0.14 & 0.86 \\ 0.86 & 0.14 \end{bmatrix}$	$\begin{bmatrix} 0.16 & 0.84 \\ 0.84 & 0.16 \end{bmatrix}$

As noted previously, the demand for purified water has little effect on the sensitivities, and consequently, little effect on the relative gains. Repeating the calculations using K_{12} and K'_{12} gives the following results:

Water Flow to Users (gpm):	0	380
Sensitivity K_{12} (psig/%)	0.0444	0.0495
Sensitivity K'_{12} (psig/%)	0.0492	0.0559
Relative gain λ_{12}	0.90	0.89
Relative gain array Λ	$\begin{bmatrix} 0.10 & 0.90 \\ 0.90 & 0.10 \end{bmatrix}$	$\begin{bmatrix} 0.11 & 0.89 \\ 0.89 & 0.11 \end{bmatrix}$

The results are not identical, but the difference is only nominal and is due to the nonlinear nature of the process.

Process Test. For a 2×2 process that responds quickly, process testing is a potentially viable approach to obtaining values for the relative gains. Such a process test will be illustrated for the recirculation pressure loop in the control configuration in Figure 7.6, for which the controlled and manipulated variables are paired as follows:

- Recirculation pressure C_1 is controlled by manipulating the recirculation valve opening M_1.
- Recirculation flow C_2 is controlled by manipulating the recirculation pump speed M_2.

While maintaining a constant user demand for purified water, the process test proceeds as follows:

- Line the process out at the desired operating conditions. For the test, the desired conditions are a recirculation pressure C_1 of 60 psig and a recirculation flow C_2 of 30 gpm.
- Place both controllers on manual.
- Change the recirculation valve opening M_1, which is the output of the recirculation pressure controller. For this example, the valve opening will be increased by 5%.
- Wait until the process attains equilibrium.
- Note the values of the recirculation pressure C_1 and, if desired, the recirculation flow C_2. (The value of C_2 is not used in computing the relative gain.)
- Adjust the recirculation pump speed M_2 until the value of the recirculation flow C_2 is the same as for the normal operating conditions. If the recirculation flow controller functions when used alone, the easiest way to accomplish this is to switch the controller to automatic, specify the desired set point for the recirculation flow C_2, and wait for the process to line-out.
- Note the values of the recirculation pressure C_1 and, if desired, the recirculation pump speed M_2 (The value of M_2 is not used in computing the relative gain.)

The trend in Figure 7.13 presents the results of this test. Usually, one gets a sense from the responses in Figure 7.13 as to the severity of the interaction. When the recirculation flow controller is switched to automatic, the pump speed is reduced so as to restore the recirculation flow to 30 gpm. But to do so, the pump speed must be reduced from 2739 rpm to 2294 rpm, which reduces the recirculation pressure from 57.2 psig to 40.1 psig. Both changes are very significant, which suggest that the flow controller has a significant effect on the recirculation pressure loop. This implies significant interaction.

To quantify the degree of interaction, the following values from the trends in Figure 7.13 are of interest:

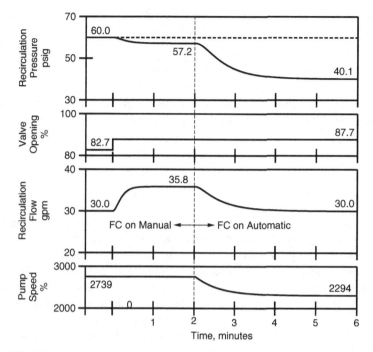

Figure 7.13 Process test to determine a value for relative gain λ_{11} for the purified water supply process.

Variable	Base Case	FC on Manual (Case 1)	FC on Auto (Case 3)
Water flow to users (gpm)	0	0	0
Recirculation valve opening M_1 (%)	82.7	87.7	87.7
Recirculation pump speed M_2 (rpm)	2739	2739	2294
Recirculation pressure C_1 (psig)	60.0	57.2	40.1
Recirculation flow C_2 (gpm)	30.0	35.8	30.0

The numbers assigned to the cases correspond to those computed in the preceding section from the steady-state model. The trends in Figure 7.13 were obtained using a dynamic simulation of the purified water supply process. The values are slightly different because the dynamic simulation is more detailed than the steady-state model based on Figure 7.12.

The sensitivity K_{11} is computed from case 1 and the base case as follows:

$$K_{11} = \left.\frac{\partial C_1}{\partial M_1}\right|_{M_2} \cong \left.\frac{\Delta C_1}{\Delta M_1}\right|_{M_2} = \frac{57.2 \text{ psig} - 60.0 \text{ psig}}{87.7\% - 82.7\%} = -0.56 \text{ psig}/\%$$

The sensitivity K'_{11} is computed from case 3 and the base case as follows:

$$K'_{11} = \left.\frac{\partial C_1}{\partial M_1}\right|_{C_2} \cong \left.\frac{\Delta C_1}{\Delta M_1}\right|_{C_2} = \frac{40.1 \text{ psig} - 60.0 \text{ psig}}{87.7\% - 82.7\%} = -3.98 \text{ psig}/\%$$

The value of the relative gain λ_{11} can be computed from the sensitivities K_{11} and K'_{11}:

$$\lambda_{11} = \frac{K'_{11}}{K_{11}} = \frac{-0.56 \text{ psig}/\%}{-3.98 \text{ psig}/\%} = 0.14$$

The remaining elements of the relative gain array can be computed from the fact that each row and each column must sum to unity:

	M_1	M_2
C_1	0.14	0.86
C_2	0.86	0.14

Process testing must never be taken lightly. The recirculation pressure and recirculation flow respond very rapidly, but even so the test requires over 5 minutes. Throughout the test period, the purified water flow to the users must be constant (any changes affect both the recirculation pressure and recirculation flow). Except during plant shutdowns, achieving this will entail some disruption to production operations. Perhaps this would be tolerable for 5 minutes or so, but should the test period be 5 hours, the disruption would be unacceptable.

For a 2×2 process, only one test of the type illustrated in Figure 7.13 is required. For a 3×3 process, four such tests are required. An alternative approach presented next requires only three tests. But even for fast processes, this makes the plant testing approach difficult to justify. As the dimensionality increases, the effort becomes even more daunting.

Calculation of Relative Gains: Approach 2. As in approach 1, the starting point is either with the process lined out at its normal operating conditions or with a model solution corresponding to the normal operating conditions. For a 2×2 process, the approach is as follows:

- Determine the sensitivities K_{11} and K_{21}. This will require one process test or one model solution (such as case 1 presented in the preceding section).
- Determine the sensitivities K_{12} and K_{22}. This will require one process test or one model solution (such as case 2 presented in the preceding section).
- Construct the process gain matrix $\mathbf{K}$.
- Compute $\mathbf{K}^{-1}$, the inverse of the process gain matrix $\mathbf{K}$.
- Compute $(\mathbf{K}^{-1})^{\mathbf{T}}$, the transpose of the inverse of the process gain matrix $\mathbf{K}$.
- In the preceding section, the following equation was derived for computing the sensitivities K'_{ij}:

$$K'_{ij} = \frac{1}{(\mathbf{K}^{-1})^{\mathbf{T}}_{ij}}$$

This permits the relative gains to be computed as follows:

$$\lambda_{ij} = \frac{K_{ij}}{K_{ij}'} = K_{ij} \times (\mathbf{K}^{-1})_{ij}^{\mathrm{T}}$$

To obtain values for the relative gains, each element of the process gain matrix $\mathbf{K}$ is multiplied by the corresponding element of $(\mathbf{K}^{-1})^{\mathrm{T}}$, the transpose of the inverse of the process gain matrix. Note that this is not matrix multiplication.

In the preceding section, values for the sensitivities K_{11}, K_{12}, K_{21}, and K_{22} were computed from the steady-state process model. The relative gain array is computed from these sensitivities as follows:

Water Flow to Users (gpm)	0	380
Sensitivity K_{11} (psig/%)	−0.546	−0.636
Sensitivity K_{12} (psig/rpm)	0.0444	0.0495
Sensitivity K_{21} (gpm/%)	1.160	1.132
Sensitivity K_{22} (gpm/rpm)	0.0109	0.0121
Process gain matrix $\mathbf{K}$	$\begin{bmatrix} -0.546 & 0.0444 \\ 1.160 & 0.0109 \end{bmatrix}$	$\begin{bmatrix} -0.636 & 0.0495 \\ 1.132 & 0.0121 \end{bmatrix}$
Inverse $\mathbf{K}^{-1}$	$\begin{bmatrix} -0.190 & 0.773 \\ 20.2 & 9.50 \end{bmatrix}$	$\begin{bmatrix} -0.190 & 0.777 \\ 17.8 & 9.98 \end{bmatrix}$
Inverse transposed $(\mathbf{K}^{-1})^{\mathrm{T}}$	$\begin{bmatrix} -0.190 & 20.2 \\ 0.773 & 9.50 \end{bmatrix}$	$\begin{bmatrix} -0.190 & 17.8 \\ 0.777 & 9.98 \end{bmatrix}$
Relative gain array $\mathbf{\Lambda}$	$\begin{bmatrix} 0.104 & 0.896 \\ 0.896 & 0.104 \end{bmatrix}$	$\begin{bmatrix} 0.121 & 0.879 \\ 0.879 & 0.121 \end{bmatrix}$

For 3×3 and higher-dimensional processes, this is usually the most appropriate approach.

7.5. LOOP PAIRING

Often, the justification for obtaining numerical values for the relative gains is to assess the loop pairing, that is, the choice of manipulated variable for each controlled variable. Ideally, this should be investigated early in the design cycle for the process. However, the more common situation is that tuning problems are being experienced with one or more loops in the current P&I diagram. A possible cause is that the loop pairing from the P&I diagram is incorrect.

Based on the relative gains, the statement for loop pairing is simple:

The manipulated variable chosen for each controlled variable must correspond to the largest value of the relative gain on the respective row (for the controlled variable) and column (for the manipulated variable).

There is one disclaimer regarding this statement. The relative gain assesses only the steady-state aspect of interaction. If there is significant dynamic separation (a factor of 5 or more) between two loops, they will function properly even in the face of very adverse steady-state interaction. Tuning must begin with the fast loop, and the fast loop must remain on automatic when the slow loop is in automatic.

The desire is for the largest relative gain to be in the vicinity of 1. For a 2×2 process, the possibilities are:

Relative Gain Array			Explanation
	M_1	M_2	No interaction; control C_1 with M_1 and C_2 with M_2.
C_1	1	0	
C_2	0	1	
	M_1	M_2	No interaction; control C_1 with M_2 and C_2 with M_1.
C_1	0	1	
C_2	1	0	
	M_1	M_2	Little interaction; control C_1 with M_1 and C_2 with M_2.
C_1	0.8	0.2	
C_2	0.2	0.8	
	M_1	M_2	Little interaction; control C_1 with M_2 and C_2 with M_1.
C_1	0.2	0.8	
C_2	0.8	0.2	
	M_1	M_2	Serious degree of interaction; neither pairing will perform properly unless significant dynamic separation is available.
C_1	0.5	0.5	
C_2	0.5	0.5	
	M_1	M_2	Serious degree of interaction. Suggested pairing is to control C_1 with M_1; however, this loop will function only if significant dynamic separation is available. Even then, switching the fast loop from automatic to manual increases the process sensitivity by a factor of 17 in the slow loop, which will probably cause that loop to become unstable.
C_1	17	-16	
C_2	-16	17	

For a 2×2 process, a relative gain of 0.5 means serious interaction. For higher-order processes, a relative gain of 0.5 suggests significant interaction, but the loop paring may function properly, especially if some dynamic separation is available. For example, suppose that a row in a 4×4 process is as follows:

	M_1	M_2	M_3	M_4
C_i	0.2	0.5	0.1	0.2

Controlling C_i with M_2 may prove to be satisfactory. It is not just the value of the relative gain; it is its value relative to other relative gains on each row and column. For example, suppose that a row in a 3×3 process is as follows:

	M_1	M_2	M_3
C_i	−1.9	1.0	0.9

The relative gain for controlling C_i with M_2 is exactly 1.0. However, the relative gain for controlling C_i with M_3 is 0.9, which is close. Without a strong preference for M_2 over M_3, performance problems are likely.

For the purpose of determining the preferable loop pairing, there is no need to compute the relative gains to a high precision. Two significant figures are adequate. That is, if the largest relative gains on a row or column are 1.1 and 1.2, performance problems are likely with the respective loops. Adding more precision to the relative gains does not provide any benefit.

In the relative gain arrays computed for the purified water supply process, the largest relative gain ranged from a low of 0.86 to a high of 0.90. But such variability has no impact on the conclusion: Control the recirculation pressure C_1 with the recirculation pump speed M_2 and the recirculation flow C_2 with the recirculation valve opening M_1. All relative gain arrays are in agreement: Use the control configuration in Figure 7.8.

This discussion has also assumed that all possible loop pairings can be considered. For the purified water supply process, there are two possible pairings (Figures 7.6 and 7.8). Provided that the two loops perform satisfactorily, either would be acceptable. But in some applications, other considerations make one of the pairings unacceptable. In a 2×2 process, there are two possible pairings, one of which may not be acceptable for other reasons. In a 3×3 process, there are six possible pairings, one or more of which may not be acceptable for other reasons.

7.6. STARCH PUMPING SYSTEM

The purpose of this example is to illustrate the following:

- That one of the possible loop pairings cannot be considered because of process issues.

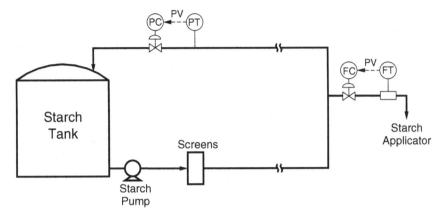

Figure 7.14 Starch pumping system.

- That the degree of interaction can change significantly with process operating conditions.
- That with adequate dynamic separation, two loops can be tuned even in the face of adverse steady-state interaction.

This example has one aspect in common with the purified water supply process—the objective is to control a flow and a pressure.

Process Description. Starch slurries have a property that must be addressed in any starch pumping system. As long as the starch slurry continues to flow, it can be pumped much as any fluid. But should the flow stop, the starch "sets up." When this occurs, the piping must be dismantled and cleaned before flow can be restored. To avoid plugging the piping, flow must continue until the lines are flushed with water.

Figure 7.14 presents the flowsheet for a starch pumping system similar to those encountered in the paper industry. In manufacturing fine papers (such as writing papers), starch is applied to the surface of the sheet to provide a smooth surface. The starch slurry is prepared in a feed tank, from which it is pumped to the starch applicator on the machine.

All paper machines occasionally experience paper breaks. When a paper break occurs, the starch applicator is shut down and the flow of starch to the applicator must be stopped. However, the starch flow through the piping between the starch tank and the starch applicator must not be allowed to stop. Since a paper break can occur with little or no warning, a recirculation loop must be provided so that the starch can flow back to the starch tank. However, the flow in this piping cannot be stopped either, so there must be some recirculation flow at all times. A typical design is as follows:

Starch Applicator:	Running	Stopped
Flow to applicator	80% of pump flow	None
Flow back to starch tank	20% of pump flow	100% of pump flow

The operation of paper machines is such that the starch flow basically switches between these two. On a paper break, the flow to the applicator stops abruptly. When the paper machine is brought online again, the starch flow to the applicator usually goes from 0% to 80% very rapidly. But for this example, the behavior at intermediate flows to the applicator will be examined. To simplify the notation, hereafter the pump flow is assumed to be 100 gpm.

Control Configuration. For the starch pumping system in Figure 7.14, the controlled and manipulated variables are as follows:

Manipulated Variable	Controlled Variable
M_1: recirculation valve opening	C_1: recirculation pressure
M_2: applicator valve opening	C_2: applicator starch flow

The starch pumping system clearly exhibits interaction. With no controls in operation, the behavior is as follows:

Manipulated Variable	Controlled Variable
Increase recirculation valve opening M_1	Decrease recirculation pressure C_1
	Decrease applicator starch flow C_2
Increase applicator valve opening M_2	Decrease recirculation pressure C_1
	Increase applicator starch flow C_2

Both manipulated variables affect both controlled variables.

The control configuration in Figure 7.14 suggests the following pairing of the controlled and manipulated variables:

- Control the recirculation pressure C_1 using the recirculation valve opening M_1.
- Control the applicator starch flow C_2 using the applicator valve opening M_2.

This configuration is consistent with what one typically sees in P&I diagrams: Control each variable with the nearest control valve.

Reversing the loop pairing gives the control configuration in Figure 7.15:

- Control the recirculation pressure C_1 using the applicator valve opening M_2.

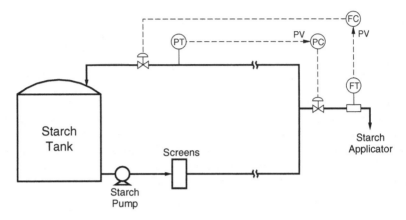

Figure 7.15 Alternative control configuration for starch pumping system.

- Control the applicator starch flow C_2 using the recirculation valve opening M_1.

This configuration may initially appear strange, but it could be considered provided that the flow to the applicator is not zero.

The problem with the configuration in Figure 7.15 arises when the starch applicator is stopped. The applicator valve is fully closed, so no control is being provided for the recirculation pressure. The configuration in Figure 7.15 is not acceptable, making the configuration in Figure 7.14 the only viable option. Reversing the pairing is not an option for the starch pumping system.

Relative Gain. For the starch pumping system, is the relative gain affected significantly by the flow of starch to the applicator? To answer this question, the sensitivities and the relative gain will be evaluated for two cases:

- An applicator flow of 80 gpm (most of the pump flow goes to the applicator).
- An applicator flow of 20 gpm (most of the pump flow is returned to the starch tank).

Using a dynamic simulation of the starch pumping system, each relative gain will be determined as follows:

- Line the process out at the desired operating conditions, that is, at the desired recirculation pressure (80 psig) and the desired flow to the applicator (20 gpm or 80 gpm).
- Place both controllers on manual.
- Change the recirculation valve opening M_1, which is the output of the recirculation pressure controller. For the test, the valve opening will be increased by 5%.

- Wait until the process attains equilibrium.
- Note the values of the recirculation pressure C_1 and, if desired, the applicator flow C_2 (the value of C_2 is not be used in computing the relative gain).
- Adjust the applicator valve opening M_2 until the value of the applicator flow C_2 is the same as for the initial operating conditions. This will be accomplished by switching the flow controller to automatic, specifying the desired set point for the applicator flow, and waiting for the process to line-out.
- Note the values of the recirculation pressure C_1 and, if desired, the applicator valve opening M_2 (the value of M_2 is not used in computing the relative gain).

Figure 7.16 presents the results of the test when the flow to the applicator is 80 gpm. The values of interest are as follows:

Variable	Base Case	FC on Manual	FC on Auto
Recirculation valve opening M_1 (%)	46.1	51.1	51.1
Applicator valve opening M_2 (%)	84.3	84.3	89.3
Recirculation pressure C_1 (psig)	80.0	73.4	53.7
Applicator flow C_2 (gpm)	80.0	76.6	80.0

Sensitivity K_{11} (psig/%)

$$\frac{73.4 \text{ psig} - 80.0 \text{ psig}}{51.1\% - 46.1\%} = -1.32 \text{ psig/\%}$$

Sensitivity K'_{11} (psig/%)

$$\frac{53.7 \text{ psig} - 80.0 \text{ psig}}{51.1\% - 46.1\%} = -5.26 \text{ psig/\%}$$

Relative gain λ_{11}

$$\frac{-1.32 \text{ psig/\%}}{-5.24 \text{ psig/\%}} = 0.25$$

Relative gain array Λ

	M_1	M_2
C_1	0.25	0.75
C_2	0.75	0.75

What does this say about the control configuration in Figure 7.14? The answer: that the interaction between the pressure and flow loops is substantial. If it were possible, the loop pairing should be reversed. With 80% of the pump flow going to the applicator, the recirculation valve has a larger influence on the flow to the applicator than that of the applicator valve. In turn, the applicator valve has a larger influence on the recirculation pressure than on the recirculation valve.

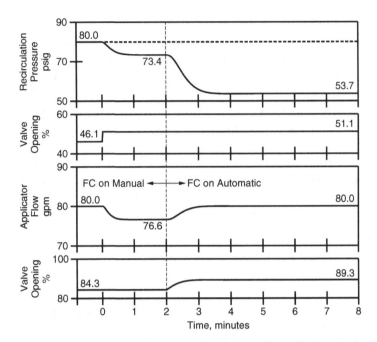

Figure 7.16 Process test to determine a value for relative gain λ_{11} for the starch pumping system with 80 gpm to starch applicator.

Figure 7.17 presents the results of the test when the flow to the applicator is 20 gpm. The values of interest are as follows:

Variable	Base Case	FC on Manual	FC on Auto
Recirculation valve opening M_1 (%)	81.6	86.6	86.6
Applicator valve opening M_2 (%)	49.6	49.6	54.2
Recirculation pressure C_1 (psig)	80.0	59.4	55.5
Applicator flow C_2 (gpm)	20.0	17.2	20.0

Sensitivity K_{11} (psig/%)
$$\frac{59.4 \text{ psig} - 80.0 \text{ psig}}{86.6\% - 81.6\%} = -4.12 \text{ psig/\%}$$

Sensitivity K'_{11} (psig/%)
$$\frac{55.5 \text{ psig} - 80.0 \text{ psig}}{86.6\% - 81.6\%} = -4.90 \text{ psig/\%}$$

Relative gain λ_{11}
$$\frac{-4.12 \text{ psig/\%}}{-4.90 \text{ psig/\%}} = 0.84$$

Relative gain array

	M_1	M_2
C_1	0.84	0.16
C_2	0.16	0.84

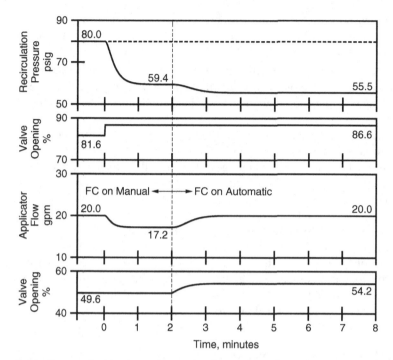

Figure 7.17   Process test to determine a value for relative gain λ_{11} for the starch pumping system with 20 gpm to starch applicator.

What does this say about the control configuration in Figure 7.14? The answer: that the interaction between the pressure and flow loops is nominal. With 20% of the pump flow going to the applicator, the recirculation pressure is most influenced by the recirculation valve and the flow to the applicator is most influenced by the applicator valve.

For the starch pumping system, the relative gain is very much influenced by the operating conditions for the process. At low flows to the applicator, the control configuration in Figure 7.14 exhibits little interaction. However, the degree of interaction increases as the flow to the applicator increases.

Loop Tuning. The conclusions for the starch pumping system are summarized as follows:

- The only viable control configuration is the one shown in Figure 7.14. Reversing the pairing in Figure 7.15 is not acceptable because of issues that arise when the flow to the applicator is stopped.
- For the configuration in Figure 7.14, the degree of interaction increases as the flow to the applicator increases.

- When 80% of the pump flow goes to the applicator, the relative gain for each loop is 0.25, which suggests a serious degree of interaction.

With such a degree of interaction, there is only one possibility for getting the two controllers to operate on automatic simultaneously: tune one loop to respond as rapidly as possible and slow down the other loop until the degree of oscillations in the two loops is acceptable.

Which should be the fast loop and which should be the slow loop? One's first impulse is that the fast loop should be the one that is most critical to process operations. For the starch pumping system, the flow to the applicator is more critical to process operations than is the recirculation pressure. As long as the flow to the applicator is on target and the recirculation flow to the starch tank is adequate, any recirculation pressure is acceptable.

Is it necessary to control the recirculation pressure? When the flow to the applicator is stopped, a recirculation valve opening of 87.7% gives a recirculation pressure of 80 psig. With this valve opening, a fully open applicator valve gives an applicator flow of only 60 gpm. As the flow to the applicator increases, it is necessary to decrease the opening of the recirculation valve. The pressure controller is responsible for doing so.

Fast Flow Loop, Slow Pressure Loop. The approach to tuning the controllers is as follows:

Flow controller. The customary settings ($K_C = 0.2\%/\%$; $T_I = 3$ sec) are used for the flow controller.

Pressure controller. Use very conservative tuning, usually by reducing the controller gain K_C, so that the pressure controller responds much more slowly than the flow controller. The final tuning is $K_C = 0.4\%/\%$ and $T_I = 54$ sec.

For this application, more aggressive tuning for the flow controller should be considered, as less conservative settings could be used for the pressure controller.

Figure 7.18 presents the response to applicator flow set point changes in increments of 20 gpm from an applicator flow of zero to an applicator flow of 80 gpm. For the step change from 0 gpm to 20 gpm, the response exhibits only nominal overshoot. The degree of overshoot and oscillations increases as the flow to the applicator increases. For the step change from 60 gpm to 80 gpm, the response exhibits approximately a quarter decay ratio and noticeable overshoot. The recirculation pressure loop also exhibits increased oscillations as the applicator flow increases. This behavior is because the degree of interaction between the pressure loop and the flow loop increases as the flow to the applicator increases.

Figure 7.19 presents the response for an applicator flow set point change from 0 gpm (applicator is stopped) to 80 gpm. Several features are worth noting:

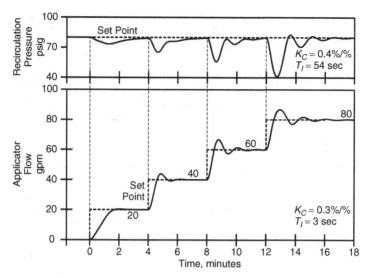

Figure 7.18 Fast flow loop, slow pressure loop; applicator flow set point changes in increments of 20 gpm.

- There is a substantial drop in the recirculation pressure: specifically, from 80 psig to 25.7 psig. The recirculation pressure is the pressure drop across the applicator control valve, so this drop in pressure reduces the driving force for starch slurry to flow to the applicator.
- The applicator flow controller drives the applicator control valve fully open, and this valve remains fully open for approximately 44 seconds. During this period of time, the recirculation pressure controller is closing the recirculation control valve in order to increase the recirculation pressure.
- The period of the cycles is relatively long, being somewhat greater than 2 min. These long periods are the result of the recirculation pressure controller being very conservatively tuned.

The applicator flow attains 80 gpm in just over 1 min, but then increases to 86.6 gpm and remains above the set point for over 1 min. Is this performance acceptable? Only those familiar with the process can answer this question. In most processes, 1 min is not considered to be a long period of time. But paper machines are far faster than most processes. If the paper machine is running at 1000 ft/min (not considered fast for today's machines), 1 min translates into 1000 ft of paper.

Fast Pressure Loop, Slow Flow Loop. The approach to tuning the controllers is as follows:

Pressure controller. The pressure controller is tuned as aggressively as possible to obtain the fastest possible response. The final tuning is $K_C = 8\%/\%$ and $T_I = 18$ sec.

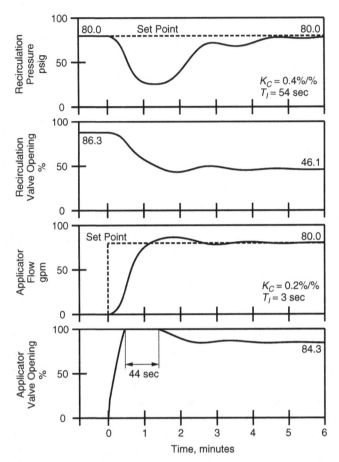

Figure 7.19 Fast flow loop, slow pressure loop; applicator flow set point change from 0 gpm to 80 gpm.

Flow controller. The customary tuning for flow controllers ($K_C = 0.2\%/\%$; $T_I = 3$ sec) gives excessive oscillations. More conservative tuning can be realized by either decreasing the controller gain or increasing the reset time. As a controller gain of $0.2\%/\%$ is very conservative, the reset time of 6 sec is used.

Figure 7.20 presents the response to applicator flow set point changes in increments of 20 gpm from an applicator flow of 0 gpm to an applicator flow of 80 gpm. For the step change from 0 gpm to 20 gpm, the response exhibits no overshoot. For the step change from 60 gpm to 80 gpm, the response exhibits a small overshoot but no oscillations. For the recirculation pressure controller, the effect of the increase in the applicator flow becomes more noticeable as the applicator flow increases. However, the departures from the set point are far less than exhibited in Figure 7.18 for the conservative tuning of the pressure controller.

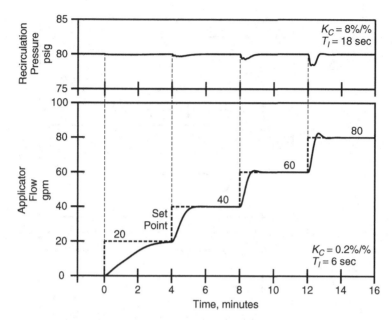

Figure 7.20 Fast pressure loop, slow flow loop; applicator flow set point changes in increments of 20 gpm.

Figure 7.21 presents the response for an applicator flow set point change from 0 gpm (applicator is stopped) to 80 gpm. Several features are worth noting:

- The drop in the recirculation pressure is nominal.
- The applicator flow controller does not drive the applicator control valve fully open.
- The period of the oscillations is much shorter than those in Figure 7.19. In general, the system appears to be more responsive.

The applicator flow attains 80 gpm in approximately 1 min, which is about the same as in Figure 7.19. The overshoot is only slightly less, but the duration is much shorter. Again only those responsible for the process can determine if this performance is acceptable.

Effect of Controller Gain. The expected effect of reducing the controller gain is to decrease the overshoot and oscillations in the response. For a controller gain of 8%/% in the recirculation pressure controller, the response in Figure 7.20 of the recirculation pressure to a change in applicator flow set point from 60 gpm to 80 gpm exhibits a very small oscillation. If the controller gain is reduced below 8%/%, the usual expectation is for this oscillation to disappear.

But in the presence of interaction with another loop, the effect can be very different. Figure 7.22 presents the same responses as in Figure 7.20, but for a

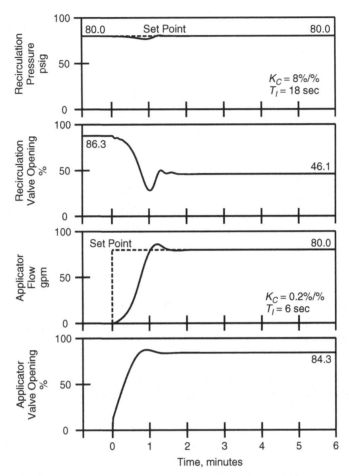

Figure 7.21 Fast pressure loop, slow flow loop; applicator flow set point change from 0 gpm to 80 gpm.

controller gain of 4%/%. For the change in applicator flow set point from 60 gpm to 80 gpm, the oscillation in the recirculation pressure is more pronounced. This is also apparent in the applicator flow response. If the controller gain is reduced to 2%/%, the oscillations in both responses become more pronounced.

Why do the oscillations become more pronounced as the controller gain is lowered? A controller gain of 8%/% in the recirculation pressure controller gives a certain separation of dynamics between the recirculation pressure loop and the applicator flow loop. Lowering the controller gain in the recirculation pressure controller reduces the separation of dynamics between the two loops. The degree of steady-state interaction between the two loops is rather severe (relative gain is about 0.25). As the dynamic separation is reduced, the steady-state interaction has a greater impact on the responses, making any oscillations more pronounced.

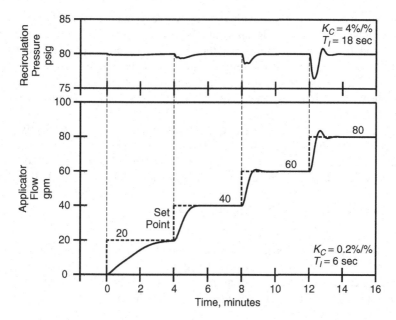

Figure 7.22   Effect of reducing the controller gain for the pressure controller.

Reducing the controller gain should reduce the degree of oscillations in the responses. If the opposite effect is observed, the next action item should be to understand why. Interaction with another loop is a likely explanation.

7.7. REDUCING THE DEGREE OF INTERACTION

In applying single-loop controllers to a multivariable control application on a process with interaction, one has two choices:

1. *Totally eliminate the interaction between the loops.* Each loop can then be tuned individually. Switching one of the loops between manual and automatic has no effect on any of the other loops. This is the objective of decouplers and model predictive controllers. This provides the best performance, but at the expense of increased complexity (and increased cost, both for implementation and subsequent support).
2. *Reduce the degree of interaction only to the point that each loop can be tuned successfully to deliver the required performance.* The objective is to attain adequate performance by relying on simple and readily available function blocks, including summers, multipliers, characterization functions, and so on.

The choice of these two approaches is often dictated by economics. As originally developed, model predictive control was used in conjunction with a process optimization endeavor that provided the incentive to pursue the total effort. What can be done in multivariable applications where some degree of performance is required from the controllers but there is little or no economic incentive to obtain the best possible performance? Often, the only option is to insert simple function blocks into the control configuration with the objective of reducing the interaction to the point where adequate performance can be obtained from the individual loops.

There are rigorous design techniques whose objective is to completely eliminate the interaction between two or more loops. But when the objective is only to reduce the degree of interaction to the point that the individual single-loop controllers will perform adequately, no rigorous methodology is available. Often, one has to rely on one's understanding of the process behavior to suggest what function block should be inserted, and where.

Starch Pumping System. For the starch pumping system, the starch flow to the applicator is the most critical of the controlled variables. Suppose that the process is currently lined-out but an increase in the starch flow is required. The configuration in Figure 7.14 attempts to accomplish this as follows:

- The flow controller increases the applicator valve opening. This increases the applicator flow, but also decreases the recirculation pressure.
- On any decrease in the recirculation pressure, the pressure controller must respond quickly by closing the recirculation valve. The recirculation pressure is the driving force for fluid flow through the applicator valve, so decreases in the recirculation pressure cause the applicator flow to decrease.

The control configuration must quickly translate a change in the applicator valve opening to a change in the recirculation valve opening. There are two ways to do this:

Simple feedback loops (Figure 7.14). The recirculation pressure controller must be tuned aggressively. Due to the interaction between the loops, the applicator flow controller must be conservatively tuned, which affects performance negatively.

Feedforward logic. Logic can be incorporated into the control configuration that establishes a fixed relationship between the applicator valve opening and the recirculation valve opening. For any increase in the applicator valve opening, the logic decreases the recirculation valve opening by an amount that offsets the effect of the increase in applicator valve opening on the recirculation pressure.

To accomplish the latter, in this section we present two configurations, one that uses a summer and one that uses a characterization function. In both configurations, the pressure controller provides the feedback trim and can be tuned

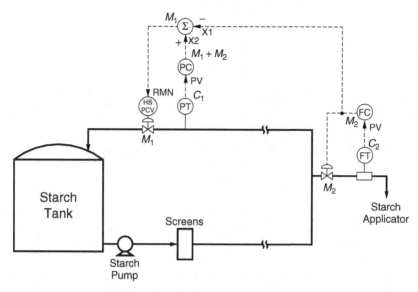

Figure 7.23   Control configuration with a summer.

conservatively, which permits the applicator flow controller to be aggressively tuned.

Summer. Figure 7.23 presents a control configuration that uses a summer to translate a change in the applicator valve opening to a change in the recirculation valve opening. The recirculation valve opening is the output of a summer whose inputs are the following:

> *The applicator valve opening (input X1).* The sign on this input to the summer is negative. If the applicator valve opening increases, the recirculation valve opening must decrease.
>
> *Output of the recirculation pressure controller (input X2).* The sign on this input to the summer is positive. If the recirculation pressure is increasing, the recirculation pressure controller should decrease its output—that is, the controller must be reverse acting. However, changing the sign on input X2 only changes the action of the controller.

With the signs on the inputs as in Figure 7.23, the relationship for the summer is as follows:

$$Y = X_2 - X_1$$

The output Y is the opening M_1 of the recirculation valve. Input $X1$ is the opening M_2 of the applicator valve. Substituting these gives the following relationship for the summer:

$$M_1 = X_2 - M_2$$
$$X_2 = M_1 + M_2$$

The output of the pressure controller is the recirculation valve opening plus the applicator valve opening.

Most implementations of summers in digital systems provide a coefficient on each input to the summer. That is, the equation for the summer is as follows:

$$Y = k_1 X_1 + k_2 X_2$$

For the summer in Figure 7.23, the coefficient k_1 would be -1.0 and coefficient k_2 would be $+1.0$. But with these coefficients, an increase of 1% in the applicator valve opening translates to a decrease of 1% in the recirculation valve opening. Is this appropriate?

The responses in Figures 7.19 and 7.21 give the valve openings for applicator flows of 0 gpm and 80 gpm. A change in the applicator valve opening M_2 from 0.0% to 84.3% translates into a recirculation valve opening M_1 from 86.3% to 46.1%. A change of $+84.3\%$ for the applicator valve opening translates to a change of -38.2% for the recirculation valve opening. For input X2, this suggests a coefficient of $-38.2/84.3 = -0.453 \cong -0.5$. The resulting summer equation is:

$$Y = X_2 - \tfrac{1}{2} X_1$$
$$M_1 = X_2 - \tfrac{1}{2} M_2$$
$$X_2 = M_1 + \tfrac{1}{2} M_2$$

The following table computes $M_1 + M_2$ and $M_1 + \tfrac{1}{2} M_2$ from the valve openings for 0 and 80 gpm:

Variable	Value	Value
Applicator starch flow C_2	0.0 gpm	80.0 gpm
Recirculation valve opening M_1	86.3%	46.1%
Applicator valve opening M_2	0.0%	84.3%
$M_1 + M_2$	86.3%	130.4%
$M_1 + \tfrac{1}{2} M_2$	86.3%	88.3%

Using $M_1 + \tfrac{1}{2} M_2$, the pressure controller output must only change from 86.3% to 88.3%. However, the results for intermediate flows could be different.

When summers are inserted as in Figure 7.23, some attention must be paid to the output range of the pressure controller. The lower and upper range values for both M_1 and M_2 are 0 to 100%. To allow the pressure controller to fully open both valves, the output range for the pressure controller must be 0 to 150%.

Most organizations have adopted recommendations for the tuning coefficients for a flow controller (in this book, $K_C = 0.2\%/\%$ and $T_I = 3$ sec are used). For the control configuration in Figure 7.23, there are two problems with the customary settings:

1. *Customary settings apply to loops where the measured flow is the flow through the flow control valve.* The flow controller in Figure 7.23 is changing the opening of both the applicator control valve and the recirculation control valve, so the customary settings may not be acceptable.
2. *Customary settings are very conservative.* To change the applicator flow from 0 to 80 gpm as rapidly as possible, more aggressive tuning will be required.

The following test assesses the effectiveness of the control configuration in Figure 7.23 in achieving this objective:

• With the flow to the applicator stopped, adjust the recirculation pressure controller output until the recirculation pressure is 80 psig.
• Place the pressure controller on manual.
• Place the flow controller on automatic.
• Change the flow controller set point to 80 gpm.

Figure 7.24 presents the results of the test. The following observations apply:

• With the recirculation pressure controller on manual, the recirculation pressure lines-out at 82.1 psig at an applicator flow of 80 gpm. Only a small adjustment is required from the feedback trim controller to obtain 80.0 psig. However, this may not be the case for intermediate flows.
• The recirculation pressure peaks at 132.5 psig. This suggests that the recirculation valve is closing too quickly. One contributor is the equal-percentage characteristics of the two valves.
• The applicator flow attains 80 gpm is less than 30 sec. The overshoot is less than 3 gpm, and the loop lines-out in less than 60 sec.

This is clearly an improvement over the responses in Figure 7.21, mainly because the flow controller is tuned to respond much more aggressively.

Assessing the Degree of Interaction. When evaluating a measure of interaction, the menu of controlled and manipulated variables must reflect the controlled and manipulated variables for the individual controllers. The following menu of controlled and manipulated variables applies to the control configuration in Figure 7.23:

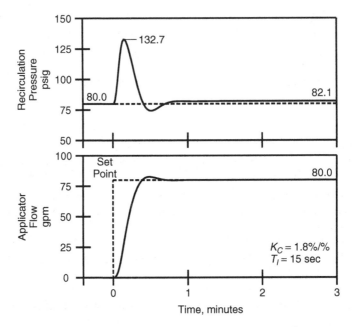

Figure 7.24 Performance of configuration using a summer.

Manipulated Variable	Controlled Variable
$M_1 + \frac{1}{2}M_2$: trim for recirculation valve opening	C_1: recirculation pressure
M_2: applicator valve opening	C_2: starch flow to applicator

The data for computing the relative gain can be obtained by either of the following approaches:

- Procedures analogous to those applied to obtain responses such as in Figure 7.16.
- Compute the values from steady-state solutions of a process model.

In either case, the relative gain for the pressure controller at an applicator flow of 80 gpm is computed as follows:

- *Initial equilibrium.* For a recirculation pressure of 80 psig and an applicator flow of 80 gpm, the recirculation valve opening M_1 must be 46.1% and the applicator valve opening M_2 must be 84.3% (same as in Figure 7.16). The output of the pressure controller is $M_1 + \frac{1}{2}M_2 = 88.2\%$.

- *Increase of 5% in pressure controller output, fixed applicator valve opening* M_2. The values are (the results are the same as in Figure 7.16):

$$M_1 + \tfrac{1}{2}M_2 = 93.2\% \text{ (5\% increase from initial equilibrium)}$$

$$M_2 = 84.3\% \text{ (constant applicator valve opening)}$$

$$M_1 = 51.1\%$$

$$C_1 = 73.4 \text{ psig (recirculation pressure)}$$

$$C_2 = 76.6 \text{ gpm (applicator flow)}$$

The gain K_{11} is computed as follows:

$$K_{11} = \frac{73.4 \text{ psig} - 80.0 \text{ psig}}{93.2\% - 88.2\%} = -1.32 \text{ psig}/\%$$

- *Increase of 5% in pressure controller output, fixed applicator flow* C_2. To increase the applicator flow from 76.6 gpm to 80.0 gpm, the applicator valve opening M_2 must increase from 84.3% to 87.6%. The results are as follows:

$$M_2 = 87.6\%$$

$$M_1 + \tfrac{1}{2}M_2 = 93.2\% \text{ (5\% increase from initial equilibrium)}$$

$$M_1 = 49.4\%$$

$$C_1 = 61.4 \text{ psig (recirculation pressure)}$$

The gain K'_{11} is computed as follows:

$$K'_{11} = \frac{61.4 \text{ psig} - 80.0 \text{ psig}}{93.2\% - 88.2\%} = -3.72 \text{ psig}/\%$$

The relative gain λ_{11} is computed as follows:

$$\lambda_{11} = \frac{K_{11}}{K'_{11}} = \frac{-1.32 \text{ psig}/\%}{-3.72 \text{ psig}/\%} = 0.35$$

The relative gain array is:

	M_1	M_2
C_1	0.35	0.65
C_2	0.65	0.35

Incorporating the summer provides only a slight improvement in the relative gain: specifically, from 0.25 to 0.35.

Characterization Function. Figures 7.18 and 7.20 present the responses at applicator flows of 20, 40, 60, and 80 gpm. The valve openings for each of these flows (and a flow of 90 gpm) are as follows:

Applicator Flow (gpm):	0	20	40	60	80	90
Applicator valve (%)	0	50	67	77	84	87
Recirculation pressure (psig)	80	80	80	80	80	80
Recirculation valve (%)	88	82	74	64	46	29

Figure 7.25 presents the recirculation valve opening as a function of the applicator valve opening. Also illustrated in Figure 7.25 is the linear approximation provided by the summer in the control configuration in Figure 7.23. Clearly, the linear approximation is not very good at intermediate flows.

The control configuration in Figure 7.26 includes a characterization function (the "PY" element) in the control configuration. The output of the applicator flow controller (or the applicator valve opening M_2) is the input to the characterization function; the output of the characterization function is the recirculation valve opening M_1 computed from the applicator valve opening.

Especially for applications such as pumping slurries, characterization functions will not be perfect. Maintaining a recirculation pressure of 80 psig requires

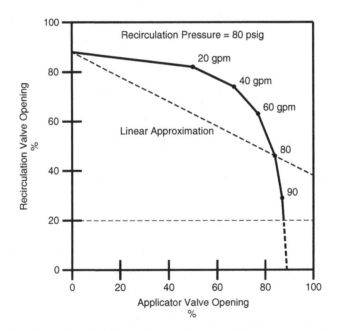

Figure 7.25   Recirculation valve opening vs. applicator valve opening.

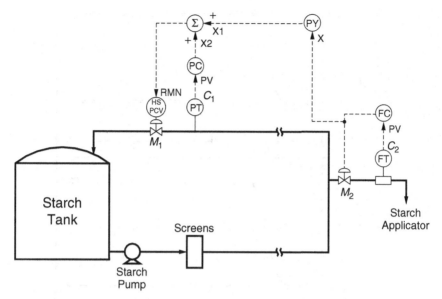

Figure 7.26 Control configuration with a characterization function for the recirculation valve opening.

feedback trim similar to that required in ratio and feedforward control applications. In the control configuration in Figure 7.26, a summer is inserted between the output of the characterization function and the recirculation control valve. The two inputs to this summer are as follows:

Input X1. Output of the characterization function, which is the recirculation valve opening that corresponds to the current applicator valve opening. The coefficient on this input is $+1.0$.

Input X2. Output of the recirculation pressure controller that is providing the feedback trim. The coefficient on this input is $+1.0$. If the recirculation pressure is increasing, the recirculation pressure controller should increase its output (direct action).

The output range for the recirculation pressure controller determines how much the recirculation valve opening can differ from the value computed by the characterization function. For example, an output range of -25 to $+25\%$ permits the recirculation valve opening to be as much as 25% above or as much as 25% below the value computed by the characterization function.

The same test that was applied to the configuration with the summer can be applied to the configuration with the characterization function. Figure 7.27 presents the results of the test. The applicator flow attains 80 gpm in less than 30 sec with inconsequential overshoot. The peak in the recirculation pressure is also smaller. The performance at intermediate flows should be comparable.

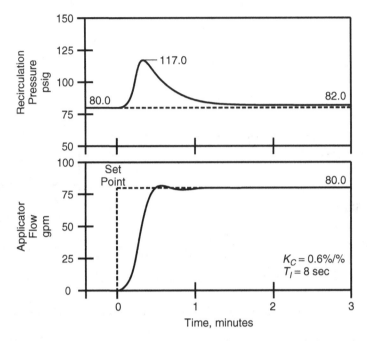

Figure 7.27 Performance of configuration using a characterization function.

Assessing the Degree of Interaction. The following menu of controlled and manipulated variables applies to the control configuration in Figure 7.26:

Manipulated Variable	Controlled Variable
$M_1 - f(M_2)$: trim for computed recirculation valve opening	C_1: recirculation pressure
M_2: applicator valve opening	C_2: starch flow to applicator

For the pressure controller in Figure 7.26, the relative gain at an applicator flow of 80 gpm is computed as follows:

- *Initial equilibrium.* For a recirculation pressure of 80 psig and an applicator flow of 80 gpm, the recirculation valve opening M_1 must be 46.1% and the applicator valve opening M_2 must be 84.3% (same as in Figure 7.16). The output of the characterization function $f(M_2)$ is 44.4%. The output of the pressure controller is

$$M_1 - f(M_2) = 46.1\% - 44.4\% = 1.7\%$$

- *Increase of 5% in pressure controller output, fixed applicator valve opening M_2.* The values are (the results are the same as in Figure 7.16):

$$M_1 - f(M_2) = 6.7\% \text{ (5\% increase from initial equilibrium)}$$

$$M_2 = 84.3\% \text{ (constant applicator valve opening)}$$

$$f(M_2) = 44.4\%$$

$$M_1 = 51.1\%$$

$$C_1 = 73.4 \text{ psig (recirculation pressure)}$$

$$C_2 = 76.6 \text{ gpm (applicator flow)}$$

The gain K_{11} is computed as follows:

$$K_{11} = \frac{73.4 \text{ psig} - 80.0 \text{ psig}}{6.7\% - 1.7\%} = -1.32 \text{ psig/\%}$$

- *Increase of 5% in pressure controller output, fixed applicator flow C_2.* To increase the applicator flow from 76.6 gpm to 80.0 gpm, the applicator valve opening M_2 must increase from 84.3% to 85.0%. The results are as follows:

$$M_2 = 85.0\%$$

$$f(M_2) = 40.1\%$$

$$M_1 - f(M_2) = 6.7\% \text{ (5\% increase from initial equilibrium)}$$

$$M_1 = 46.8\%$$

$$C_1 = 75.3 \text{ psig (recirculation pressure)}$$

The gain K'_{11} is computed as follows:

$$K'_{11} = \frac{75.3 \text{ psig} - 80.0 \text{ psig}}{6.7\% - 1.7\%} = -0.94 \text{ psig/\%}$$

The relative gain λ_{11} is computed as follows:

$$\lambda_{11} = \frac{K_{11}}{K'_{11}} = \frac{-1.32 \text{ psig/\%}}{-0.94 \text{ psig/\%}} = 1.40$$

The relative gain array is

	M_1	M_2
C_1	1.40	−0.40
C_2	−0.40	1.40

Although a value of 1.40 for the relative gain indicates a modest degree of interaction, this is tolerable even in applications where the pressure controller cannot be tuned conservatively.

Manual Control of Pressure. In the process industries, the operators are usually permitted to specify the output to any final control element. In most loops (including the flow loop in Figures 7.23 and 7.26), the controller can be switched to manual and its output adjusted by the process operator. Since the output of the flow controller is the opening of the applicator valve, the operator can specify the applicator valve opening.

But this is not the case for the pressure controller in Figures 7.23 and 7.26. Switching the pressure controller to manual permits the operator to change the controller output. But this has two shortcomings:

- Operators want to specify the valve opening directly, not some controller output that affects the valve opening.
- A constant value of the pressure controller output does not give a constant recirculation valve opening. The summer or characterization function translates changes in the applicator valve opening to changes in the recirculation valve opening.

The purpose of the hand station in Figures 7.23 and 7.26 is to permit the process operator to directly specify the recirculation valve opening.

For bumpless transfer from local to remote, output tracking must be configured in the pressure controller as follows.

Condition for output tracking to be active. Output tracking should be active whenever the hand station is in local. Input TRKMN is true when output RMT from the hand station is false.

Value for output tracking. The computations performed by the summer that outputs to the hand station must be "inverted." For the configuration in Figure 7.26, input MNI must be the output of the hand station HSPCV.MN less the output of the characterization function PY.Y.

This is expressed by the following logic:

```
PC.TRKMN = !HSPCV.RMT
PC.MNI = HSPCV.MN - PY.Y
```

Limits on Recirculation Valve Opening. A mechanical stop is fitted to the recirculation valve to assure that the valve will never close. A value must be specified for the lower output limit for the hand station that corresponds to the position of the mechanical stop, or perhaps slightly less to provide some overrange. The value of the upper output limit is the customary 102%, which assures that the recirculation valve can fully open.

Should the hand station be driven to either output limit, windup protection must be activated in the pressure controller. The options for preventing windup are as follows:

Integral tracking. Preventing windup using integral tracking involves the following inputs:

- *Input TRKMN.* Integral tracking must be activated if the output of the hand station is either at its upper limit (output QH is true) or at its lower limit (output QL is true).
- *Input MRI.* The appropriate value for the controller output bias is the output of the hand station less the output of the characterization function.

The following logic implements output tracking and integral tracking in the recirculation pressure controller in Figure 7.26:

```
PC.TRKMN = !HSPCV.RMT
PC.MNI = HSPCV.MN - PY.Y
PC.TRKMR = HSPCV.QH | HSPCV.QL
PC.MRI = HSPCV.MN - PY.Y
```

External reset. Preventing windup using external reset involves only one input:

- *Input XRS.* The appropriate value for the input to the reset mode is the output of the hand station less the output of the characterization function.

The following logic implements output tracking and external reset in the recirculation pressure controller in Figure 7.26:

```
PC.TRKMN = !HSPCV.RMT
PC.MNI = HSPCV.MN - PY.Y
PC.XRS = HSPCV.MN - PY.Y
```

Inhibit increase/inhibit decrease. Preventing windup using inhibit increase/inhibit decrease involves the following two inputs:

- *Input NOINC.* If the output of the hand station has been driven to its upper output limit (output QH is true), the pressure controller must not increase its output further.
- *Input NODEC.* If the output of the hand station has been driven to its lower output limit (output QL is true), the pressure controller must not decrease its output further.

The following logic implements output tracking and inhibit increase/inhibit decrease in the recirculation pressure controller in Figure 7.26:

```
PC.TRKMN = !HSPCV.RMT
PC.MNI = HSPCV.MN - PY.Y
```

```
PC.NOINC = HSPCV.QH
PC.NODEC = HSPCV.QL
```

LITERATURE CITED

1. Nisenfeld, A. E., and H. M. Schultz, "Interaction Analysis Applied to Control System Design," *InTech*, Vol. 18, No. 4 (April 1971), pp. 52–57.
2. Bristol, E. H., "On a New Measure of Interaction for Chemical Process Control," *IEEE Transactions on Automatic Control*, Vol. 11 (January 1966), pp. 133–134.

8

MULTIVARIABLE CONTROL

The starting point for most multivariable control applications is a configuration consisting of the necessary number of single-loop controllers. If these can be successfully tuned and if their performance meets the requirements of process operations, no further time or effort will be devoted to this configuration.

But success is not assured. Untunable controllers are common within process control configurations. Sometimes the controllers can be on auto if the process is operating smoothly, but on any upset, one or more controllers must be switched to manual. Such controllers are of dubious value. Sometimes one or more controllers are tuned so conservatively that their performance is inadequate.

A potential cause is interaction with other loops. If interaction is the culprit, the solutions include the following:

- *Leave one or more controllers on manual.* Although not really a "solution," too often this is the final result. With digital controls, this is not appropriate.
- *Check the loop pairing.* This was discussed in Chapter 7. This should always be examined, but even if the pairing is incorrect, changing the pairing may not be a viable option, due to other process considerations.
- *Tune one loop to respond rapidly, the other to respond slowly.* This approach was presented in Chapter 7 for a 2×2 process. For higher-dimensional processes, adequately separating the dynamics becomes more challenging. The slow loop is often the one that most affects process operations.
- *Reduce the degree of interaction.* Usually, this involves judiciously inserting simple function blocks (summers, multipliers, characterization functions,

Advanced Process Control: Beyond Single-Loop Control   By Cecil L. Smith
Copyright © 2010 John Wiley & Sons, Inc.

etc.) into the control configuration to reduce the degree of interaction to the point that the single-loop controllers can be tuned to deliver adequate performance. An example was presented in Chapter 7.

- *Install a decoupler.* The decoupler is inserted between the outputs of the single-loop controllers and the inputs to the process, the objective being to incorporate interaction within the decoupler that offsets the interaction within the process.

- *Apply a multivariable control technique.* The army of single-loop PID controllers is replaced by a multivariable controller that is designed to cope with the interaction within the process. Within the process industries, the only such technique that is widely applied is model predictive control.

The latter two are examined in this chapter.

Predictive Controllers. Using the derivative mode of the PID controller causes the control action to be based on a predicted value of the control error instead of the current value of the control error. The derivative mode relies on a simple straight-line predictor:

$$\widehat{PV} = PV + T_D \frac{dPV}{dt}$$

where

PV = current value of the process variable
$\widehat{PV}$ = predicted value of the PV at one derivative time in the future
T_D = derivative time (min)
t = time (min)

The predicted control error $\hat{E}$ is computed from $\widehat{PV}$. The proportional mode is always based on $\hat{E}$. The basis for the integral mode depends on the form of the PID control equation:

Parallel (noninteracting; ideal). Integral mode is based on the actual control error E.

Series (interacting; nonideal). Integral mode is based on the predicted control error $\hat{E}$.

Model Predictive Controller. A model predictive controller bases its control actions on predicted values of the control error that are computed using a model for the process. On the plus side, the predictions are both better and can be projected farther into the future, which improves the control performance. On the negative side, a plant test must be conducted to develop the process model, and as noted previously, plant testing is never easy.

Model predictive control encompasses all control configurations that

- Base control actions on predicted values of the process variable.

- Compute the predicted values using a process model of some form.

The following three control methodologies meet these criteria:

Dead-time compensation. With industrial applications dating from the late 1960s, this was the earliest form of model predictive control to be applied to industrial processes. However, it is applicable only to processes that are dominated by dead time or transportation lag, such as Fourdrinier paper machines. This technology will be explained shortly.

Internal model control. This technology has received considerable attention in academia. However, few applications to industrial processes have been reported. Consequently, this technology is not presented herein.

Dynamic matrix control (DMC). The term "MPC" or even "advanced control" is often used within the industry to refer to some form of this technology. Basically, the process is characterized by its step response or impulse response. The principle of superposition is then applied to predict the effect of past control actions as well as proposed future control actions on future values of the process variable. The basic technology is explained herein, but a thorough examination is the subject of an entire book.

8.1. DECOUPLER

As illustrated for a 2×2 process in Figure 8.1, decouplers are inserted between the single-loop PID controllers and the process. The interaction within the process is determined primarily by the nature of the process. However, the interaction within the decoupler is the result of its design, and is completely at the discretion of the control engineers.

The following notation will be used for the decoupler in Figure 8.1:

C_i = process variable for PID controller i; also output i from the process

M_i = input i to the process; also output i from the decoupler

X_i = output of PID controller i (process variable for controller i is C_i)

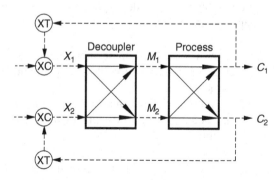

Figure 8.1 Decoupler for a 2×2 process.

In the decoupler designs presented herein, the relationship for the "straight-through" paths (X_1 to M_1 and X_2 to M_2) of the decoupler will be unity. Consequently:

- When the PID controller for controlled variable C_1 makes a change in its output X_1, the decoupler makes this same change in manipulated variable M_1.
- When the PID controller for controlled variable C_2 makes a change in its output X_2, the decoupler makes this same change in manipulated variable M_2.

The design of the decoupler determines the relationships for the diagonal paths (X_1 to M_2 and X_2 to M_1). The design objectives are as follows:

1. *A change in X_1 affects only C_1.* A change in M_1 affects both C_1 and C_2. For a change in M_1, the decoupler will be designed to change M_2 in such a manner as to offset the effect of M_1 on C_2.
2. *A change in X_2 affects only C_2.* A change in M_2 affects both C_1 and C_2. For a change in M_2, the decoupler is designed to change M_1 in such a manner as to offset the effect of M_2 on C_1.

To the PID controllers, the decoupler-process combination appears to have no interaction. This concept is easily extended to higher-dimensional processes; disturbance inputs can also be incorporated.

Implementing a Decoupler. With modern control systems, the following two options are available for implementing a decoupler:

Programmed implementation. The advantage of this approach is that routines are available for the matrix computations. As the dimensionality of the decoupler increases, the advantages of this approach mount.

Function blocks. Commercial control systems do not generally provide a "decoupler function block." The decoupler has to be constructed using summers, multipliers, and so on.

The function block approach will be used in the subsequent examples for a 2×2 process. The issues discussed for the function block implementation also pertain to the programmed implementations.

Numerous articles have presented the avenues for designing a decoupler. However, far less attention has been directed to the practical issues pertaining to the implementation of a decoupler:

Manual operation. Especially for high-dimensional decouplers, an "all on auto" or "all on manual" is unacceptable. Manual and automatic must apply to each individual decoupler output, not to the decoupler as a whole. Specifically, the plant operators must be able to switch any decoupler output to manual and specify a value for that output. The functions provided by the decoupler for the remaining outputs must continue. All decoupler control configurations presented herein will include a hand station (as described in Chapter 1) for each decoupler output, giving the following meaning for manual and automatic:

- *Manual.* The hand station is on local and the operator is specifying the value for the output.
- *Automatic.* The hand station is on remote and the value for this output is the value from the decoupler.

Bumpless transfer. When any decoupler output is switched from manual to automatic or from automatic to manual, the transition must be bumpless. There must be no abrupt change in:

- The output of the hand station being switched.
- Any other output from the decoupler.

Limits on outputs. Limits apply to all final control elements installed in process plants. Should any output of the decoupler attain a limit, the decoupler must react in an appropriate fashion, including preventing reset windup in the PID controllers.

In many respects, these issues are more challenging than the design of the decoupler.

Issues Pertaining to Auto/Manual. Suppose that the process variable C_1 for the controller whose output is X_1 becomes invalid. The PID calculations must always be suspended. For continuous processes that respond slowly (and most do), the normal action in the short term is to hold the last value of the controller output. The operator is alerted and is expected to assume manual control.

In a decoupler application, how does the operator assume control for one of the outputs? There are two possibilities:

Specify X_1. The operator switches the controller to manual and specifies its output, which is an input to the decoupler. By taking this approach, the decoupler will continue to provide its functions.

Specify M_1. This capability is provided by the hand station or its equivalent. The operator switches the hand station to local and specifies the value of the output to the final control element.

Process operators prefer the latter, which permits them directly to specify control valve openings, pump speeds, and so on. Although arguments can be made for retaining the functions of the decoupler, most operators have difficulties with the consequences, which include the following:

- The operator specifies a value for X_1, but the value of M_1 is different (because of the compensation for X_2).

- Changing X_1 affects both M_1 and M_2.
- Holding X_1 constant does not provide a constant value of M_1 (the decoupler is compensating for changes in X_2).

Failures such as the process variable for a controller will occur, so the controls must provide the necessary features to enable the operators to cope. However, such events should be infrequent, which reduces the incentive to retain the functions provided by the decoupler during such events.

For a decoupler, the operator must be able to place any hand station on local but leave the other hand station(s) on remote. For each hand station, the transition between local and remote must be smooth. Specifically, the following is required:

- When a hand station is switched to local, the output of the hand station must not change abruptly. The hand station is designed to behave in this manner.
- Switching a hand station to local must not cause an abrupt change in the output of any other hand station.
- No PID controller must wind up because a hand station is on local.
- When a hand station is switched to remote, the output of the hand station must not change abruptly.

To meet these requirements, appropriate initialization calculations must be executed when a hand station is on local. Output tracking is normally activated for a PID controller that provides one of the inputs to the decoupler. But when output tracking is activated in a PID controller, the controller output usually changes abruptly, which can cause the values calculated by the decoupler for other outputs to change abruptly.

Design Objectives. Basically, the objective of the design of a decoupler is to incorporate interaction within the decoupler that offsets or cancels the interaction within the process. The description of the process, the description of the decoupler, and the design objective for the decoupler can be expressed as follows (an alternative expression will be presented shortly):

Dimensionality:	2×2	$n \times n$
Process equations	$C_1 = f_1(M_1, M_2)$ $C_2 = f_2(M_1, M_2)$	$C_i = f_i(M_1, \ldots, M_n),$ $i = 1, \ldots, n$
Decoupler equations	$M_1 = g_1(X_1, X_2)$ $M_2 = g_2(X_1, X_2)$	$M_i = g_i(X_1, \ldots, X_n),$ $i = 1, \ldots, n$
Design objectives	$C_1 = h_1(X_1)$ $C_2 = h_2(X_2)$	$C_i = h_i(X_i),$ $i = 1, \ldots, n$

The decoupler–process combination must appear as having no interaction between the loops. For a 2×2 multivariable process, the explanation of the design objective is as follows:

Design Objective	Explanation
$C_1 = h_1(X_1)$	C_1 is a function of X_1 only; otherwise, no specification is imposed on the nature of the function h_1
$C_2 = h_2(X_2)$	C_2 is a function of X_2 only; otherwise, no specification is imposed on the nature of the function h_2

With each controlled variable being a function of only the output of the respective PID controller, the decoupler–process combination appears to the controller as having no interaction. Certainly, this is the ideal result. But to achieve this, the process must be characterized in great detail. Consequently, most decouplers fall short of the ideal result.

Decouplers and Feedforward Control. In a sense, a decoupler is a logical extension of feedforward control. Consider the control configuration presented in Figure 7.26 for the starch pumping system. The applicator valve opening can be viewed as a disturbance to the recirculation pressure loop. The objective of adding the characterization function is to compensate for the effect of this disturbance on the recirculation pressure. Ideally, the result would be that changes in the applicator valve opening have no effect on the recirculation pressure; in practice, the effect of changes in the applicator valve on the recirculation pressure are greatly reduced.

Figure 8.2(a) presents an alternative representation of the control configuration in Figure 7.26. The "PY" element encompasses the computations required to compensate the recirculation valve opening for changes in the applicator valve opening. In the control configuration in Figure 7.26, the PY element is a characterization function. But in the general case, the PY element includes whatever is required to compensate the recirculation valve opening for changes in the applicator valve opening to minimize the effect of these changes on the recirculation pressure.

A similar approach can be taken to the applicator flow. Changes in the recirculation valve opening affect the applicator flow. The "FY" element in Figure 8.2(b) encompasses the computations required to compensate the applicator valve opening for changes in the recirculation valve opening to minimize the effect of these changes on the applicator flow.

The two feedforward control configurations are as follows:

Figure 8.2(a): compensates the recirculation valve opening for changes in the applicator valve opening to minimize the effect of these changes on the recirculation pressure

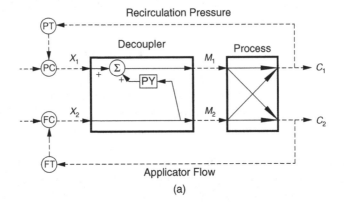

(a)

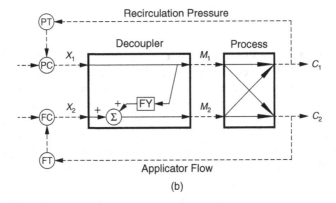

(b)

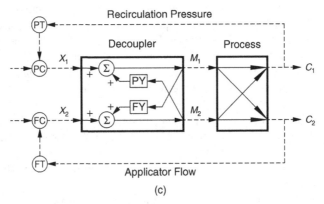

(c)

Figure 8.2 Relationship of decoupler to feedforward control: (a) feedforward control of applicator valve opening; (b) feedforward control of recirculation valve opening; (c) decoupler for starch pumping system.

Figure 8.2(b): compensates the applicator valve opening for changes in the recirculation valve opening to minimize the effect of these changes on the applicator flow

These are not mutually exclusive; both can be implemented, the result being the control configuration for a decoupler in Figure 8.2(c).

Decoupler Structures. The control configuration in Figure 8.2(c) is a decoupler. However, its structure differs from that of the decoupler in Figure 8.1. The relationships for the decoupler are expressed as follows:

	2×2	3×3
Figure 8.1	$M_1 = g_1(X_1, X_2)$	$M_1 = g_1(X_1, X_2, X_3)$
	$M_2 = g_2(X_1, X_2)$	$M_2 = g_2(X_1, X_2, X_3)$
		$M_3 = g_2(X_1, X_2, X_3)$
Figure 8.2(c)	$M_1 = g_1'(X_1, M_2)$	$M_1 = g_1'(X_1, M_2, M_3)$
	$M_2 = g_2'(X_2, M_1)$	$M_2 = g_2'(X_2, M_1, M_3)$
		$M_3 = g_3'(X_3, M_1, M_2)$

Most classical presentations start with the configuration in Figure 8.1 for a decoupler. But for reasons to be explained subsequently, a structure analogous to that in Figure 8.2(c) is usually preferable for process applications.

Figure 8.3 presents two structures, designated as the V-canonical form and the P-canonical form, for implementing a decoupler. The expression for each output from the decoupler is summarized as follows:

V-canonical. Each decoupler output is determined from all inputs to the decoupler. For $i \neq j$, relationship G_{ij} compensates decoupler output i for changes in decoupler input j.

P-canonical. Each decoupler output is determined from the respective input to the decoupler and all other outputs from the decoupler. For $i \neq j$, relationship G_{ij}' compensates decoupler output i for changes in decoupler output j.

For some applications, steady-state relationships suffice for G_{ij} and G_{ij}'. The simplest possibility is to use only a gain, but this assumes linear (or nearly linear) process behavior. To compensate for nonlinear behavior, algebraic relationships and/or characterization functions must be incorporated. When dynamics are required, the options are similar to those presented in Chapter 6. Process tests can be conducted to obtain values for the dynamic coefficients. Alternatively, lead-lag elements possibly coupled with a dead time can be incorporated and the coefficients adjusted based on operational experience. Regardless of the approach taken, the undertaking becomes more ambitious as the dimensionality of the system increases.

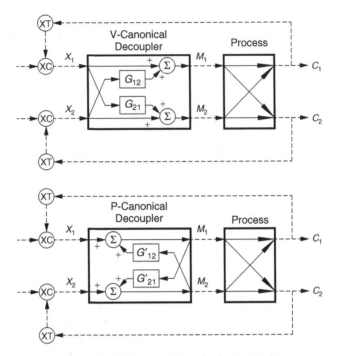

Figure 8.3 Alternative forms for the decoupler.

V-Canonical Steady-State Decoupler. When linear behavior is assumed for the process, the design of a steady-state decoupler is straightforward. As explained in Chapter 7, the process can be represented by the following equation:

$$\begin{bmatrix} \Delta C_1 \\ \Delta C_2 \end{bmatrix} = \begin{bmatrix} K_{11} & K_{12} \\ K_{21} & K_{22} \end{bmatrix} \begin{bmatrix} \Delta M_1 \\ \Delta M_2 \end{bmatrix}$$
$$\mathbf{c} \quad = \quad \mathbf{K} \qquad \mathbf{m}$$

where

$$\mathbf{c} = \begin{bmatrix} \Delta C_1 \\ \Delta C_2 \end{bmatrix} = \text{vector of changes in the controlled variables}$$

$$\mathbf{m} = \begin{bmatrix} \Delta M_1 \\ \Delta M_2 \end{bmatrix} = \text{vector of changes in the manipulated variables}$$

$$\mathbf{K} = \begin{bmatrix} K_{11} & K_{12} \\ K_{21} & K_{22} \end{bmatrix} = \text{process gain matrix}$$

The V-canonical form of the decoupler is represented by the following equation:

$$\begin{bmatrix} \Delta M_1 \\ \Delta M_2 \end{bmatrix} = \begin{bmatrix} G_{11} & G_{12} \\ G_{21} & G_{22} \end{bmatrix} \begin{bmatrix} \Delta X_1 \\ \Delta X_2 \end{bmatrix}$$
$$\mathbf{m} \quad = \quad \mathbf{G} \qquad \mathbf{x}$$

where

$$\mathbf{x} = \begin{bmatrix} \Delta X_1 \\ \Delta X_2 \end{bmatrix} = \text{vector of changes in the decoupler inputs}$$

$$\mathbf{G} = \begin{bmatrix} G_{11} & G_{12} \\ G_{21} & G_{22} \end{bmatrix} = \text{decoupler matrix}$$

The decoupler-process combination is represented by the following equation:

$$\mathbf{c} = \mathbf{Km} = \mathbf{KGx}$$

The interaction between the loops is eliminated completely if the decoupler matrix **G** satisfies the following relationship:

$$\mathbf{KG} = \mathbf{I}$$

where

$$\mathbf{I} = \begin{bmatrix} 1 & 0 \\ 0 & 1 \end{bmatrix} = \text{identity matrix}$$

The design of the decoupler requires the inverse $\mathbf{K}^{-1}$ of the process gain matrix $\mathbf{K}$:

$$\mathbf{K}^{-1}\mathbf{KG} = \mathbf{K}^{-1}$$

$$\mathbf{G} = \mathbf{K}^{-1}$$

Using this approach, the decoupler matrix **G** is the inverse $\mathbf{K}^{-1}$ of the process gain matrix **K**.

In practice, there is some incentive to simplify the control configuration as much as possible. For the V-canonical form of the decoupler in Figure 8.3, the equations for the decoupler are as follows:

$$\begin{bmatrix} \Delta M_1 \\ \Delta M_2 \end{bmatrix} = \begin{bmatrix} 1 & G_{12} \\ G_{21} & 1 \end{bmatrix} \begin{bmatrix} \Delta X_1 \\ \Delta X_2 \end{bmatrix}$$

The interaction between the two loops is eliminated completely if the following equation is satisfied:

$$\begin{bmatrix} K_{11} & K_{12} \\ K_{21} & K_{22} \end{bmatrix} \begin{bmatrix} 1 & G_{12} \\ G_{21} & 1 \end{bmatrix} = \begin{bmatrix} Z_{11} & 0 \\ 0 & Z_{22} \end{bmatrix}$$

There are four equations and four unknowns (G_{12}, G_{21}, Z_{11}, and Z_{22}). However, any values are acceptable for Z_{11} and Z_{22}.

These equations can be satisfied by proceeding as follows:

1. Compute the inverse $\mathbf{K}^{-1}$ of the process gain matrix $\mathbf{K}$.

2. To obtain the decoupler matrix **G**, divide each element of the inverse $\mathbf{K}^{-1}$ by the diagonal element on that column. That is,

$$G_{ij} = \frac{(\mathbf{K}^{-1})_{ij}}{(\mathbf{K}^{-1})_{jj}}$$

3. The decoupler-process sensitivity or gain Z_{ii} is

$$Z_{ii} = \frac{1}{(\mathbf{K}^{-1})_{ii}}$$

Any value is acceptable for Z_{ii}; it affects the value of the controller gain only for the respective controller.

Herein the decoupler will always be designed so that the elements on the diagonal are unity. To most, this would seem natural, but it is somewhat arbitrary. A more general design would be to force one element on each row and each column to be unity. The alternative is to change the subscripts assigned to each manipulated variable and each controlled variable and retain the requirement that the unity elements be on the diagonal. Only this approach is used herein.

P-Canonical Steady-State Decoupler. The P-canonical form of the decoupler is represented by the following equation:

$$\begin{bmatrix} \Delta M_1 \\ \Delta M_2 \end{bmatrix} = \begin{bmatrix} \Delta X_1 \\ \Delta X_2 \end{bmatrix} + \begin{bmatrix} 0 & G_{12}' \\ G_{21}' & 0 \end{bmatrix} \begin{bmatrix} \Delta M_1 \\ \Delta M_2 \end{bmatrix}$$
$$\mathbf{m} \quad = \quad \mathbf{x} \quad + \quad \mathbf{G}' \quad \mathbf{m}$$

The P-canonical decoupler matrix $\mathbf{G}'$ has zeros on the diagonal, but where interaction is present, the off-diagonal elements will be nonzero.

The design of the V-canonical decoupler led to the following relationship for the decoupler matrix **G**:

$$\mathbf{G} = \mathbf{K}^{-1}$$

The equation for the decoupler can also be expressed as follows:

$$\mathbf{m} = \mathbf{G}\mathbf{x} = \mathbf{K}^{-1}\mathbf{x}$$

or

$$\mathbf{K}\mathbf{m} = \mathbf{x}$$

The process gain matrix can be divided into two matrices, one containing the diagonal elements and the other containing the off-diagonal elements:

$$\mathbf{K}_0\mathbf{m} + \mathbf{K}_1\mathbf{m} = \mathbf{x}$$

where

$K_0 = KI$ = diagonal elements of the process gain matrix K

$K_1 = K - KI$ = off-diagonal elements of the process gain matrix K

This equation can be rearranged as follows:

$$K_0 m = x - K_1 m$$

Next, each row of this equation will be divided by the diagonal element of the process gain matrix K. The results are defined as follows:

- x' = vector x with each element divided by the diagonal element of the process gain matrix K:

$$X_i' = \frac{X_i}{K_{ii}}$$

- K_1' = matrix K_1 with each element divided by the diagonal element of the process gain matrix K:

$$(K_1')_{ij} = \frac{(K_1)_{ij}}{K_{ii}}$$

$$= \frac{K_{ij}}{K_{ii}} \quad \text{if } i \neq j$$

$$= 0 \quad \text{if } i = j$$

When each row of matrix K_0 is divided by the diagonal element of the process gain matrix K, the result is the identity matrix I. The resulting equation is the following:

$$m = x' - K_1' m$$

$$= x' + G' m$$

Technically, the controller outputs should be designated as X_i'. But the elements of vectors x and x' differ only by a coefficient. This affects only the value of the controller gain for the respective controller. As these must be established by the usual tuning procedures, the distinction between x and x' is of no significance. Hereafter, x will be used.

The decoupler matrix G' for the P-canonical decoupler is the negative of matrix K_1'. The elements of G' are defined by the following equations:

$$G_{ij}' = -\frac{K_{ij}}{K_{ii}} \quad \text{if } i \neq j$$

$$= 0 \quad \text{if } i = j$$

For the P-canonical decoupler, the elements for the decoupler can easily be computed from the elements of the process gain matrix $\mathbf{K}$. Unlike the V-canonical decoupler, no matrix inversion is required. However, two requirements are imposed on the process gain matrix $\mathbf{K}$:

- No diagonal element is zero ($K_{ii} \neq 0$ for all i).
- The determinant must be nonzero (all variables are independent).

Pairing Controlled and Manipulated Variables. When single-loop controllers are applied to a 2×2 multivariable process, the proper pairing for the controlled and manipulated variables is crucial. But for a decoupler, this is not essential, at least for the normal control functions.

One's usual interpretation of either of the decoupler structures in Figure 8.3 is as follows:

- PID controller 1 responds to changes in C_1 by changing input X_1 to the decoupler.
- The decoupler changes M_1 in exactly the same manner as input X_1 changes.
- To offset the effect of the changes in M_1 on C_2, the decoupler makes small changes in M_2.
- The net result is that the change in X_1 primarily affects C_1, with little or no effect on C_2.

Intuitively, it seems that the straight-through paths in the decoupler should be more significant that the diagonal paths. This is not necessarily correct. If necessary to offset the effect of the changes in M_1 on C_2, the decoupler will make larger changes in M_2.

Suppose that the effect of M_2 on C_1 is larger than the effect of M_1 on C_1. For a change in X_1, the decoupler will make larger changes in M_2 than it makes in M_1. The primary control path is via the diagonal elements of both the decoupler and the process. That is, the principal control path is from X_1 to M_2 to C_1, not X_1 to M_1 to C_1. However, if the manipulated and controlled variables are paired properly, the principal control path will be via the straight-through paths instead of the diagonal paths. Although the normal control functions do not depend on the proper pairing, other issues must be considered:

- Impact of switching one of the decoupler outputs to manual
- Impact of a decoupler output being driven to a limit

These make the proper pairing very desirable, and possibly mandatory.

V-Canonical Decoupler for the Purified Water Process. In accordance with the control configuration in Figure 7.6 that was originally proposed for the purified water process, the manipulated and controlled variables were designated as follows:

Manipulated Variable	Controlled Variable
M_1: recirculation valve opening (%)	C_1: recirculation pressure (psig)
M_2: recirculation pump speed (rpm)	C_2: recirculation flow (gpm)

For a user demand of 380 gpm of purified water, the values for the elements in the process gain matrix $\mathbf{K}$ are as follows:

$$\mathbf{K} = \begin{bmatrix} -0.636 \text{ psig/\%} & 0.0495 \text{ psig/rpm} \\ 1.132 \text{ gpm/\%} & 0.0121 \text{ gpm/rpm} \end{bmatrix}$$

The inverse $\mathbf{K}^{-1}$ of the process gain matrix $\mathbf{K}$ is

$$\mathbf{K}^{-1} = \begin{bmatrix} -0.190\%/\text{psig} & 0.777\%/\text{gpm} \\ 17.8 \text{ rpm/psig} & 9.98 \text{ rpm/gpm} \end{bmatrix}$$

To obtain a decoupler matrix $\mathbf{G}$ with unity diagonal elements, each column of the inverse $\mathbf{K}^{-1}$ is divided by the diagonal element on that column:

$$\mathbf{G} = \begin{bmatrix} 1 & 0.0778\%/\text{rpm} \\ -93.6 \text{ rpm/\%} & 1 \end{bmatrix}$$

When the elements of the decoupler matrix $\mathbf{G}$ are in engineering units, wide variations can occur for the magnitudes of the elements, as is the case for this example. In order to attach any significance to the magnitudes of the elements, each must be converted to %/%. Basically, the issues are the same as for K_C (in %/%) vs. $K_{C,\text{EU}}$ (in engineering units).

For the decoupler, ranges are only required for the manipulated variables. For this example, the following will be used:

Manipulated Variable	Range
Recirculation valve opening	0 to 100%
Recirculation pump speed	0 to 4000 rpm

For those decoupler elements whose units are %/rpm, multiply by 40 to obtain the value in %/%. For those decoupler elements whose units are rpm/%, divide by 40 to obtain the value in %/%. The resulting decoupler matrix $\mathbf{G}$ is as follows:

$$\mathbf{G} = \begin{bmatrix} 1 & 3.11\%/\% \\ -2.34\%/\% & 1 \end{bmatrix}$$

The magnitudes of the off-diagonal elements (G_{12} and G_{21}) are larger than the straight-through elements. For an increase of 1% in input X_1, the decoupler increases M_1 by 1% but decreases M_2 by 2.34%. For an increase of 1% in input

X_2, the decoupler increases M_2 by 1% but increases M_1 by 3.11%. Basically, the decoupler is relying primarily on M_2 to control C_1 and on M_1 to control C_2.

For each loop the gain of the decoupler-process combination is the reciprocal of the diagonal element of the inverse $\mathbf{K}^{-1}$ of the process gain matrix $\mathbf{K}$:

$$Z_{11} = \frac{1}{-0.190\%/\text{psig}} = -5.26 \text{ psig}/\%$$

$$Z_{22} = \frac{1}{9.98 \text{ rpm/gpm}} = 0.100 \text{ gpm/rpm} = 4.00 \text{ gpm}/\%$$

These results are repeated as option 1 in Table 8.1. To obtain option 2, the assignments for M_1 and M_2 are swapped to give the following designations:

Controlled Variable	Manipulated Variable
C_1: recirculation pressure	M_1: recirculation pump speed
C_2: recirculation flow	M_2: recirculation valve

Table 8.1   V-canonical decoupler for the purified water process

	Option 1	Option 2
Controlled variables **c**	C_1: recirculation pressure (psig) C_2: recirculation flow (gpm)	C_1: recirculation pressure (psig) C_2: recirculation flow (gpm)
Manipulated variables **m**	M_1: valve opening (%) M_2: pump speed (rpm)	M_1: pump speed (rpm) M_2: valve opening (%)
Process gain matrix **K**	$\begin{bmatrix} -0.636 \text{ psig}/\% & 0.0495 \text{ psig/rpm} \\ 1.132 \text{ gpm}/\% & 0.0121 \text{ gpm/rpm} \end{bmatrix}$	$\begin{bmatrix} 0.0495 \text{ psig/rpm} & -0.636 \text{ psig}/\% \\ 0.0121 \text{ gpm/rpm} & 1.132 \text{ gpm}/\% \end{bmatrix}$
Inverse **K**$^{-1}$	$\begin{bmatrix} -0.190\%/\text{psig} & 0.777\%/\text{gpm} \\ 17.8 \text{ rpm/psig} & 9.98 \text{ rpm/gpm} \end{bmatrix}$	$\begin{bmatrix} 17.8 \text{ rpm/psig} & 9.98 \text{ rpm/gpm} \\ -0.190\%/\text{psig} & 0.777\%/\text{gpm} \end{bmatrix}$
Decoupler inputs **x**	X_1: valve opening (%) X_2: pump speed (rpm)	X_1: pump speed (rpm) X_2: valve opening (%)
Decoupler matrix **G**	$\begin{bmatrix} 1 & 0.0778\%/\text{rpm} \\ -93.6 \text{ rpm}/\% & 1 \end{bmatrix}$ $\begin{bmatrix} 1 & 3.11\%/\% \\ -2.34\%/\% & 1 \end{bmatrix}$	$\begin{bmatrix} 1 & 12.8 \text{ rpm/gpm} \\ -0.0107\%/\text{rpm} & 1 \end{bmatrix}$ $\begin{bmatrix} 1 & 0.320\%/\% \\ -0.428\%/\% & 1 \end{bmatrix}$
Decoupler-process gain	$Z_{11} = -5.26 \text{ psig}/\%$ $Z_{22} = 0.100 \text{ gpm/rpm} = 4.00 \text{ gpm}/\%$	$Z_{11} = 0.0562 \text{ psig/rpm} = 2.25 \text{ psig}/\%$ $Z_{22} = 1.29 \text{ gpm}/\%$

The entries for option 2 are obtained as follows:

Process gain matrix **K**. To reflect swapping M_1 and M_2, the columns of the process gain matrix **K** for option 1 must be swapped to obtain the process gain matrix **K** for option 2.

Inverse $\mathbf{K}^{-1}$ *of the process gain matrix* **K**. Swapping columns in the process gain matrix **K** results in the rows being swapped in the inverse $\mathbf{K}^{-1}$.

Decoupler matrix **G**. Dividing each column of the inverse $\mathbf{K}^{-1}$ by the diagonal element on that column gives the decoupler matrix **G**.

Gains of decoupler-process combination. Each gain is the reciprocal of the diagonal element of the inverse $\mathbf{K}^{-1}$ of the process gain matrix **K**.

For option 2, the magnitudes (in %/%) of the off-diagonal elements (G_{12} and G_{21}) are smaller than the straight-through elements. For an increase of 1% in input X_1, the decoupler increases M_1 by 1% and decreases M_2 by 0.428%. For an increase of 1% in input X_2, the decoupler increases M_2 by 1% and increases M_1 by 0.320%. The decoupler is relying primarily on M_1 to control C_1 and on M_2 to control C_2. Intuitively, this seems more logical.

The values of the decoupler-process gains Z_{11} and Z_{22} are different for the two options. Consequently, the appropriate values for the controller gains will be different. However, if the product $K_{C,\text{EU}} Z_{ii}$ for each loop is the same for the two options, the performance will be exactly the same. However, there are two caveats to this statement:

- Both decoupler outputs are on automatic.
- No limiting condition has been encountered.

These issues will be examined shortly.

P-Canonical Decoupler for the Purified Water Process. In accordance with the control configuration in Figure 7.6 proposed originally for the purified water process, the manipulated and controlled variables were designated as follows:

Manipulated Variable	Controlled Variable
M_1: recirculation valve opening (%)	C_1: recirculation pressure (psig)
M_2: recirculation pump speed (rpm)	C_2: recirculation flow (gpm)

The process gain matrix **K** is as follows:

$$\mathbf{K} = \begin{bmatrix} -0.636 \text{ psig}/\% & 0.0495 \text{ psig/rpm} \\ 1.132 \text{ gpm}/\% & 0.0121 \text{gpm/rpm} \end{bmatrix}$$

The decoupler matrix **G**′ is computed from the process gain matrix **K** as follows:

$$\mathbf{G}' = \begin{bmatrix} 0 & -K_{12}/K_{11} = 0.0778\%/\text{rpm} \\ -K_{21}/K_{22} = -93.6 \text{ rpm}/\% & 0 \end{bmatrix}$$

For an output range of 0 to 4000 rpm for the pump speed, the decoupler matrix $\mathbf{G}'$ in %/% is as follows:

$$\mathbf{G}' = \begin{bmatrix} 0 & 3.11\%/\% \\ -2.34\%/\% & 0 \end{bmatrix}$$

The magnitudes of the off-diagonal elements (G'_{12} and G'_{21}) are greater than 1. For an increase of 1% in input X_1, the decoupler increases M_1 by 1%. This change in M_1 is multiplied by G'_{21}, which decreases M_2 by 2.34%. For an increase of 1% in input X_2, the decoupler increases M_2 by 1%. This change in M_2 is multiplied by G'_{12}, which increases M_1 by 3.11%. Again, the decoupler is relying primarily on M_2 to control C_1 and on M_1 to control C_2.

These results are repeated as option 1 in Table 8.2. To obtain option 2, the assignments for M_1 and M_2 are swapped to give the following designations:

Manipulated Variable	Controlled Variable
M_1: recirculation pump speed	C_1: recirculation pressure
M_2: recirculation valve	C_2: recirculation flow

The entries for option 2 are obtained as follows:

Process gain matrix $\mathbf{K}$. To reflect swapping M_1 and M_2, the columns of the process gain matrix $\mathbf{K}$ for option 1 must be swapped to obtain the process gain matrix $\mathbf{K}$ for option 2.

Decoupler matrix $\mathbf{G}'$. The off-diagonal elements are computed from the elements of the process gain matrix $\mathbf{K}$.

Table 8.2 P-canonical decoupler for the purified water process

	Option 1	Option 2
Controlled variables **c**	C_1: recirculation pressure (psig) C_2: recirculation flow (gpm)	C_1: recirculation pressure (psig) C_2: recirculation flow (gpm)
Manipulated variables **m**	M_1: valve opening (%) M_2: pump speed (rpm)	M_1: pump speed (rpm) M_2: valve opening (%)
Process gain matrix **K**	$\begin{bmatrix} -0.636 \text{ psig}/\% & 0.0495 \text{ psig/rpm} \\ 1.132 \text{ gpm}/\% & 0.0121 \text{ gpm/rpm} \end{bmatrix}$	$\begin{bmatrix} 0.0495 \text{ psig/rpm} & -0.636 \text{ psig}/\% \\ 0.0121 \text{ gpm/rpm} & 1.132 \text{ gpm}/\% \end{bmatrix}$
Decoupler inputs **x**	X_1: valve opening (%) X_2: pump speed (rpm)	X_1: pump speed (rpm) X_2: valve opening (%)
Decoupler matrix **G**′	$\begin{bmatrix} 0 & 0.0778\%/\text{rpm} \\ -93.6 \text{ rpm}/\% & 1 \end{bmatrix}$ $\begin{bmatrix} 0 & 3.11\%/\% \\ -2.34\%/\% & 0 \end{bmatrix}$	$\begin{bmatrix} 0 & 12.8 \text{ rpm}/\% \\ -0.0107\%/\text{rpm} & 0 \end{bmatrix}$ $\begin{bmatrix} 0 & 0.320\%/\% \\ -0.428\%/\% & 0 \end{bmatrix}$

For option 2, the magnitudes (in %/%) of the off-diagonal elements (G'_{12} and G'_{21}) are less than 1. For an increase of 1% in input X_1, the decoupler increases M_1 by 1%, which is then multiplied by G'_{21} to decrease M_2 by 0.428%. For an increase of 1% in input X_2, the decoupler increases M_2 by 1%, which is then multiplied by G'_{12} to increase M_1 by 0.320%. The decoupler is relying primarily on M_1 to control C_1 and on M_2 to control C_2.

The performance of the V-canonical decoupler and the P-canonical decoupler are exactly the same, but again with the following two caveats:

- Both decoupler outputs are on automatic.
- No limiting condition has been encountered.

These issues will be examined shortly.

Implementation of a V-Canonical Decoupler. For a 2×2 process the decoupler matrix **G** is as follows:

$$\mathbf{G} = \begin{bmatrix} 1 & G_{12} \\ G_{21} & 1 \end{bmatrix}$$

The implementation will be based on the following equations for the decoupler:

$$\Delta M_1 = \Delta X_1 + G_{12}\Delta X_2$$

$$\Delta M_2 = G_{21}\Delta X_1 + \Delta X_2$$

The changes ΔM_i and ΔX_i can be interpreted either as an incremental change or as a change from their respective initial values. The decoupler implementation can be based on either, but only the latter is presented here.

When interpreted as changes from their respective initial values, ΔM_i and ΔX_i are the current values of M_i and X_i minus their respective reference or initial values. When the decoupler output is switched from manual to automatic, the values of M_i and X_i are retained and designated as $M_{0,i}$ and $X_{0,i}$. With this approach, ΔM_i and ΔX_i are defined as follows:

$$\Delta M_i = M_i - M_{0,i}$$

$$\Delta X_i = X_i - X_{0,i}$$

Substituting these into the decoupler equations expressed in terms of ΔM_i and ΔX_i gives the following equations expressed in terms of M_i and X_i:

$$M_1 - M_{0,1} = (X_1 - X_{0,1}) + G_{12}(X_2 - X_{0,2})$$

$$M_2 - M_{0,2} = G_{21}(X_1 - X_{0,1}) + (X_2 - X_{0,2})$$

These equations can be further rearranged to the following:

$$M_1 = X_1 + G_{12}X_2 + (M_{0,1} - X_{0,1} - G_{12}X_{0,2})$$
$$M_2 = G_{21}X_1 + X_2 + (M_{0,2} - G_{21}X_{0,1} - X_{0,2})$$

The term $M_{0,1} - X_{0,1} - G_{12}X_{0,2}$ can be viewed as a bias on the decoupler output for M_1. This bias will be designated as $M_{R,1}$, and the corresponding bias for M_2 as $M_{R,2}$. The equations can be written as follows:

$$M_1 = X_1 + G_{12}X_2 + M_{R,1}$$
$$M_2 = G_{21}X_1 + X_2 + M_{R,2}$$

Bias $M_{R,1}$ is only recomputed when the output M_1 is on manual and the controller for X_1 is being initialized. Consequently, $M_{0,1} = M_1$, $X_{0,1} = X_1$, and $X_{0,2} = X_2$. Similar statements apply to the bias $M_{R,2}$, so each bias can be computed from current values of M_1, M_2, X_1, and X_2:

$$M_{R,1} = M_1 - X_1 - G_{12}X_2$$
$$M_{R,2} = M_2 - G_{21}X_1 - X_2$$

The equation for most summer function blocks includes a bias. However, the value of the bias in the summer equation is often a configuration coefficient whose value cannot be changed except through the configuration tools. For the decoupler, the value of both $M_{R,1}$ and $M_{R,2}$ must be determined as part of the initialization logic. Consequently, an input to each summer must usually be configured for the respective bias.

Figure 8.4 presents the P&I diagram for an implementation of a V-canonical decoupler for the purified water process. The configuration is based on option 2 in Table 8.1. In addition to the two controllers, the following additional blocks are provided:

- Hand stations that permit the operator to switch either or both decoupler outputs to manual.
- Summers to compute the decoupler outputs. To compute the decoupler output for M_1, a summer with three inputs is configured as follows:
  - *Input X1*: Output X_1 of the pressure controller; coefficient is 1.0.
  - *Input X2*: Output X_2 of the flow controller; coefficient is G_{12}.
  - *Input X3*: Bias $M_{R,1}$; coefficient is 1.0.

  A similar summer is required to compute output M_2.

Computing values for the biases $M_{R,1}$ and $M_{R,2}$ will be discussed shortly.

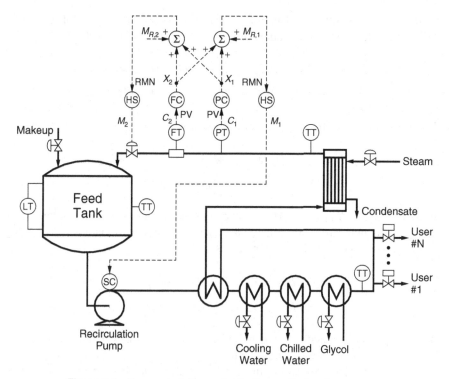

Figure 8.4 V-canonical decoupler for the purified water process.

Implementation of a P-Canonical Decoupler. For a 2×2 process the equation for the decoupler matrix $\mathbf{G}'$ is as follows:

$$\begin{bmatrix} \Delta M_1 \\ \Delta M_2 \end{bmatrix} = \begin{bmatrix} \Delta X_1 \\ \Delta X_2 \end{bmatrix} + \begin{bmatrix} 0 & G'_{12} \\ G'_{21} & 0 \end{bmatrix} \begin{bmatrix} \Delta M_1 \\ \Delta M_2 \end{bmatrix}$$

The corresponding equations for the decoupler are as follows:

$$\Delta M_1 = \Delta X_1 + G'_{12} \Delta M_2$$
$$\Delta M_2 = \Delta X_2 + G'_{21} \Delta M_1$$

As for the V-canonical form of the decoupler, ΔM_i and ΔX_i can be interpreted either as an incremental change or as a change from their respective initial values. When interpreted as changes from their respective initial values, ΔM_i and ΔX_i are the current values of M_i and X_i minus their respective reference or initial values $M_{0,i}$ and $X_{0,i}$:

$$\Delta M_i = M_i - M_{0,i}$$
$$\Delta X_i = X_i - X_{0,i}$$

Substituting these into the decoupler equations expressed in terms of ΔM_i and ΔX_i gives the following equations expressed in terms of M_i and X_i:

$$M_1 - M_{0,1} = (X_1 - X_{0,1}) + G'_{12}(M_2 - M_{0,2})$$
$$M_2 - M_{0,2} = (X_2 - X_{0,2}) + G'_{21}(M_1 - M_{0,1})$$

These equations can be rearranged to the following:

$$M_1 = X_1 + G'_{12}M_2 + (M_{0,1} - X_{0,1} - G'_{12}M_{0,2})$$
$$M_2 = X_2 + G'_{21}M_1 + (M_{0,2} - G'_{21}M_{0,1} - X_{0,2})$$

The term $M_{0,1} - X_{0,1} - G'_{12}M_{0,2}$ can be viewed as a bias on the decoupler output for M_1. This bias will be designated as $M'_{R,1}$, and the corresponding bias for M_2 as $M'_{R,2}$. The equations can be written as follows:

$$M_1 = X_1 + G'_{12}M_2 + M'_{R,1}$$
$$M_2 = X_2 + G'_{21}M_1 + M'_{R,2}$$

Bias $M'_{R,1}$ is recomputed only when the output M_1 is on manual and the controller for X_1 is being initialized. Consequently, $M_{0,1} = M_1$, $X_{0,1} = X_1$, and $X_{0,2} = X_2$. Similar statements apply to the bias $M'_{R,2}$, so each bias can be computed from current values of M_1, M_2, X_1, and X_2:

$$M'_{R,1} = M_1 - X_1 - G'_{12}M_2$$
$$M'_{R,2} = M_2 - G'_{21}M_1 - X_2$$

Figure 8.5 presents the P&I diagram for implementation of a P-canonical decoupler for the purified water process. The configuration is based on option 2 in Table 8.2. In addition to the two controllers, the following additional blocks are provided:

- Hand stations that permit the operator to switch either or both decoupler outputs to manual.
- Summers to compute the decoupler outputs. To compute the decoupler output M_1, a summer with three inputs is configured as follows:
  - *Input X1*: Output X_1 of the pressure controller; coefficient is 1.0.
  - *Input X2*: Output X_2 of the hand station for M_2; coefficient is G'_{12}.
  - *Input X3*: Bias $M'_{R,1}$; coefficient is 1.0.

  A similar summer is required to compute output M_2.

Computing values for the biases $M'_{R,1}$ and $M'_{R,2}$ will be discussed shortly.

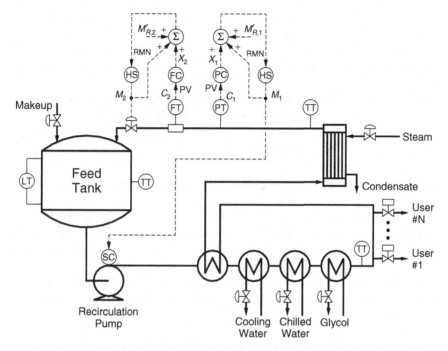

Figure 8.5 P-canonical decoupler for the purified water process.

Single-Loop Configurations. Suppose that one of the following occurs:

- Decoupler output M_2 is switched to manual.
- Decoupler output M_2 is driven to a limit.

Only the value of output M_1 will be at the discretion of the decoupler. Either C_1 or C_2, but not both, can be maintained at its respective target.

Using the purified water process as an example, suppose that the process operator switches the recirculation pump hand station to manual. The decoupler can now only change the recirculation valve opening. Either the recirculation pressure can be controlled or the recirculation flow can be controlled, but not both. When the hand station is switched to manual, the configuration of the tracking logic determines which variable will be controlled:

- Tracking is activated in the recirculation pressure controller. The recirculation flow controller continues to function.
- Tracking is activated in the recirculation flow controller. The recirculation pressure controller continues to function.

Applying the relative gain to the purified water process suggested the following pairing:

- Control the recirculation pressure with the recirculation pump speed.
- Control the recirculation flow with the recirculation valve opening.

In keeping with this pairing, the behavior of the decoupler is to be as follows:

- If the recirculation pump hand station is switched to local, the recirculation flow is to be controlled with the recirculation valve opening.
- If the recirculation valve hand station is switched to local, the recirculation pressure is to be controlled with the recirculation pump speed.

Pump Speed on Manual, V-Canonical Decoupler. Figure 8.6(a) presents the V-canonical decoupler for the original designations (option 1 in Table 8.1) for the controlled and manipulated variables. With the pump speed output on manual, the single-loop configuration would be represented as in Figure 8.6(a). The configuration works perfectly well, but it seems more logical to control via the straight-through paths rather than via the diagonal paths.

Whether the control is via the straight-through paths or the diagonal paths depends on the subscripts assigned to the controlled and manipulated variables. There are two possibilities:

Swap designations for manipulated variables. For Figure 8.6(b), the subscripts assigned to the manipulated variables have been swapped (option 2 in Table 8.1).

Swap designations for controlled variables. For Figure 8.6(c), the subscripts assigned to the controlled variables have been swapped.

In both cases, the same numerical subscript is used for the recirculation valve opening and the recirculation flow. Control is via the straight-through paths, not via the diagonal paths.

Pump Speed on Manual, P-Canonical Decoupler. The P-canonical decoupler is normally implemented as in Figure 8.7. The following details are significant:

- The input to decoupler element G'_{12} is the output of the hand station for M_2, not the corresponding output of the decoupler.
- The input to decoupler element G'_{21} is the output of the hand station for M_1, not the corresponding output of the decoupler.

The P&I diagram in Figure 8.5 for the purified water process is also configured in this manner.

When implemented in this manner, the P-canonical decoupler is not so forgiving with regard to pairing. The structure in Figure 8.7 imposes the following limitations:

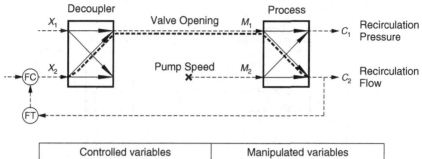

Controlled variables	Manipulated variables
C_1 – Recirculation pressure, psig	M_1 – Valve opening, %
C_2 – Recirculation flow, gpm	M_2 – Pump speed, rpm

(a)

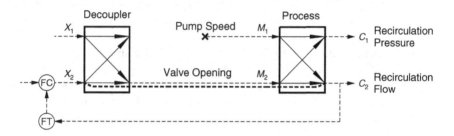

Controlled variables	Manipulated variables
C_1 – Recirculation pressure, psig	M_1 – Pump speed, rpm
C_2 – Recirculation flow, gpm	M_2 – Valve opening, %

(b)

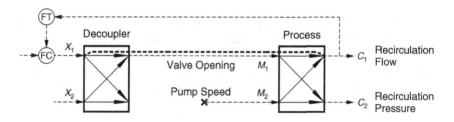

Controlled variables	Manipulated variables
C_1 – Recirculation flow, gpm	M_1 – Valve opening, %
C_2 – Recirculation pressure, psig	M_2 – Pump speed, rpm

(c)

Figure 8.6 Single-loop options when the pump speed is not at the discretion of the decoupler: (a) Option 1; (b) Option 2 (manipulated variables swapped); (c) Option 3 (controlled variables swapped).

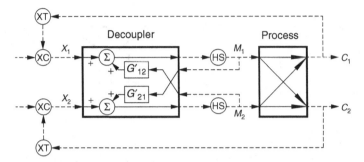

Figure 8.7   Hand stations with the P-canonical form of the decoupler.

- If the hand station for M_1 is switched to local, decoupler input X_1 has no effect on either M_1 or M_2. Consequently, tracking must be activated in the controller for C_1. Control of C_2 continues, and should the operator change the value of M_1, the decoupler will compensate the value of M_2 for this change.
- If the hand station for M_2 is switched to local, decoupler input X_2 has no effect on either M_1 or M_2. Consequently, tracking must be activated in the controller for C_2. Control of C_1 continues, and should the operator change the value of M_2, the decoupler will compensate the value of M_1 for this change.

Consider the options in Table 8.2 for the P-canonical decoupler in light of the requirement that the recirculation flow is to be controlled with the recirculation valve opening when the pump speed hand station is switched to local. This is not possible with option 1 (C_1 for the recirculation pressure and M_2 for the recirculation pump speed). For the P-canonical decoupler to meet this requirement, proper pairing is mandatory, the options being as follows:

1. *Swap designations for manipulated variables.* Use C_1 for the recirculation pressure and M_1 for the recirculation pump speed. This is option 2 in Table 8.2.
2. *Swap designations for controlled variables.* Use C_2 for the recirculation pressure and M_2 for the recirculation pump speed.

Initialization Calculations, V-Canonical Decoupler. When the hand station for M_1 is on local, what should be the value of X_1? Two possible answers are:

- $X_1 = M_1$. The controller is initialized so that its output X_1 is the same as the output M_1 to the process. To provide a smooth transfer from local to remote, the value of the bias $M_{R,1}$ must be computed as follows:

$$M_{R,1} = M_1 - X_1 - G_{12}X_2$$

- $X_1 = M_1 - G_{12}X_2$. When initialized in this manner, the value of the bias $M_{R,1}$ will be zero, which permits the input to the summer for the bias to be removed. However, X_1 could be significantly different from M_1, which would require some explanation for the process engineers, technicians, and operators.

To most, initializing using $X_1 = M_1$ (and $X_2 = M_2$) seems more reasonable, and this approach will be used hereafter.

With the hand station on remote, the value of output M_1 is computed using the following equation:

$$M_1 = X_1 + G_{12}X_2 + M_{R,1}$$

The difference between X_1 and M_1 depends on the term $G_{12}X_2$. As the value of X_2 changes from the value used in the computation of $M_{R,1}$, the difference between X_1 and M_1 increases.

The initialization occurs when the hand station for M_1 is on local. Output tracking is activated in the controller for C_1 as follows:

Input TRKMN: True when output RMT from the hand station for M_1 is false.

Input MNI: Output of the hand station for M_1.

When input TRKMN changes from false to true, the value of X_1 changes abruptly from its current value to the value of M_1.

This raises an issue with regard to the output for M_2. If the hand station for M_2 is on remote, the value for M_2 is being computed by the following equation:

$$M_2 = G_{21}X_1 + X_2 + M_{R,2}$$

If there is an abrupt change in X_1, the value for M_2 will also change abruptly, the change in M_2 being G_{21} times the change in X_1. Consequently, switching the hand station for M_1 from remote to local could cause an abrupt change in the output for M_2. This is not acceptable and must be prevented.

In a similar manner, M_1 would be affected should the hand station for M_2 be switched from remote to manual. To prevent an abrupt change in M_1 on the sampling instant that the hand station for M_2 is switched from remote to local, the value of the bias $M_{R,1}$ must be recomputed using the following equation:

$$M_{R,1} = M_1 - X_1 - G_{21}X_2$$

This computation must be performed under two conditions:

- All sampling instants on which the hand station for M_1 is on local. On these sampling instants, output tracking is active in the controller whose output is X_1, which causes the controller output X_1 to be set equal to the current value of M_1. The bias calculation could be simplified to

$$M_{R,1} = -G_{21}X_2$$

- On the sampling instant that the hand station for M_2 is switched from remote to local. In this case, X_1 is not necessarily equal to M_1, so the equation for the bias cannot be simplified.

Analogous logic is required for X_2 and $M_{R,2}$.

Initialization Calculations, P-Canonical Decoupler. Suppose that the process operator assumes manual control of output M_1 by switching the hand station for M_1 to local. What should be the value of X_1? Two possible answers are:

- $X_1 = M_1$. The value of the bias $M_{R,1}$ must be computed as follows:

$$M_{R,1} = M_1 - X_1 - G'_{12}M_2$$
$$= -G'_{12}M_2 \text{ since } X_1 = M_1$$

- $X_1 = M_1 - G'_{12}M_2$. The value of the bias $M'_{R,1}$ will be zero, but X_1 will probably be significantly different from M_1.

Most prefer initializing using $X_1 = M_1$ (and $X_2 = M_2$), so this approach will be used hereafter. This is achieved by configuring output tracking as follows for the controller whose output is X_1:

Input TRKMN: True when output RMT from the hand station for M_1 is false.

Input MNI: Output of the hand station for M_1.

When input TRKMN changes from false to true, the value of X_1 changes abruptly from its current value to the value of M_1. The issues regarding the computation of decoupler output M_2 are as follows:

V-canonical form of the decoupler. The decoupler calculations are based on X_1, so an abrupt change in X_1 will cause an abrupt change in M_2. To avoid this "bump," the value of the bias $M_{R,2}$ must be recomputed on the sampling instant that output tracking is activated in the controller for C_1.

P-canonical form of the decoupler. The decoupler calculations are based on M_1, so an abrupt change in X_1 has no effect on M_2. There is no need to recompute the bias $M'_{R,2}$.

In this respect, the P-canonical form of the decoupler is simpler to implement than the V-canonical form. The bias must be computed only when the hand station for the respective output is on local.

Computing the Bias. Only computing the bias for output M_1 will be described; the requirements for computing the bias for output M_2 are analogous.

The equations for computing the bias for output M_1 are as follows:

V-canonical form of the decoupler:

$$M_{R,1} = M_1 - X_1 - G_{12}X_2$$

P-canonical form of the decoupler:

$$M'_{R,1} = M_1 - X_1 - G'_{12}M_2$$
$$= -G'_{12}M_2$$

This computation is performed only when output tracking is active ($X_1 = M_1$), which permits the simplified form of the equation to be used.

One approach to computing the value for the bias is to use a summer. This raises the following issues:

- The value of the bias needs to be recomputed only when the hand station for M_1 is on local or (for the V-canonical decoupler) when the hand station for M_2 is switched from remote to local.
- The summer block normally executes continuously, which means that its output is recomputed on every sampling instant.

The possibilities depend on the features provided by the control system. If the bias is to be recomputed only then the hand station for M_1 is on local (as for the P-canonical decoupler), the possibilities include the following:

- Some systems allow a control block to be "turned on" and "turned off." If this capability is available, the summer should be "turned on" only when the hand station for M_1 is on local.
- A sample-and-hold block (or its equivalent) must be configured. There are two inputs to the sample-and-hold:
  - *Input X.* This is the output of the summer that is computing the equation for the bias.
  - *Input TRK.* This input must be true when the hand station for M_1 is not on remote (output RMT is false). The output Y is set equal to input X only when input TRK is true.

  The output of the sample-and-hold is the value of the bias.

For the V-canonical decoupler, the bias must also be recomputed on the sampling instant that tracking is activated in the controller for C_2. This occurs when the hand station for M_2 is switched from remote to local. The one-shot described in Chapter 1 can be used to do this. The input is output RMT of the hand station for M_2. If the one-shot is configured to detect a 1-to-0 transition, the output of the one-shot will be true for the scan on which the mode of the hand station changes from remote (1) to local (0).

The purpose of a one-shot is to cause certain calculations to be performed on one sampling instant only. But a word of caution: The sequence of the calculations is often crucial. For the initialization calculations required on the instant that the hand station for M_2 is switched from remote to local, the calculations must occur in the following order:

PID calculations for the flow controller (output is X_2). On the sampling instant that the hand station for M_2 is switched to local, output tracking is activated, which causes the value of X_2 to change abruptly to the current value of M_2.

Summer calculations for the bias $M_{R,1}$. The output of these calculations will reflect the change in the value of X_2 that occurs on the sampling instant that output tracking is activated in the flow controller.

Sample-and-hold for bias $M_{R,1}$. On the sampling instant that the hand station for M_2 switches from remote to local, the output of the one-shot will be true, making input TRK true. The output of the sample-and-hold will change to the value computed by the summer for the bias $M_{R,1}$.

Summer to calculate M_1. To produce the desired results, the following actions must previously occur on the same sampling instant:

- Change in X_2 as a result of output tracking being activated in the flow controller
- Change in bias $M_{R,1}$

These two changes must offset, resulting in no change for the value computed for M_1.

These issues do not arise for the P-canonical form of the decoupler, which makes its implementation much simpler.

Output Range and Limits for the Controllers. Each hand station must be configured with an output range and output limits that are appropriate for the final control element (pump speed or recirculation valve opening for the purified water process). For the controllers, one's first impulse is to use the output range and output limits of the hand station as the output range and output limits for the controller. However, the output limits for the hand station cannot be translated to equivalent output limits for the controller (terms such as $G'_{12}M_2$ do not have fixed values). Basically, the output limits must be applied by the hand station, not by the controller. The solution:

- Specify wide output limits for the controller so that they are never encountered.

- Activate windup protection in the controller upon attaining an output limit for the hand station.

For controllers that require an output range (most do, so K_C can be in $\%/\%$), whether or not the output range for the controller can be the same as the output range for the hand station depends on the restrictions imposed on the values for the output limits:

> *No restrictions are imposed.* The output range for the controller can be the same as the output range for the hand station.
>
> *Maximum overrange is imposed.* The output limits are normally used to provide an overrange on the controller output. Some systems impose a maximum on the overrange that can be provided. If so, the specifications for the output range must permit the required values for the output limits to be specified. The simplest is to make the output range the same as the output limits.

For this application, it would be preferable to express the controller gain in engineering units and dispense with the output range and output limits for the controller. However, this approach is incompatible with the PID block as implemented in most control systems.

Output Limits, V-Canonical Decoupler. For the purified water loop, the maximum flow that can be delivered is when the recirculation pump is running at full speed. Suppose that users request more than this amount of water? The recirculation pressure will drop below its set point, causing the recirculation pressure controller to attempt to increase its output. If this is allowed to happen, there are two consequences:

- Windup occurs in the recirculation pressure controller.
- The increases in output X_2 are not implemented (the pump speed is at the upper limit), but the decoupler compensates output M_1 for all changes in X_2.

The upper output limit on the pump speed must be imposed at the hand station for M_1. Let X_{M1} be the value of X_1 that corresponds to the current value of M_1. This value can be back-calculated from the current value of output M_1, the current value of X_2, and the bias $M_{R,1}$:

$$X_{M1} = M_1 - G_{12}X_2 - M_{R,1}$$

When no limit is being imposed on M_1, X_{M1} will equal X_1. But if a limit is being imposing on M_1, X_{M1} will not equal X_1.

Unless protection is provided, windup in the recirculation pressure controller will occur when a limiting condition is encountered on M_1. Normally, such windup is prevented using either integral tracking, external reset, or inhibit increase/inhibit decrease:

	Input	Configuration
Integral tracking	TRKMR	Logical OR of output QH and output QL of the hand station for M_1. Integral tracking is active when M_1 is at either output limit.
	MRI	Value of X_{M1}.
External reset	XRS	Value of X_{M1}.
Inhibit increase/inhibit decrease	NOINC	Output QH of the hand station for M_1. When M_1 is at the upper output limit, the pressure controller is not permitted to increase its output.
	NODEC	Output QL of the hand station for M_1. When M_1 is at the lower output limit, the pressure controller is not permitted to decrease its output.

There is one other issue to be addressed: the compensation of output M_2 for changes in X_1. In Figure 8.5 this compensation is based on the output X_1 of the recirculation pressure controller. But when a limit is being imposed on M_1, the value of X_1 does not correspond to the current value of M_1. Instead, the compensation of M_2 should be based on X_{M1}. If no limit is being applied, X_1 and X_{M1} are equal, but if a limit is being applied, X_{M1} corresponds to the current value of output M_1.

As more of the practical issues are addressed, the complexity of the implementation of the V-canonical decoupler is increasing rapidly. The P-canonical form of the decoupler addresses limits imposed on the outputs in a far simpler manner than the V-canonical form of the decoupler. When limits are imposed on the outputs (and they usually are), the P-canonical form is usually preferable.

Output Limits, P-Canonical Decoupler. As noted in the discussion on the V-canonical decoupler, the maximum flow that can be delivered to the users is when the recirculation pump is running at full speed. Should users request more than this amount of water, the recirculation pressure will drop below its set point, causing the recirculation pressure controller to attempt to increase its output. For the P-canonical decoupler, there is only one consequence: that windup occurs in the recirculation pressure controller. Even if this controller is permitted to increase its output X_1, the limits applied in the hand station do not permit the output M_1

to increase. For the P-canonical decoupler, the equation for compensating M_2 is based on M_1 instead of X_1. Consequently, the compensation provided by the decoupler is not affected even if output X_1 continues to increase.

The upper output limit on the pump speed must still be imposed at the hand station for M_1. Let X'_{M1} be the value of X_1 that corresponds to the current value of M_1. This value can be back-calculated from the current value of output M_1, the current value of M_2, and the bias $M'_{R,1}$:

$$X'_{M1} = M_1 - G'_{12}M_2 - M'_{R,1}$$

When no limit is being imposed on M_1, the values of X_1 and X'_{M1} will be the same. But if a limit is being imposed, the two values will not be equal.

Windup in the recirculation pressure controller is prevented using either integral tracking, external reset, or inhibit increase/inhibit decrease:

	Input	Configuration
Integral tracking	TRKMR	Logical OR of output QH and output QL of the hand station for M_1. Integral tracking is active when M_1 is at either output limit.
	MRI	Value of X'_{M1}.
External reset	XRS	Value of X'_{M1}.
Inhibit increase/inhibit decrease	NOINC	Output QH of the hand station for M_1. When M_1 is at the upper output limit, the pressure controller is not permitted to increase its output.
	NODEC	Output QL of the hand station for M_1. When M_1 is at the lower output limit, the pressure controller is not permitted to decrease its output.

With the P-canonical decoupler, the compensation of output M_2 is based on M_1. Consequently, neither X_1 nor X_{M1} has any effect on the compensation of output M_2. This is a major advantage of the P-canonical decoupler over the V-canonical decoupler.

Outputs to the Controller Set Points. An output from a decoupler can be to either of the following:

Final control element. These are usually via a hand station.

Set point of a PID controller (usually, a flow controller). If the controller is external (such as the speed controller for the recirculation pump), a hand station is usually included. But if the output is to a flow controller configured within the digital control system, the hand station is omitted and the remote/local switch within the PID controller is used in lieu of the hand station.

Although the decoupler can be implemented with either approach, a strong argument can be made for the decoupler to output to the set point of a flow controller whenever possible. The coefficients within the decoupler must reflect the characteristics of the process. When the decoupler outputs to a final control element (especially a control valve), the characteristics of the final control element are part of the process characteristics. Whenever maintenance work is performed on the final control element, the characteristics can change abruptly. This would degrade the performance of the decoupler. But when the decoupler outputs to a set point of a flow controller, the flow controller isolates the decoupler from the characteristics of the final control element.

Dynamics. The issues pertaining to dynamics for a decoupler are comparable to the issues pertaining to dynamics in a feedforward control configuration. One should always begin with a steady-state decoupler and add dynamics only if warranted. Dynamics are required only for the diagonal paths of the decoupler, not in the straight-thorough paths. The usual approach is to insert a lead-lag element, although occasionally, a dead-time element is also appropriate. The tuning approach is analogous to that for a feedforward control configuration. Obviously, as the dimension of the decoupler increases, the tuning endeavor becomes increasingly more challenging. Fortunately, some of the diagonal paths do not require dynamic compensation.

8.2. DEAD-TIME COMPENSATION

In the mid-to-late 1960s, every paper company was actively pursuing the application of dead-time compensation to the control of sheet weight and moisture. At the time, this was unproven technology, so why were they all in hot pursuit? For commodity papers (newsprint, linerboard for cardboard boxes, tissue paper, etc.), the margins are small. The potential improvement by a successful application of dead-time compensation could exceed the profit margin. If your competitor succeeds and you do not, you have a big, big problem! Most efforts encountered difficulties along the way but eventually became both technical and economic successes.

Paper Machine. Figure 8.8 presents a simplified representation of a Fourdrinier paper machine. Dating from the early 1800s, variations of this design account for

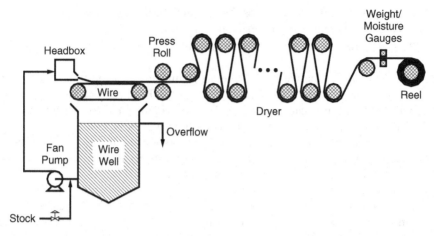

Figure 8.8 Simplified representation of a Fourdrinier paper machine.

almost all paper manufactured today. Dating from 1858, the Beloit Corporation, formerly the Beloit Iron Works, is the leading manufacturer of papermaking machinery. The paper machine is an impressive piece of mechanical equipment, but the process aspects are rather simple. The key attributes are as follows:

Fan pump. Fresh stock (wood fiber in water slurry) consisting of about 2% fiber is fed into the suction of the fan pump, where it is diluted to about 0.25% fiber before entering the headbox. The wire well, the fan pump, the headbox, and the moving wire constitute a large recirculation loop for the water.

Headbox. The headbox uniformly distributes the large flow of diluted stock over the moving wire, which is no simple task (most paper machines are over 20 ft in width).

Moving wire. The dilute slurry jets out of the headbox lip at approximately the same speed as the moving wire (the difference in speeds is the rush/drag ratio). Most of the fiber is retained on the wire to form the sheet; most of the water drains through and into the wire well.

Press rolls. Pressure applied by the press rolls removes even more water from the sheet. However, the sheet leaving the press rolls is still mostly water (by weight).

Dryer section(s). Most of the remaining water is removed by applying heat through the steam-heated rotating cylinders that comprise the dryer section of the paper machine.

Gauges. The final sheet passes through gauges that sense moisture and sheet weight (or possibly sheet thickness). The gauges scan across the sheet and report both an instantaneous value and a scan average value for the moisture and sheet weight. The data can be processed to provide both a machine-direction profile (weight as a function of time) and a cross-direction profile (weight as a function of position on the sheet) for moisture and weight. Attention herein is on the machine-direction profile only.

Reel. The final sheet is rolled onto the reel.

The dynamics of the paper machine are dominated by dead time. The length of the paper path through the machine is known. The speed of the machine is known. The dead time can be computed as the length of the paper path (in feet) divided by the speed (in ft/min). The dead time will be 1 min or so. The time constant will be 10 sec or so. The dead time dominates. In the paper machine, the dry sheet weight (or fiber per unit area of the sheet) at the press rolls will be the dry sheet weight at the reel one dead time later.

A model consisting of a time constant and a dead time is sufficient. The dead time can be computed from the length of the paper path and the machine speed; the time constant is so much shorter that its effect on control system performance is nominal. The weight of the sheet can be computed as

$$W = \frac{FC}{DV}$$

where

W = dry sheet weight (lb fiber/ft^2); the basis weight is the weight of a specified area, such as 1000 ft^2 or 240 ft^2 (a ream)

F = stock flow (normally sensed by a magnetic flow meter) (lb/min)

C = stock consistency (lb fiber/lb slurry) (normally expressed in %)

D = width of the machine (ft)

V = machine speed (actually, the sheet velocity) (ft/min)

All quantities in this formula are accurately known except one—the stock consistency. The measurement is based on sensing a pseudoviscosity that depends on the stock consistency; unfortunately, the pseudoviscosity is heavily influenced by the nature of the fiber. The consistency of the stock feed to the machine is measured and controlled (usually by diluting with water). However, maintaining the measured value for the consistency at 2% does not assure that the actual consistency is maintained at 2%.

PID Control. It is well known that dead time degrades the performance of a PID controller. The first to suffer is the derivative mode; in the presence of significant dead time, rates of change are of little value. To obtain a stable loop, the controller gain must be lowered to the point where the proportional mode is contributing very little. The end result is effectively integral-only control, which is a slow-responding mode of control. In applications where the process is dominated by dead time, the usual result is that the PID controller responds so slowly that it is of no value.

Despite this, attempts were made to control basis weight using a PID controller. These date back to the 1960 time frame and were implemented entirely with analog controls. The effort was further complicated by the fact that the control actions had to be based on the scan average value for the basis weight. Prior to computers, the data from the gauges could not be resolved into machine-direction (time) and cross-direction (position) profiles. The scans were on the order of 30 sec, which is slow by today's standards. Nevertheless, the scan average provided the input to a sample-and-hold, which provided a continuous value for the PV input to the PID controller.

If sufficient incentive exists, a gauge can be installed at a fixed position, which would give a continuous measurement of the instantaneous basis weight at that location on the sheet. With such a measurement for the PV for a PID controller, Figure 8.9 illustrates the performance of PID control of basis weight for two different sets of tuning coefficients. The responses are to an increase in the actual consistency from 2% to 2.2% (the measured value for the consistency remains at 2%). Time zero is when this change occurs.

The two sets of tuning coefficients for the responses in Figure 8.9 are based on the following process model:

- *Process dead time* θ: 1.2 min
- *Process time constant* τ: 10 sec (0.17 min)
- *Process gain K:* 0.67%/%

The rationale for each set of tuning coefficients is as follows:

Figure 8.9(a). For this model, the ultimate gain K_u is 1.6%/% and the ultimate period P_u is 2.7 min. To obtain a quarter decay ratio, the controller gain K_C should approximately equal one-half the ultimate gain (or $K_C = 0.8\%/\%$) and the reset time T_I should approximately equal the ultimate period (or $T_I = 2.7$ min). For reasons to be discussed shortly, the response in Figure 8.9(a) is not exactly a smooth damped sinusoid, but its decay ratio is on the order of one-fourth. Perhaps the most obvious deficiency is that a reset time of 2.7 min is clearly too long.

Figure 8.9(b). In the paper industry, the *Lambda tuning method* [1] is popular. Using the value of the dead time θ for the closed loop time constant τ_{CL}, the tuning equations are applied as follows:

$$K_C = \frac{\tau}{\tau_{CL} + \theta} \frac{1}{K} = \frac{0.17\,\text{min}}{1.2\,\text{min} + 1.2\,\text{min}} \frac{1}{0.67\%/\%} = 0.11\%/\%$$

$$T_I = \tau = 0.17\,\text{min}$$

The objective of the Lambda method is to provide a response with little or no overshoot, and the response in Figure 8.9(b) clearly is consistent with this. This response is also quite smooth, especially as compared to the response in Figure 8.9(a).

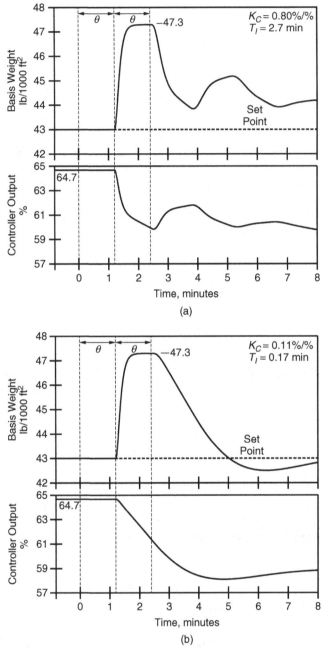

Figure 8.9 Performance of PID control for basis weight: (a) quarter decay ratio tuning; (b) Lambda tuning.

Using Lambda method tuning, the line-out time for the response is approximately 8 min (by shortening the reset time, comparable results could be obtained for the $\frac{1}{4}$ decay ratio tuning). For most processes, lining-out within 8 min would be very acceptable. But for a paper machine, 8 min is a long time. For a machine running 1600 ft/min, 8 min translates into 12,800 ft or over 2 miles of paper!

All feedback control strategies exhibit a certain behavior when responding to a disturbance to a process with a large dead time. The machine for Figure 8.9 has a dead time of 1.2 min, and the first two dead times are indicated on the responses. Observe the following:

- When the consistency changes, there is no effect on the measured value of the basis weight until one dead time has elapsed. Of course, until the value of the PV changes, the PID controller will not respond.
- As the remaining dynamics of the paper machine are very short, the basis weight rapidly changes once the dead time has elapsed. The PID controller can now respond. However, the results of its control actions will not be observed until another dead time has elapsed.

Following the occurrence of a disturbance, two dead times must elapse before any corrective action will be observed. This type of behavior continues for successive dead times, leading to the somewhat erratic appearance of the response in Figure 8.9(a). The only way to avoid this is to reduce the controller gain and rely primarily on the reset mode for control action, the result being the smoother response in Figure 8.9(b).

In the days prior to computer controls, the process operators were rather effective at controlling basis weight using the following approach:

- Obtain the last scan average of the basis weight.
- Based on the difference between the scan average value and the desired value, make an adjustment in the stock valve opening. Most operators had the stock valve "calibrated"; to change the basis weight by one unit, they knew approximately how many turns to make on the stock valve.
- Wait until the effect of this change had worked its way entirely through the machine. From experience, they knew the dead time, and also knew to wait for another complete scan of the weight gauge.

If necessary, this procedure could be repeated, but usually one try got them close enough. Their performance best PID hands-down!

Moving Dead Time Outside the Loop. Figure 8.10 is a simplified version of the PID control loop for basis weight. The dynamics of the paper machine are separated into dead-time and the non-dead-time dynamics, which consist of one or more time constants. The time constants are associated primarily with the headbox and other equipment at the wet end. These precede the dead time associated with the transport of the sheet through the machine.

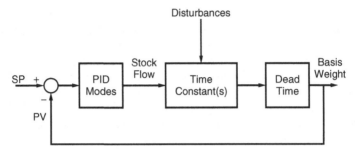

Figure 8.10   PID control of basis weight, gauge at reel.

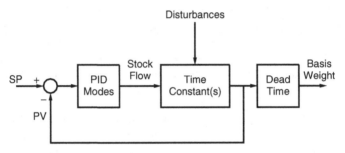

Figure 8.11   PID control of basis weight, gauge at press rolls.

At least in the abstract, there is a simple solution to this control problem. As illustrated in Figure 8.11, use the signal between the time constant(s) and the dead time as the PV for the PID controller. This moves the dead time outside the loop, which permits the controller to be tuned more aggressively to achieve a faster response to the disturbances.

For a paper machine, it is possible to at least consider doing this in practice. Instead of sensing the basis weight at the reel, sense the basis weight just downstream of the press rolls. This moves most of the dead time outside the feedback loop, and the controller can be more aggressively tuned. But there are issues:

- The gauge must be installed at a fixed location, but this is acceptable for the purposes of Figure 8.11 (only the machine-direction profile is needed).
- The weight gauge senses total weight, that is, fiber plus moisture. To determine fiber-only (the dry basis weight), both a weight and a moisture gauge must be installed.
- The sheet is largely moisture. Subtracting the moisture from the total means subtracting two large numbers to obtain a small number. This amplifies the errors in both measurements.

For a paper machine, implementing Figure 8.11 is possible, but is not popular. For other dead-time processes, the dynamics are not as cleanly separated. For such processes, Figure 8.11 cannot be implemented in practice.

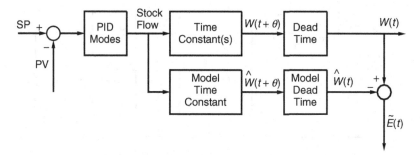

Figure 8.12 Model of the paper machine.

Model-Based Control. For processes that are dominated by dead time, the remaining dynamics can usually be approximated by a time constant. Figure 8.12 illustrates running a process model (consisting of a single time constant and a dead time) in parallel with the paper machine itself. The input to the model is the stock flow; the output of the model is the computed value for the basis weight. The notation is as follows:

$W(t) =$ basis weight at time t

$\hat{W}(t) =$ value for basis weight at time t computed using the model

$\tilde{E}(t) = W(t) - \hat{W}(t) =$ model error at time t

If $W(t)$ is the basis weight at time t, what is the output of the dynamic element for the time constant(s) in Figure 8.12? If the remaining dynamics are a dead time of θ, the output of the dynamic element for the time constant(s) is the value of the basis weight at time θ in the future, which would be designated as $W(t + \theta)$.

A similar statement applies to the output of the dynamic element for the time constant for the model. Specifically, the output of this element is $\hat{W}(t + \theta)$. In a sense, this is a predictor: The output of the time constant element is the predicted value of $W(t)$ at time θ in the future. A model is being used to predict the value of the controlled variable $W(t)$ at some time in the future. Hence, the term *model predictive* is clearly applicable.

In Figure 8.13, the value of $\hat{W}(t + \theta)$ is used as the PV for the PID controller. As the "process" now consists of only the model time constant, the controller gain can be set to a very large value, giving a very responsive control loop. But there is clearly an issue: The system is controlling the model, not the process. However, the value of the model error $\tilde{E}(t)$ is known. One could argue that if the model error is small, then by controlling the model, the system is effectively controlling the process. This raises questions as to what is considered a "small" model error. Fortunately, there is a way to incorporate the model error into the control configuration.

Finally, a word about the control error. The control error $E(t)$ should be defined as the difference between the basis weight set point and the measured

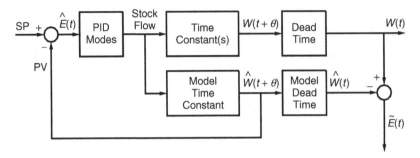

Figure 8.13   Model-based control of sheet weight.

value of the basis weight:

$$E(t) = SP - W(t)$$

But for the control configuration in Figure 8.13, the control error is based on the value predicted for the basis weight:

$$\hat{E}(t) = SP - \hat{W}(t + \theta)$$

The standard features of most control systems display values for the values associated with a PID controller (set point, process variable, etc.). For the configuration in Figure 8.13, these displays present the predicted value of the basis weight $\hat{W}(t + \theta)$ and the control error $\hat{E}(t)$, not the measured basis weight $W(t)$ and the control error $E(t)$.

Dead-Time Compensation.   Consider the following expressions for the model error:

$$\tilde{E}(t) = W(t) - \hat{W}(t) = \text{model error at time } t$$

$$\tilde{E}(t + \theta) = W(t + \theta) - \hat{W}(t + \theta) = \text{model error at time } t + \theta$$

The value of $\tilde{E}(t)$ is known but the value of $\tilde{E}(t + \theta)$ is not. If it were, the value of the basis weight one dead time in the future could be computed as follows:

$$W(t + \theta) = \hat{W}(t + \theta) + \tilde{E}(t + \theta)$$

If the process model is a reasonable approximation to the process, abrupt changes in the model error would not be expected. If so, the value of the current model error $\tilde{E}(t)$ can be used for the model error $\tilde{E}(t + \theta)$ at one dead time in the future. This permits a value to be computed as follows for the basis weight at one dead time in the future:

$$\tilde{W}(t + \theta) = \hat{W}(t + \theta) + \tilde{E}(t)$$

The value of $\tilde{W}(t+\theta)$ is the value $\hat{W}(t+\theta)$ predicted by the model for the basis weight one dead-time in the future plus the current model error $\tilde{E}(t)$. This logic is incorporated into the control configuration in Figure 8.14. The result is the dead-time compensator as proposed by O. J. M. Smith in 1957 [2].

Predicted Effect of Control Actions Taken Within the Last Dead Time.
The block diagram in Figure 8.14 can be rearranged in several different ways, one of which is presented in Figure 8.15. In this form, the following quantities are computed using the model:

$$\hat{W}(t) = \text{expected value for the basis weight at time } t$$

$$\hat{W}(t+\theta) = \text{expected value for the basis weight at one dead} \\ \text{time in the future}$$

$$\hat{W}(t+\theta) - \hat{W}(t) = \text{expected change in the basis weight over the} \\ \text{next dead time}$$

The PV for the PID controller is $\tilde{W}(t+\theta)$ and is exactly the same as in Figure 8.14. But in Figure 8.15, it is computed as follows:

$$\tilde{W}(t+\theta) = W(t) + [\hat{W}(t+\theta) - \hat{W}(t)]$$

In a sense, the role of $\hat{W}(t+\theta) - \hat{W}(t)$ within the control configuration is to provide a memory capability to account for the expected effect of control actions taken within the past dead time. The behavior is as follows:

- The appearance of a control error (difference between the set point and the measured basis weight) causes the controller to take corrective action.
- The expected effect of this corrective action on the basis weight quickly appears in the output $\hat{W}(t+\theta)$ of the non-dead-time elements of the process model.

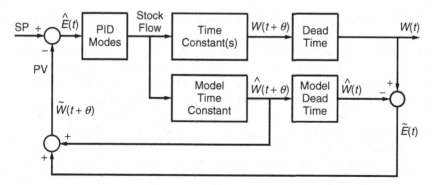

Figure 8.14 Dead-time compensator.

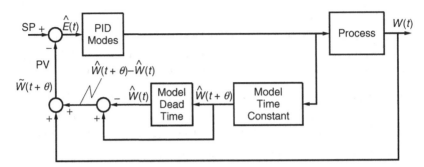

Figure 8.15   Alternative structure for the dead-time compensator.

- By adding $\hat{W}(t + \theta)$ to the current value of the basis weight $W(t)$, the PV to the controller reflects the expected result of the corrective action.
- Adding $\hat{W}(t + \theta)$ to the PV is equivalent to subtracting $\hat{W}(t + \theta)$ from the control error, the result being that the controller does not take further corrective actions.
- By subtracting $\hat{W}(t)$ from the current value of the basis weight $W(t)$, the following two effects should largely cancel:
  1. One dead time later, the result of the corrective action will appear in the measured basis weight $W(t)$.
  2. One dead time later, the expected result of the corrective action will appear in the output $\hat{W}(t)$ of the process model.

The change in the basis weight that occurs between time t and $t + \theta$ (one dead time in the future) depends solely on the corrective actions taken between time $t - \theta$ and t (one dead time in the past). The difference $\hat{W}(t + \theta) - \hat{W}(t)$ can be thought of as the expected results of all corrective actions taken within the last dead time. If no further corrective actions are taken, the basis weight will change by $\hat{W}(t + \theta) - \hat{W}(t)$ because of the corrective actions taken within the past dead time.

Dead-Time Compensation Function Block. If the control system provides a dead-time function block, the configuration in Figure 8.15 can be implemented easily. However, there is one drawback. The PV for the PID controller is not the current measured value for the basis weight. Instead, it is the expected value of the basis weight one dead time in the future. When displaying the PV and control error for the PID controller, it would be preferable for the operators to see the actual PV and the actual control error, not the anticipated values. The standard loop displays would not display the actual PV and the actual control error, but developing a custom display for this loop is always an option.

Further rearrangement of the block diagram for dead-time compensation gives the control configuration in Figure 8.16. Two summers precede the PID mode calculations:

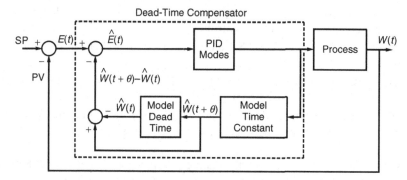

Figure 8.16 Dead-time compensator function block.

- The first summer subtracts the measured basis weight from the set point to give the current control error $E(t)$. This summer is the counterpart to the comparator in the customary implementations of the PID function block.
- The second summer subtracts $\hat{W}(t + \theta) - \hat{W}(t)$ from the current control error $E(t)$ to give the expected control error $\hat{E}(t)$.
- The input to the PID model calculations is the expected control error $\hat{E}(t)$.

This structure is the basis for a dead-time compensation control block provided by some systems. This makes it easy to incorporate dead-time compensation into a control configuration. A further advantage is that the PV to this function block is the actual basis weight, and the output of the first summer is the actual control error.

For most processes that are dominated by dead time, a process model consisting of a process gain, a time constant, and a dead time is adequate. Such a model can easily be incorporated into the dead-time compensation function block. But as will be discussed shortly, the performance of dead-time compensation requires an accurate value of the dead time. In applications such as paper machines, the dead time depends on the machine speed, which is not necessarily constant. In Chapter 1 we presented a dead-time function block and discussed the simulation of a variable dead time. These issues also apply to the dead-time compensation function block.

Bias for the Control Error. The configuration in Figure 8.15 can always be implemented using function blocks, assuming that a dead-time function block is available (and it usually is). But to implement the configuration in Figure 8.16 using function blocks, the PID function block must have a capability referred to as *control error bias*. This feature requires an input, say EBIAS, to provide a value for the control error bias. This input is used as follows:

- The control error $E(t)$ is computed in the usual manner: as the difference between the value of the set point and the value of the PV input.

Consequently, the standard loop displays present the actual value of the PV and the actual value of the control error.

- The value of input EBIAS can be applied in either of the following ways:
  1. Subtracted (or added) to the control error $E(t)$ to give $\hat{E}(t)$. The control calculations are based on $\hat{E}(t)$.
  2. Added (or subtracted) from the PV to give $\widehat{PV}$. The control error $\hat{E}(t)$ for the control calculations is computed from $\widehat{PV}$.

As an example of how to use this feature advantageously, consider controlling the megawatts being generated by a steam-driven turbo-generator. The controls must address the following issues:

- The actual megawatts are sensed and compared to the target for the megawatts, the difference being the actual control error.
- Separate controls are responsible for the turbine throttle pressure. If there is an error in the turbine throttle pressure, the actions taken by the throttle pressure controller to drive the throttle pressure to its target will also affect the megawatts.

Equations can be developed that relate the change in the megawatts that will occur for a change in the turbine throttle pressure. This leads to the following control configuration:

- A PID controller is configured with the PV being the sensed value for the megawatts.
- Input EBIAS is the value computed for the change in megawatts from the turbine throttle pressure error.

Adding the bias to the actual megawatts gives the zero-throttle-pressure error megawatts, which should be the basis for the control actions.

To implement dead-time compensation using input EBIAS, the configuration is as follows:

- A PID controller is configured with the PV being the measured value of the basis weight.
- The non-dead-time dynamic elements are simulated, the output being $\hat{W}(t + \theta)$. In most applications these elements comprise a single time constant, so a single lead-lag function block suffices.
- The dead time is simulated using the dead-time function block, the output being $\hat{W}(t)$.
- A summer is configured to compute $\hat{W}(t + \theta) - \hat{W}(t)$.
- Input EBIAS to the PID controller is the output of the summer block.

As compared to using a dead-time compensation function block, this approach permits the non-dead-time dynamic elements to be something other than a single time constant. However, this is required infrequently.

Performance of Dead-Time Compensation. Figure 8.17 illustrates the performance of dead-time compensation. The response is to an increase in the actual stock consistency from 2.0% to 2.2% that occurs at time zero. This is the same as for the responses in Figure 8.9 for PID control. Also included in Figure 8.17 are trends for the following basis weight errors:

Actual. This is the difference between set point and the measured value of the basis weight.

Net. This is the actual basis weight error less the value of $\hat{W}(t + \theta) - \hat{W}(t)$ from the dead-time compensator.

The control logic is applied to the net basis weight error, not the actual basis weight error. As control action is taken, the anticipated effect is subtracted from the actual basis weight error, giving the net error. The results of the control actions quickly appear in the net basis weight error, which permits a much larger value to be used for the controller gain K_C along with a short value for the reset time T_I. This enables the control logic to maintain the net basis weight error close to zero.

For the responses in Figure 8.17, there are two sources for error:

Process gain. The value of the process gain for the model is about 10% less than the true process gain.

Dynamics. The dynamics of the true process are not exactly represented by a single time constant.

Of these two, errors in the process gain have the greatest effect; errors in the dead time will be discussed shortly.

After the second dead time has elapsed, the controller has applied too much corrective action, as evidenced by the following characteristics of the responses in Figure 8.17:

- The basis weight has dropped below its set point.
- The controller output is less than its final line-out value.

For a given actual basis weight error, the lower the process gain in the model, the greater the corrective action required to drive the net basis weight error to zero. The effect is as follows:

- If the process gain for the model is low, the controller over-corrects for errors in the basis weight (as in Figure 8.17).

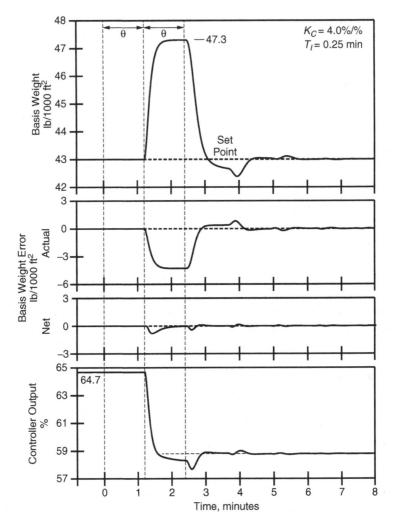

Figure 8.17 Performance of dead-time compensation.

- If the process gain for the model is high, the controller undercorrects for errors in the basis weight.

If the process gain for the model closely agrees with the true process gain, the dead time compensator would line-out in two dead times. But errors in the process gain are likely. Unless the errors are excessive, a realistic value for the line-out time is three dead times, which is the case for the response in Figure 8.17.

Errors in the Dead Time. For the same disturbance as Figure 8.17, Figure 8.18 presents the performance of dead-time compensation when the model dead time

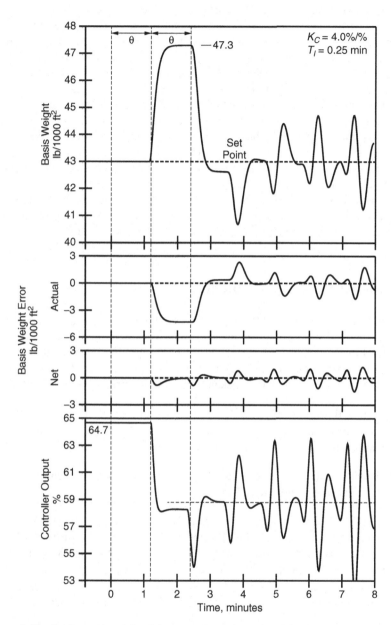

Figure 8.18 Performance of dead-time compensation, model dead time 10% too short.

is 10% shorter than the actual process dead time. Any error in the dead time degrades performance; a model dead time that is too short degrades performance in about the same manner as a model dead time that is too long. The performance of dead-time compensation degrades rapidly as the error in the model dead time increases. The response in Figure 8.18 is unstable. Stability can be restored by reducing the controller gain, but with a sacrifice in performance.

Dead-time compensation is applied routinely with great success in the sheet-processing industries, which includes paper machines. In other industries, applying dead-time compensation has yielded mixed results. The key is the value of the process dead time used in the model. In the paper industry, the dead time can be computed form the length of the paper path and the machine speed (actually, the paper velocity). Both are known accurately, so the value for the dead time in the model is also accurate. In applications where the paper machine speed changes, the value of the dead time in the model is adjusted for changes in machine speed. In other industries, the value of the dead time is not always known with sufficient accuracy for the potential benefits of dead-time compensation to be realized.

8.3. MODEL PREDICTIVE CONTROL

Model predictive control (MPC) has enjoyed both widespread industrial applications and theoretical interest in academia. The focus of this section reflects industrial users of MPC technology. Users do not program their own, but instead use software supplied by others. But whereas PID is a standard feature of industrial control systems, an MPC package must be acquired separately, either:

- From a third-party supplier.
- From the manufacturer of the control system. The package supplied by some manufacturers is from a third-party supplier.

Separate versions of MPC technology were developed independently within two organizations in the late 1970s:

IDCOM (identification and command). Developed within ADERSA (a French company), Richelet et al. [3] relied on an impulse response model of the process as the basis for their model predictive control implementation.

DMC (dynamic matrix control). Developed within Shell Oil in the United States, Cutler and Ramaker [4] relied on a step response model of the process as the basis for their model predictive control implementation.

Both have gone through subsequent evolutions, such as QDMC (quadratic dynamic matrix control) and ADMC (adaptive dynamic matrix control).

Several technologies belong to the broad classification of MPC. But as the term is used in industry, MPC refers to some version of the foregoing technologies, and indeed most industrial applications of MPC rely on some version of one of the above. The term is used similarly herein.

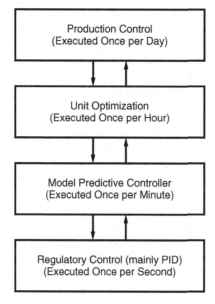

Figure 8.19   Role for MPC.

Role for MPC. Figure 8.19 illustrates the role for which MPC technology was originally developed. In commodity production facilities, significant incentives exist for optimization at both the plant level (setting targets for production rates for the various products) and at the unit level. The unit-level optimization calculations are normally based on steady-state relationships, and are performed on intervals such as 1 hr.

As implemented prior to MPC, the optimization routines implemented changes by adjusting the set points of selected PID controllers at the regulatory level. Due to factors such as interaction between the various loops at the regulatory level, these adjustments were not implemented as expeditiously or as smoothly as desired. Adjusting the various set points was essentially an upset process. The time required for the various loops to line-out proved to be not much shorter than the interval on which the optimization calculations were performed. The process was almost in a continual upset state, which offset much of the benefits of process optimization.

MPC was developed to provide a more effective mechanism for implementing the decisions from the optimization routines. As illustrated in Figure 8.19, MPC is inserted between the unit optimization and regulatory levels. The optimization routines adjust the targets for the model predictive controller, which in turn adjusts the targets for PID controllers at the regulatory level. When implemented in this manner, the MPC routines can be executed on a slower interval than the regulatory controllers. Whereas the regulatory routines are typically executed on an interval of 1 sec more or less, the MPC routines are executed on an interval of 1 min more or less.

In the example to be presented shortly, the MPC outputs to the set point of a flow controller. Most flow controllers are so fast that the flow through the control valve is very close to the flow set point; consequently, the terms *flow* and *flow set point* will essentially be used interchangeably. After the basics of MPC are presented, the issues pertaining to outputting directly to a valve or outputting to a PID controller set point are examined in more detail.

Forms for Process Models. Great advances have been made in developing process models from basic principles (material balances, energy balances, rate expressions, etc). Most recent process designs are based on a steady-state process model, and clearly the technology is available to develop such a model for any process. Unlike industries such as aerospace, dynamic models of industrial processes are not developed routinely. The commissioning (startup) of most processes can be accomplished with most if not all of the loops on manual; the aerospace industry does not have this option, which mandates models.

There are two alternatives to developing a dynamic model from process test data:

Lumped-parameter model. Although one occasionally encounters an exception, the dynamics of most processes can be characterized by one or more time constants and a dead time. The values of the model parameters are determined by analyzing data obtained from a process test. Mathematical routines can determine the values for the parameters, but the user must supply the structure of the model (number of time constants, etc).

Free-form model. The behavior of the process is characterized by either its step response or its impulse response. These models are also developed by analyzing the data from a process test. However, the user is freed from the responsibility of supplying the structure of the model.

Both approaches have one aspect in common: The model is linear. Most processes are nonlinear, especially with regard to changes in throughput. Fortunately, many commodity production facilities are operated within a narrow range of operating conditions, and linear models have proven to be adequate for approximating the behavior of such processes.

Figure 8.20 presents alternatives for the process model. The lumped-parameter model in Figure 8.20(a) consists of two time constants and a dead time. Figure 8.20(b) illustrates the equivalent free-form models: specifically, the impulse response $g(t)$ and the step response $s(t)$. The impulse and step responses are related very simply:

$$g(t) = \frac{ds(t)}{dt}$$

Whether a model is based on the step response or the impulse response is mainly a matter of preference.

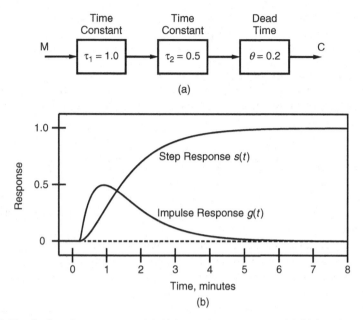

Figure 8.20 Options for process model: (a) lumped-parameter model; (b) free-form models.

Inpulse Function. The unit step function $u(t)$ is defined as follows:

$$u(t) = \begin{cases} 0 & \text{for } t < 0 \\ 1 & \text{for } t \geq 0 \end{cases}$$

Although not always practical, at least conceptually the step response $s(t)$ of a process to a unit step function $u(t)$ can be obtained by direct process testing.

The impulse function is also referred to as the *Dirac delta function*, $\delta(t)$. It can be thought of as the derivative of the unit step function:

$$\delta(t) = \frac{du(t)}{dt} = \begin{cases} 0 & \text{for } t \neq 0 \\ \text{undefined} & \text{for } t = 0 \end{cases}$$

Sometimes the value of $\delta(t)$ at $t = 0$ is said to be infinity. In a sense, the impulse function is a pulse of infinite height and infinitesimal width that occurs at $t = 0$. But the result of integrating the impulse function must be the unit step function. Consequently, the pulse must have an area of 1 unit. A finite pulse with a width w and height $1/w$ has an area of 1 unit. Therefore, the impulse function can be thought of as the pulse defined by the following limit:

$$\delta(t) = \lim_{w \to 0} \begin{cases} 0 & \text{for } t < 0 \text{ and } t \geq w \\ 1/w & \text{for } 0 \leq t < w \end{cases}$$

Such a function cannot be generated in practice, and thus direct process testing using the impulse function is not practical.

Finite Impulse Response. The interest in the impulse response stems from the convolution integral. For a linear process, the response $c(t)$ to any input $m(t)$ can be computed from the impulse response $g(t)$ by applying the convolution integral:

$$c(t) = \int_0^\infty m(\tau)g(t - \tau)d\tau = \int_0^\infty m(t - \tau)g(\tau)d\tau$$

Normally, an integral with an upper limit of infinity presents a problem for numerical computations. But for process applications, the problem is easily circumvented.

For nonintegrating processes, there is a cutoff time T_C for which $g(t) = 0$ for $t \geq T_C$. For the impulse response $g(t)$ illustrated in Figure 8.20(b), the value of T_C is approximately 7 min. This permits the convolution integral to be written as follows:

$$c(t) = \int_0^{T_C} m(t - \tau)g(\tau)d\tau$$

An impulse response whose value is zero outside the interval 0 to T_C is referred to as a *finite impulse response* (FIR).

Integrating processes present a minor complication. The counterpart for integrating processes is that the value of the impulse response is a constant for time greater than the cutoff time T_C; that is, $g(t) = C$ for $t \geq T_C$, with C being a constant. Although the details are not presented here, the numerical computations can be extended to accommodate this behavior.

There is also a counterpart for the step response $s(t)$. For a nonintegrating process, the value of the step response is a constant for time greater than the cutoff time T_C; that is, $s(t) = C$ for $t \geq T_C$, with C being a constant. Furthermore, the value of the constant C is the steady-state gain K of the process.

Example. Figure 8.21 presents a reactor with a once-through jacket. The cooling water enters the jacket, makes a single pass, and then exits the jacket. A flow controller is provided for the cooling water flow, and the model predictive controller will output to the set point of this controller. The objective is to control the reactor temperature. The measurement of the reactor temperature has a resolution of $0.1°F$.

A finite step response model will be developed for this process. Conceptually, the simplest approach is to conduct a direct step test on the process. Subsequently in this section, other testing approaches will be discussed along with methods to analyze the data. Figure 8.22 presents the results of a step test. Initially, the reactor is lined-out at a reactor temperature of $150°F$ and a cooling water flow

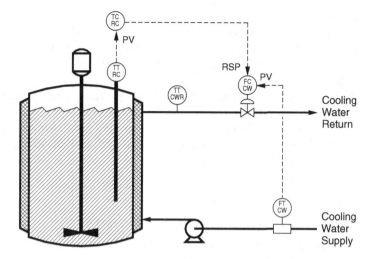

Figure 8.21   Reactor with a once-through jacket.

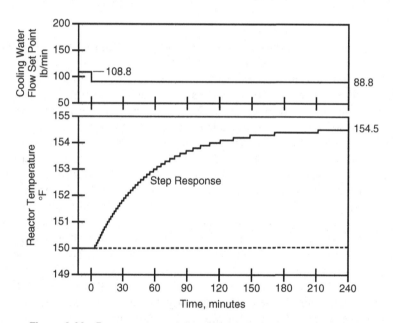

Figure 8.22   Response to a cooling water flow decrease of 20 lb/min.

of 108.8 lb/min. To initiate the step test, the set point of the cooling water flow controller is decreased by 20 lb/min. The process is reverse acting, so the reactor temperature increases, ultimately lining out at 154.5°F.

A linear model does not directly relate the actual values of the cooling water flow and the actual values of the reactor temperature. Instead, a linear model relates a change in the reactor temperature to a change in the cooling water flow.

In this context, *change* is not a rate of change; it is the change (or difference) between the actual value and an equilibrium or reference value. For the reactor test, the initial equilibrium or steady-state is a reactor temperature of 150°F and a cooling water flow of 108.8 lb/min. These values are designated as T_{eq} and F_{eq}, respectively.

The generic notation of C for the controlled variable (reactor temperature) and M for the manipulated variable (cooling water flow) will be used. Although the practice is by no means universal, uppercase letters are used here to designate the actual values, and lowercase letters are used to designate the changes. This is the basis for the following nomenclature:

$$C(t) = \text{actual reactor temperature (°F)}$$

$$c(t) = C(t) - T_{eq} = \text{change in reactor temperature (°F)}$$

$$M(t) = \text{actual cooling water flow (lb/min)}$$

$$m(t) = M(t) - F_{eq} = \text{change in cooling water flow (lb/min)}$$

A linear model relates $c(t)$ to $m(t)$.

It is also common practice to drop "actual" and "change," so both $C(t)$ and $c(t)$ are referred to as the controlled variable or reactor temperature. When these continuous functions are approximated by a discrete function (sequence of values with a fixed time interval between values), the notation will be $C(k)$, $c(k)$, $M(k)$, and $m(k)$ (an alternative notation is C_k, c_k, M_k, and m_k). In keeping with this notation, the left five columns in Table 8.3 present the data from the step response test illustrated in Figure 8.22. The time interval or sampling time Δt of 15 min between data points is much longer than would be used in practice. It is used here to reduce the number of points in the finite step response model to a manageable number so that tables such as Table 8.3 will not be excessively long. When the data are processed by computers, a sampling time of 1.0 min is more typical.

Since the test was a direct step test, $c(k)$ in Table 8.3 is the step response, but with one caveat. The test involved a change of −20 lb/min in cooling water flow. The step response $s(k)$ must be to a unit change in the cooling water flow, that is, to a change of +1 lb/min. To obtain the step response $s(k)$, the values of $c(k)$ in Table 8.3 are divided by −20 lb/min. The values of the points for $s(k)$ cease to change after 13 points. Consequently, the finite step response model will consist of 13 points. The value of the point at index 0 is understood to be zero. The values of all points beyond 13 are understood to be the same as the value of the point at index 13. The following parameters apply:

$$N = \text{number of elements in the finite step response model}$$

$$\Delta t = \text{sampling interval or sampling time}$$

$$T_N = N \Delta t = \text{time span of the finite step response model}$$

Table 8.3 Finite step response model for a reactor with a once-through jacket

Time, T (min)	Actual Cooling Water Flow, $M(k)$ (lb/min)	Change in Cooling Water Flow, $m(k)$ (lb/min)	Actual Reactor Temperature, $C(k)$ (°F)	Change in Reactor Temperature, $c(k)$ (°F)	Step Response, $s(k)$ [°F/(lb/min)]	Index, k	Impulse Response, $g(k)$ [°F/(lb/min)]
−15	108.8	0	150.0	0.0			
0	88.8	−20	150.0	0.0	0.0	0	0.0
15	88.8	−20	151.0	1.0	−0.050	1	−0.050
30	88.8	−20	151.9	1.9	−0.095	2	−0.045
45	88.8	−20	152.5	2.5	−0.125	3	−0.030
60	88.8	−20	153.0	3.0	−0.150	4	−0.025
75	88.8	−20	153.4	3.4	−0.170	5	−0.020
90	88.8	−20	153.7	3.7	−0.185	6	−0.015
105	88.8	−20	153.9	3.9	−0.195	7	−0.010
120	88.8	−20	154.1	4.1	−0.205	8	−0.010
135	88.8	−20	154.2	4.2	−0.210	9	−0.005
150	88.8	−20	154.3	4.3	−0.215	10	−0.005
165	88.8	−20	154.4	4.4	−0.220	11	−0.005
180	88.8	−20	154.4	4.4	−0.220	12	−0.000
195	88.8	−20	154.5	4.5	−0.225	13	−0.005
210	88.8	−20	154.5	4.5			
225	88.8	−20	154.5	4.5			
240	88.8	−20	154.5	4.5			

For the reactor with a once-through jacket, $N = 13$, $\Delta t = 15$ min, and $T_N = 3$ hr 15 min.

The final column in Table 8.3 presents the finite impulse response model $g(k)$, which is computed as follows:

$$g(k) = s(k) - s(k - 1)$$

The finite impulse response model also consists of 13 points. The value of the point at index 0 is understood to be zero. The values of all points beyond point 13 are understood to be zero.

Principle of Superposition. For continuous functions, the principle of superposition is stated as follows:

- Let $c_1(t)$ be the response to input $m_1(t)$.
- Let $c_2(t)$ be the response to input $m_2(t)$.
- Let $m(t) = m_1(t) + m_2(t)$.
- Then $c(t) = c_1(t) + c_2(t)$ is the response to input $m(t)$.

The principle of superposition is valid only for linear systems. For nonlinear systems, it is an approximation that is usually acceptable for small changes.

For the reactor in Figure 8.21, the response of the reactor temperature to a pulse or bump in the cooling water flow will be computed using the finite step response model in Table 8.3. The pulse is an increase in the cooling water flow of 40 lb/min for 30 min, as illustrated in Figure 8.23. This pulse is the input $m(t)$, and as illustrated in Figure 8.24(a), can be expressed as the sum of two step changes:

$$m_1(t) = 40u(t) = \begin{cases} 0 & \text{for } t < 0 \\ 40 & \text{for } t \geq 0 \end{cases}$$

$$m_2(t) = -40u(t - 30) = \begin{cases} 0 & \text{for } t < 30 \\ -40 & \text{for } t \geq 30 \end{cases}$$

The response to each of these inputs can be computed using the finite step response model. Table 8.4 presents the computations, and Figure 8.24(b) presents the responses for $c_1(k), c_2(k)$, and $C(k)$. The computations in Figure 8.24(b) proceed as follows:

- $c_1(k)$: Response to input $m_1(t)$ obtained by multiplying the finite step response model by 40
- $c_2(k)$: Response to input $m_2(t)$ obtained by multiplying the finite step response model by -40 and delaying by 30 min (two sampling times)

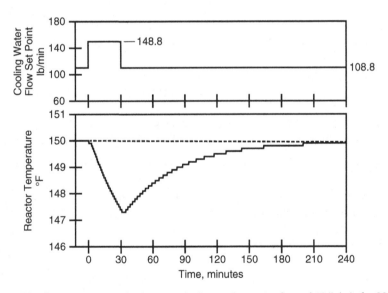

Figure 8.23 Response to a pulse increase in the cooling water flow of 40 lb/min for 30 min.

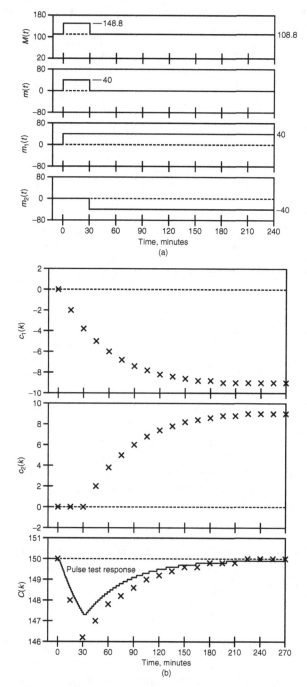

Figure 8.24 Applying the principle of superposition: (a) representation of pulse input as the sum of two step inputs; (b) $C(k)$ as the sum of the responses to the two step changes.

Table 8.4 Computing response to pulse input using the principle of superposition

t	k	$c_1(k)$	$c_2(k)$	$c(k)$	$C(k)$	Test Results
−15	−1	0.0	0.0	0.0	150.0	150.0
0	0	0.0	0.0	0.0	150.0	150.0
15	1	−2.0	0.0	−2.0	148.0	148.6
30	2	−3.8	0.0	−3.8	146.2	147.4
45	3	−5.0	2.0	−3.0	147.0	147.9
60	4	−6.0	3.8	−2.2	147.8	148.4
75	5	−6.8	5.0	−1.8	148.2	148.8
90	6	−7.4	6.0	−1.4	148.6	149.1
105	7	−7.8	6.8	−1.0	149.0	149.4
120	8	−8.2	7.4	−0.8	149.2	149.5
135	9	−8.4	7.8	−0.6	149.4	149.6
150	10	−8.6	8.2	−0.4	149.6	149.7
165	11	−8.8	8.4	−0.4	149.6	149.8
180	12	−8.8	8.6	−0.2	149.8	149.9
195	13	−9.0	8.8	−0.2	149.8	149.9
210	14	−9.0	8.8	−0.2	149.8	149.9
225	15	−9.0	9.0	0.0	150.0	149.9
240	16	−9.0	9.0	0.0	150.0	149.9
255	17	−9.0	9.0	0.0	150.0	150.0
270	18	−9.0	9.0	0.0	150.0	150.0

- $c(k)$: Sum of $c_1(k)$ and $c_2(k)$
- $C(k)$: Sum of $c(k)$ and $150.0°F$

In Figure 8.24(b), the response in reactor temperature from Figure 8.23 is included so that the response $C(k)$ computed using the principle of superposition can be compared to the response from the pulse test. If the reactor with the once-through jacket were linear, the agreement would be exact. But this is not the case for nonlinear processes.

Recursive Algorithm. The output of a digital controller changes only at discrete increments of time. The controller output can be envisioned as a sequence of step changes that are defined as follows:

$$\Delta m(i) = \Delta M(i) = M(i) - M(i-1)$$

$$= \text{change in controller output at sampling instant } i$$

Time zero corresponds to the first sampling instant on which a change in the controller output occurs. The value for index i on this sampling instant is zero. The value of time corresponding to each sampling instant is $i\Delta t$.

For the pulse test illustrated in Figure 8.23, the value of $\Delta m(i)$ is known for each sampling instant within the duration of the test, the values being as follows:

$$\Delta m(0) = +40$$
$$\Delta m(1) = 0$$
$$\Delta m(2) = -40$$
$$\Delta m(i) = 0, \qquad i > 2$$

The output of a digital controller can also be expressed as a sequence of changes $\Delta m(i)$.

To formulate a recursive algorithm, an array of expected future values of the controlled variable will be maintained. The values will be initialized on the sampling instant that corresponds to time zero. On each subsequent sampling instant, the predicted values from the preceding sampling instant will be updated to reflect the control action $\Delta m(i)$. The following notation will be used:

$i = $ index of the current sampling instant, with $i = 0$ designating the sampling instant at time zero

$k = $ index relative to the current sampling instant

$\hat{C}(k) = $ expected value of the controlled variable at sampling instant k in the future

$C(i) = $ measured value of controlled variable on sampling instant i

A storage array must be allocated for the expected values $\hat{C}(k)$ for N sampling instants in the future. Although not necessary, reserving a location for $\hat{C}(0)$ makes the algorithm easier to comprehend. The following statements apply to the storage array:

- The value of $\hat{C}(k)$ is the expected value of the controlled variable at k sampling instants in the future.
- If i is the current sampling instant, $\hat{C}(k)$ is the expected value of $C(i + k)$.
- The expected value of the controlled variable on the current sampling instant is $\hat{C}(0)$.
- The expected value of the controlled variable at the next sampling instant is $\hat{C}(1)$.
- For $k > N$, the predicted value of the controlled variable is the value of $\hat{C}(N)$.

To initialize the storage array for $\hat{C}(k)$ prior to sampling instant $i = 0$, the current value of the reactor temperature is written to all elements of the storage array. Basically, this assumes that the process is lined-out (at steady-state) and no changes have been made to the controller output within time T_N prior to the present. If this is not true, the predicted values $\hat{C}(k)$ will not be valid until the recursive algorithm has been executed N times (time T_N has elapsed).

On each sampling instant, the following two values are available:

$$C(i) = \text{measured value for the controlled variable}$$

$$\hat{C}(0) = \text{expected value for the controlled variable}$$

The difference between these is the model error $E_M(i)$ on that sampling instant:

$$E_M(i) = \hat{C}(0) - C(i)$$

If the expected value $\hat{C}(0)$ for the current sampling instant is in error by some amount, what does this suggest about the remaining values of $\hat{C}(k)$? In process applications, there are two possible sources of the error:

- *The model itself.* This is some combination of inaccuracies in the values in the finite step response model and the consequences of nonlinearities on the principle of superposition.
- *Process disturbances.* All processes are subject to disturbances from various sources that will cause the controlled variable to deviate from its expected value. Some are of short-term duration; others persist for hours or days.

One option is to assume that the error in the value for $\hat{C}(0)$ will also appear in the remaining values of $\hat{C}(k)$. With this assumption, the error in the value for $\hat{C}(0)$ should be subtracted from the remaining values of $\hat{C}(k)$.

For sampling instant i, the computations to be performed are as follows:

- Advance the time for the predicted values in the storage array by one sampling instant. The value of $\hat{C}(1)$ from the previous sampling instant becomes the expected value for $C(i)$; the value of $\hat{C}(2)$ from the previous sampling instant becomes the expected value for $C(i + 1)$; and so on. As this applies to all elements in storage array $\hat{C}(k)$, the elements must be shifted as follows:

$$\hat{C}(k) = \hat{C}(k + 1), \qquad k = 0, 1, 2, \ldots, N - 1$$

The value of $\hat{C}(N)$ remains unchanged.
- Compute the model error $E_M(i)$ for the sampling instant i:

$$E_M(i) = \hat{C}(0) - C(i)$$

- Subtract the model error $E_M(i)$ from the remaining values of $\hat{C}(k)$:

$$\hat{C}(k) = \hat{C}(k) - E_M(i), \qquad k = 1, 2, 3, \ldots, N$$

- Update the predicted values $\hat{C}(k)$ for the change $\Delta m(i)$ in the controller output on the current sampling instant. Using the finite step response model, the computation is as follows:

$$\hat{C}(k) = \hat{C}(k) + \Delta m(i)s(k), \qquad k = 1, 2, 3, \ldots, N$$

Table 8.5 presents the results of using the recursive algorithm to compute the expected response for the pulse test presented in Figure 8.23. The explanation for each column is as follows:

Initialization:
- All storage elements for $\hat{C}(k)$ are initialized to the initial reactor temperature of $150.0°F$.

Sampling instant $i = 0$:
- The values in the storage array $\hat{C}(k)$ are shifted ahead by one sampling instant (all values are 150.0, so this is not apparent from Table 8.5).
- The predicted value $\hat{C}(0)$ is $150.0°F$.
- The measured value $C(0)$ is $150.0°F$.

Table 8.5   Computations for the recursive algorithm

i:	Init	0	1	2	3	4	5	6	7
Time	—	0	15	30	45	60	75	90	105
$C(i)$	150.0	150.0	148.6	147.4	147.9	148.4	148.8	149.1	149.4
$\hat{C}(0)$	—	150.0	148.0	146.8	148.2	148.7	148.8	149.2	149.5
$E_M(i)$	—	0.0	−0.6	−0.6	+0.3	+0.3	0.0	+0.1	+0.1
$\Delta m(i)$	—	+40	0.0	−40	0.0	0.0	0.0	0.0	0.0
$\hat{C}(1)$	150.0	148.0	146.8	148.2	148.7	148.8	149.2	149.5	149.6
$\hat{C}(2)$	150.0	146.2	145.6	149.0	149.1	149.2	149.6	149.7	149.8
$\hat{C}(3)$	150.0	145.0	144.6	149.4	149.5	149.6	149.8	149.9	150.0
$\hat{C}(4)$	150.0	144.0	143.8	149.8	149.9	149.8	150.0	150.1	150.0
$\hat{C}(5)$	150.0	143.2	143.2	150.2	150.1	150.0	150.2	150.1	150.2
$\hat{C}(6)$	150.0	142.6	142.8	150.4	150.3	150.2	150.2	150.3	150.2
$\hat{C}(7)$	150.0	142.2	142.4	150.6	150.5	150.2	150.4	150.3	150.2
$\hat{C}(8)$	150.0	141.8	142.2	150.8	150.5	150.4	150.4	150.3	150.4
$\hat{C}(9)$	150.0	141.6	142.0	150.8	150.7	150.4	150.4	150.5	150.4
$\hat{C}(10)$	150.0	141.4	141.8	151.0	150.7	150.4	150.6	150.5	150.4
$\hat{C}(11)$	150.0	141.2	141.8	151.0	150.7	150.6	150.6	150.5	150.4
$\hat{C}(12)$	150.0	141.2	141.6	151.0	150.9	150.6	150.6	150.5	150.4
$\hat{C}(13)$	150.0	141.0	141.6	151.2	150.9	150.6	150.6	150.5	150.4

- The model error $E_M(0)$ is zero.
- The model error $E_M(0)$ is subtracted from the future values of $\hat{C}(k)$, but since the model error is zero, no future values are changed.
- As $\Delta m(0)$ is $+40$, the product $\Delta m(0)s(k)$ is added to each $\hat{C}(k)$ for $k = 1$ through N [recall that $s(k) < 0$ for all values of k].

Sampling instant i = 1:

- The values in the storage array $\hat{C}(k)$ are shifted ahead by one sampling instant.
- The predicted value $\hat{C}(0)$ is $148.0°$F.
- The measured value $C(1)$ is $148.6°$F.
- The model error $E_M(1)$ is $-0.6°$F.
- The model error $E_M(1)$ is subtracted from the future values of $\hat{C}(k)$, so all are increased by $0.6°$F.
- As $\Delta m(1)$ is zero, adding the product $\Delta m(1)s(k)$ to each $\hat{C}(k)$ has no effect.

Sampling instant i = 2:

- The values in the storage array $\hat{C}(k)$ are shifted ahead by one sampling instant.
- The predicted value $\hat{C}(0)$ is $146.8°$F.
- The measured value $C(2)$ is $147.4°$F.
- The model error $E_M(2)$ is $-0.6°$F.
- The model error $E_M(2)$ is subtracted from the future values of $\hat{C}(k)$, so all are increased by $0.6°$F.
- As $\Delta m(2)$ is -40, the product $\Delta m(2)s(k)$ is added to each $\hat{C}(k)$ for $k = 1$ through N.

Sampling instant i = 3:

- The values in the storage array $\hat{C}(k)$ are shifted ahead by one sampling instant.
- The predicted value $\hat{C}(0)$ is $148.2°$F.
- The measured value $C(3)$ is $147.9°$F.
- The model error $E_M(3)$ is $+0.3°$F.
- The model error $E_M(3)$ is subtracted from the future values of $\hat{C}(k)$, so all are decreased by $0.3°$F.
- As $\Delta m(3)$ is zero, adding the product $\Delta m(3)s(k)$ to each $\hat{C}(k)$ has no effect.

Sampling instant i = 4:

- The values in the storage array $\hat{C}(k)$ are shifted ahead by one sampling instant.
- The predicted value $\hat{C}(0)$ is $148.7°$F.
- The measured value $C(4)$ is $148.4°$F.

- The model error $E_M(4)$ is $+0.3°F$.
- The model error $E_M(4)$ is subtracted from the future values of $\hat{C}(k)$, so all are decreased by $0.3°F$.
- As $\Delta m(4)$ is zero, adding the product $\Delta m(4)s(k)$ to each $\hat{C}(k)$ has no effect.

Table 8.5 provides the results through $i = 7$. However, these calculations can be continued for as long as desired. If the future values were not adjusted for the model error $E_M(i)$, the results from the recursive algorithm would be the same as in Table 8.4. Observe that all model errors in Table 8.4 are negative. With the adjustment for the $E_M(i)$, Table 8.5 contains both positive and negative values for the model error. If recursive algorithm is repeated for 18 sampling instants, the standard deviations for the model error are as follows:

Without adjustment for model error (Table 8.4): $\sigma = 0.483$
With adjustment for model error (Table 8.5): $\sigma = 0.245$

At least for this example, adjusting for the model error improves the values predicted by the model.

Dynamic Matrix. Note the distinction between the following:

- $\hat{C}(k)$: Expected values of the controlled variable that result from past control actions: specifically, $\Delta m(i-1)$, $\Delta m(i-2)$, $\Delta m(i-3)$, and so on.
- $x(k)$: Changes in the controlled variable expected as the result of the current control action $\Delta m(i)$ and all proposed future control actions $\Delta m(i+1)$, $\Delta m(i+2)$, and so on.

As $\hat{C}(k)$ is the result of past control actions, the controller has no way to influence the values of $\hat{C}(k)$. But the values of $x(k)$ depend on the current and future proposed control actions, so the values of $x(k)$ are basically at the discretion of the controller. The procedure is as follows:

- Determine the desired values for $x(k)$.
- Determine the control actions $\Delta m(i)$, $\Delta m(i+1)$, $\Delta m(i+2)$, and so on, that will give the desired values for $x(k)$.

The solution to the latter problem involves the dynamic matrix.
To develop the dynamic matrix, note the following:

- $x(0)$. This will be zero. The controller has no way to influence the value of the controlled variable on the current sampling instant.
- $x(1)$. The value of $x(1)$ is only influenced by $\Delta m(i)$.

- $x(2)$. The value of $x(2)$ is only influenced by $\Delta m(i)$ and $\Delta m(i + 1)$.
- $x(k)$. The value of $x(k)$ is only influenced by $\Delta m(i), \Delta m(i + 1), \ldots,$ $\Delta m(i + k - 1)$.

The values for $x(k)$ can be related to the control actions using the finite step response model $s(k)$:

$$x(1) = s(1)\Delta m(i)$$

$$x(2) = s(2)\Delta m(i) + s(1)\Delta m(i + 1)$$

$$x(3) = s(3)\Delta m(i) + s(2)\Delta m(i + 1) + s(1)\Delta m(i + 2)$$

The general equation is

$$x(k) = s(k)\Delta m(i) + s(k - 1)\Delta m(i + 1) + s(k - 2)\Delta m(i + 2)$$
$$+ \cdots + s(1)\Delta m(i + k - 1), \qquad k = 1, 2, \ldots, N$$

This equation can also be expressed as a summation, but this is not useful for current purposes.

The dynamic matrix is formulated by expressing these relationships using vectors and matrices:

$$
\begin{bmatrix} x(1) \\ x(2) \\ x(3) \\ \vdots \\ x(N) \end{bmatrix}
=
\begin{bmatrix}
s(1) & 0 & 0 & \cdots & 0 \\
s(2) & s(1) & 0 & \cdots & 0 \\
s(3) & s(2) & s(1) & \cdots & 0 \\
\vdots & \vdots & \vdots & \vdots & \vdots \\
s(N) & s(N-1) & s(N-2) & \cdots & s(1)
\end{bmatrix}
\begin{bmatrix} \Delta m(i) \\ \Delta m(i+1) \\ \Delta m(i+2) \\ \vdots \\ \Delta m(i+N-1) \end{bmatrix}
$$

or

$$\mathbf{x} = \mathbf{A}\Delta\mathbf{m}$$

where

$$
\Delta\mathbf{m} =
\begin{bmatrix}
\Delta m(i) \\
\Delta m(i+1) \\
\Delta m(i+2) \\
\vdots \\
m(i+N-1)
\end{bmatrix}
= \text{vector of proposed current and future control actions}
$$

$$
\mathbf{x} =
\begin{bmatrix}
x(1) \\
x(2) \\
x(3) \\
\vdots \\
x(N)
\end{bmatrix}
= \text{vector of expected results of control actions } \Delta\mathbf{m}
$$

$$\mathbf{A} = \begin{bmatrix} s(1) & 0 & 0 & \cdots & 0 \\ s(2) & s(1) & 0 & \cdots & 0 \\ s(3) & s(2) & s(1) & \cdots & 0 \\ \vdots & \vdots & \vdots & \vdots & \vdots \\ s(N) & s(N-1) & s(N-2) & \cdots & s(1) \end{bmatrix} = \text{dynamic matrix}$$

Controller Formulation. The recursive algorithm presented previously can be used to compute the expected values $\hat{C}(k)$ that result from the previous control actions $\Delta m(i)$ taken by the controller. These can also be expressed as a vector:

$$\hat{\mathbf{C}} = \begin{bmatrix} \hat{C}(1) \\ \hat{C}(2) \\ \hat{C}(3) \\ \vdots \\ \hat{C}(N) \end{bmatrix}$$

The following errors must be distinguished:

$e(i) = \text{SP} - C(i) = $ control error on sampling instant i.

$\hat{e}(k) = \text{SP} - \hat{C}(k) = $ expected control error at sampling instant k in the future.

When computing the expected control errors $\hat{e}(k)$, the current value of the set point is normally used. For continuous processes, this is reasonable—changes in the set point are usually small and infrequent. In some batch applications, changes in the set point are expected to occur at a specific time in the future, and such changes can be incorporated into the computation of the expected control errors $\hat{e}(k)$.

Subtracting the expected values $\hat{C}(k)$ from the set point gives the vector of expected control errors $\hat{\mathbf{e}}$:

$$\hat{\mathbf{e}} = \begin{bmatrix} \hat{e}(1) \\ \hat{e}(2) \\ \hat{e}(3) \\ \vdots \\ \hat{e}(N) \end{bmatrix}$$

These are the expected control errors if no control actions are taken now or in the future. However, control actions $\Delta m(i), \Delta m(i+1), \ldots, \Delta m(i+k-1)$ would affect the future errors $\hat{e}(k)$. Suppose that the desire is for all future control errors $\hat{e}(k)$ to be zero. To achieve this, the result $x(k)$ of the current and future control actions must cancel these errors. Consequently,

$$\mathbf{x} = \hat{\mathbf{e}}$$

The control actions $\Delta\mathbf{m}$ required to give this result can be computed as follows:

$$\Delta\mathbf{m} = \mathbf{A}^{-1}\mathbf{x} = \mathbf{A}^{-1}\hat{\mathbf{e}}$$

Herein the controller based on this equation will be referred to as the *square DMC controller*. The dynamic matrix A is an $N \times N$ square matrix. The next N control moves will be computed so as to drive the next N predicted control errors $\hat{e}(k)$ to zero.

To implement the controller, the characteristics of the process are approximated using a finite step response model $s(k)$ consisting of N elements. The dynamic matrix $\mathbf{A}$ is composed from the elements of the finite step response model, and the inverse $\mathbf{A}^{-1}$ is computed. On each sampling instant, the following computations are executed:

1. Using the recursive algorithm presented previously, update the prediction of N future values of the controlled variable $\hat{C}(k)$.
2. Subtract each future value of the controlled variable $\hat{C}(k)$ from the set point to obtain the predicted values of the control error $\hat{e}(k)$.
3. Using the inverse $\mathbf{A}^{-1}$ of the dynamic matrix, compute the control actions $\Delta m(i + k - 1)$ that will drive the predicted control errors to zero.
4. Implement the control move $\Delta m(i)$.

Although N future values for the controller output are computed, only the value for the current sampling instant is implemented. On the next sampling instant, the computations are repeated and again only the value of the controller output for that sampling instant is implemented.

Actually, a simpler formulation for the controller is possible. The first row of the matrix equation is equivalent to the following equation:

$$x(1) = s(1)\Delta m(i)$$

Replacing $x(1)$ by $\hat{e}(1)$ and solving for $\Delta m(i)$ gives the following equation for the controller:

$$\Delta m(i) = \frac{\hat{e}(1)}{s(1)}$$

Especially when the value of $s(1)$ is small, the values computed for $\Delta m(i)$ will be unreasonably large. For a process with dead time $\theta > \Delta t$, the value of $s(1)$ will be zero. The square DMC controller can be reformulated to address these issues, but the preferred approach is to use the QDMC controller that will be presented shortly.

Performance of the Square DMC Controller. In most cases, controllers formulated to remove all of the projected error in one time interval are too aggressive. This is the case for the reactor in Figure 8.20. Figure 8.25 presents the response to a change in the reactor temperature set point from 150.0°F to 155.0°F.

Table 8.6 presents the details of the control calculations for the first two hours (iterations 0 through 8). The data in Table 8.6 are as follows:

- $\hat{C}(0)$ *through* $\hat{C}(13)$, in °F. Expected values for the process variable computed from the finite step response model and the changes in the controller output. $\hat{C}(0)$ is the expected value of the PV on the current sampling instant; $\hat{C}(1)$ is the expected value of the PV on the next sampling instant; and so on. On each sampling instant, these values are the expected values for the PV on the following basis:
  - Before being corrected for the model error on the current sampling instant
  - Before incorporating the effect of the control action, if any, that is taken on the current sampling instant.

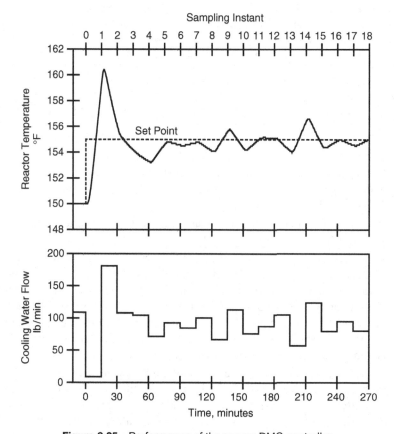

Figure 8.25 Performance of the square DMC controller.

Table 8.6 Computations for the square DMC controller

	i: 0	1	2	3	4	5	6	7	8
Time	0	15	30	45	60	75	90	105	120
$\hat{C}(0)$	150.00	155.00	155.00	155.00	155.00	155.00	155.00	155.00	155.00
$\hat{C}(1)$	150.00	159.50	150.26	155.62	155.05	156.48	154.99	155.99	154.22
$\hat{C}(2)$	150.00	162.50	147.60	155.50	155.04	157.43	155.62	155.93	154.41
$\hat{C}(3)$	150.00	165.00	145.30	155.38	155.01	158.70	155.30	156.60	154.23
$\hat{C}(4)$	150.00	167.00	143.36	155.26	155.45	158.93	155.77	156.81	153.67
$\hat{C}(5)$	150.00	168.50	141.78	155.63	155.02	159.83	155.82	156.57	154.35
$\hat{C}(6)$	150.00	169.50	141.06	155.14	155.42	160.20	155.46	157.49	153.49
$\hat{C}(7)$	150.00	170.50	139.84	155.50	155.46	160.05	156.29	156.78	153.79
$\hat{C}(8)$	150.00	171.00	139.48	150.51	154.98	161.10	155.50	157.25	153.66
$\hat{C}(9)$	150.00	171.50	139.12	150.01	155.87	160.42	155.93	157.20	153.79
$\hat{C}(10)$	150.00	172.00	138.26	150.88	155.03	160.95	155.84	157.41	153.60
$\hat{C}(11)$	150.00	172.00	138.79	150.02	155.39	160.97	156.00	157.30	153.64
$\hat{C}(12)$	150.00	172.50	137.90	150.38	155.41	161.13	155.89	157.34	153.56
$\hat{C}(13)$	150.00	172.50	137.90	150.38	155.41	161.13	155.89	157.34	153.56
SP(i)	155.00	155.00	155.00	155.00	155.00	155.00	155.00	155.00	155.00
PV(i)	150.00	159.10	156.10	154.20	153.30	154.60	154.60	154.80	154.10
$E_M(i)$	0.00	+4.10	+1.10	−0.80	−1.70	−0.40	−0.40	−0.20	−0.90
$\hat{C}(1)$ + $E_M(i)$	150.00	163.60	151.36	158.82	153.35	156.08	154.59	155.79	153.32
$\hat{e}(1)$	5.00	−8.60	+3.64	0.18	1.65	−1.08	0.41	−0.79	1.68
$\Delta m(i)$	−100.0	+172.0	−72.8	−3.6	−33.0	+21.6	−8.2	+15.8	−33.6
$M(i)$	8.8	180.8	108.0	104.4	71.4	93.0	84.8	100.6	67.0

- *SP(i)*, in °F. Set point on sampling instant i. The step change in the set point to 155°F occurs on sampling instant 0 (or time zero). The value is 155°F on all sampling instants.

- *PV(i)*, in °F. Process variable on sampling instant i. The temperature measurement has a resolution of 0.1°F.

- $E_M(i)$, in °F. Model error on sampling instant i. The model error is the value of the PV less the value of $\hat{C}(0)$ on sampling instant i. This controller always attempts to drive the PV to the set point in one sampling instant. Consequently, the value of $\hat{C}(0)$ on the current sampling instant is the value of the set point on the previous sampling instant. For this example, the value of the set point is always 155.0°F, so for this controller the value of $\hat{C}(0)$ is 155.0°F on every sampling instant, except $i = 0$.

- $\hat{C}(1) + E_M(i)$, in °F. Expected value of the PV on the next sampling instant, assuming that no control action is taken on the current sampling instant. Actually, the model error $E_M(i)$ is added to all elements of $\hat{C}(k)$ to adjust the expected values for the error in the current values. However, these values are not included in Table 8.6.

- $\hat{e}(1)$, in °F. Expected value of the control error on the next sampling instant, assuming that no control action is taken on the current sampling instant. This is the set point minus the value computed for $\hat{C}(1) + E_M(i)$. The objective of the control action will be to make the appropriate change in the controller output that will offset this error.
- $\Delta m(i)$, in lb/min. Change in the controller output made on the current sampling instant. The change in the controller output is the expected error $\hat{e}(1)$ divided by $s(1)$. As $s(1)$ is -0.05°F/(lb/min), the change in the controller output is the expected error multiplied by -20.
- $M(i)$, in lb/min. Value of the controller output after the change $\Delta m(i)$ is implemented. This is included for information purposes only; it is not used in the calculations for the current controller. However, any constraints on the controller output apply to $M(i)$. For example, the cooling water flow cannot be negative. Issues pertaining to constraints will be discussed shortly.

Two aspects pertaining to the finite step response model are illustrated by the response in Figure 8.25:

- Model accuracy
- Noise on the step response coefficients

Both have a much larger impact on aggressive controllers (such as the square DMC controller), and in practice these two issues often limit how aggressive the controller can be.

The impact of model accuracy is illustrated at the outset of the response in Figure 8.25. On sampling instant 0, the set point is changed from 150.0°F to 155°F. The controller attempts to implement this change over one sampling time. Since $s(1)$ is -0.05°F/(lb/min), the controller concludes that reducing the cooling water flow by 100 lb/min (from 108.8 lb/min to 8.8 lb/min) will increase the reactor temperature by 5.0°F over the next sampling time. However, the process sensitivity increases as the cooling water flow decreases. This explains the large overshoot exhibited in Figure 8.25 that immediately follows the change in the set point. Instead of increasing the reactor temperature from 150.0°F to 155.0°F, the reactor temperature increases from 150.0°F to 159.1°F. The actual peak is 160.4°F, but this occurs about 2 min after the sampling instant. To offset this, the initial change must be followed by an increase of 172 lb/min in cooling water flow on the next sampling instant.

The large overshoot is followed by an undershoot of 1.7°F (to 153.3°F) on sampling instant 4. But on sampling instants 5, 6, and 7, the reactor temperatures are acceptably close to 150.0°F. The values at the sampling instants are 154.6, 154.6, and 154.8°F. The trend in Figure 8.25 suggests slightly more variability, the peaks being 154.8, 154.5, and 154.8°F. But as for the initial overshoots, the peaks occur at a time slightly later than the sampling instant. But in either case, temperatures within 0.5°F of the set point are very acceptable in most applications.

But on sampling instant 7, the controller output is increased by 16 lb/min. This causes the reactor temperature on iteration 8 to drop to 154.1°F, which is 0.9°F below the set point. The reactor temperature appears to be lining-out at the set point, so why does the controller make such a large change that essentially drives the reactor temperature below its set point? In this example, there are no external disturbances to the process, so the explanation must somehow pertain to the controller.

First observe an aberration in the expected values $\hat{C}(k)$ that appears in iteration 2:

$$\hat{C}(7) = 139.84$$

$$\hat{C}(8) = 139.48$$

$$\hat{C}(9) = 139.12$$

$$\hat{C}(10) = 138.26$$

$$\hat{C}(11) = 138.79$$

$$\hat{C}(12) = 137.90$$

The first three values suggest a gradual decrease in the expected values of the PV. But the last three elements exhibit much larger changes, with both an increase as well as a decrease.

Figure 8.26 presents a plot of the expected values $\hat{C}(k)$ of the PV at the outset of sampling instant 6. The plot suggests a gradual increase with time, but with considerable scatter akin to noise. The source is the finite step response model. The values of the PV in the data from the step test illustrated in Figure 8.22 have a resolution of 0.1°F. Consequently, the values of the PV used to compute the finite step response model in Table 8.3 also have a resolution of 0.1°F. When

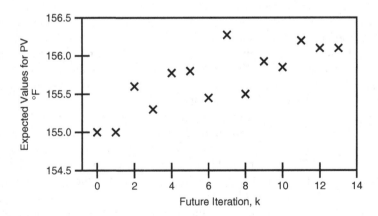

Figure 8.26 Noise on expected values of PV.

divided by the magnitude of the step change (20 lb/min), the values in the finite step response model have a resolution of 0.005°F/(lb/min). In fact, $s(11)$ and $s(12)$ are equal.

The errors introduced by a finite resolution are sometimes referred to as *quantization errors*. Often, the consequences are similar to those of noise, so the term *quantization noise* arises. A change of 0.005 seems like a small change, but this must be compared to the magnitude of 0.225 for $s(13)$. In this context, the resolution is 1 part in 45 (0.225/0.005). This is a very poor resolution, especially when multiplied by the large changes in the controller output from an aggressive controller.

Issues Pertaining to the Model. How aggressive the controller can be is often limited by the quality of the model. For the finite step response model, there are two aspects:

- *Accuracy of the predicted values.* For process applications, the degree of nonlinearity exhibited by the process is always a valid concern.
- *Noise in the values for $s(k)$.* In simple tests like that in Figure 8.22, noise in the test data translate directly to noise in the values for $s(k)$. Finite resolution in the test data is merely another source of noise. Shortly, test procedures will be discussed that use regression techniques that minimize the effect of the noise in the test data.

When faced with these issues, there are two alternatives:

1. Make the controller less aggressive.
2. Improve the quality of the model.

Shortly, the QDMC version of the DMC controller will be presented. QDMC provides additional features that can be used to make the controller less aggressive, so the discussion of this option will be deferred. But this is a good time to discuss issues pertaining to the model.

What can be done about noise in values of $s(k)$ for the finite step response model? The simple answer is to "smooth" the values of $s(k)$. For measured inputs, smoothing is usually applied using filters such as the exponential moving average. Such "one-sided" filters impart lag into the smoothed values and are not appropriate for smoothing the finite step response model.

Although exceptions definitely exist, most processes can be described adequately by a two-time-constant-plus-dead-time model. Consider the following approach:

1. Fit a two-time-constant-plus-dead-time model to the values of $s(k)$.
2. Compute the response of the two-time-constant-plus-dead-time model to a unit step change in its input.

3. For $s(k)$ use the values at the appropriate times from the step response. These values for $s(k)$ have high resolution, even if the original values of $s(k)$ have poor resolution.

Applying nonlinear regression techniques to the coefficients of $s(k)$ in Table 8.3 gives the following values for the model coefficients that minimize the sum of squares of the error:

Model Coefficient	Value
Gain K	$-0.2312°$F/(lb/min)
Minor time constant τ_1	55.49 min
Major time constant τ_2	0.0 min
Dead time θ	1.353 min

For this example, the second time constant is not required.

Table 8.7 presents the values for $s(k)$ computed from this model. In addition to smoothing the values, the number of elements for $s(k)$ has been increased from 13 to 20. Figure 8.27 presents the response of the DMC controller to a step change in reactor temperature set point from $150.0°$F to $155.0°$F. As compared to the response in Figure 8.25 for the original values of $s(k)$, the large initial overshoot is still present, but the erratic behavior following the initial cycle has been eliminated.

Quadratic Dynamic Matrix Controller. The square DMC controller was formulated as follows:

- Number of future control errors to drive to zero: N
- Number of control moves available to the controller: N

The dynamic matrix A will be a square $N \times N$ matrix, and provided the inverse $\mathbf{A}^{-1}$ exists, values can be computed for the control modes that will drive all future control errors to zero.

For QDMC the formulation is as follows:

- Number of future control errors to drive to zero: R (often called the *horizon*)
- Number of control moves available to the controller: L

All formulations presented herein assume that $L < R$; that is, the number of control moves available to the controller is less than the number of future control errors to drive to zero.

The dynamic matrix A is a nonsquare $R \times L$ matrix, which means that the inverse $\mathbf{A}^{-1}$ cannot be computed. If written as a set of equations, the number of equations (R) exceeds the number of unknowns (L). The controller cannot

Table 8.7   Smoothed finite step response model

k	Original $s(k)$	Smoothed $s(k)$
0	0.0	0.0
1	−0.050	−0.0504
2	−0.095	−0.0932
3	−0.125	−0.1259
4	−0.150	−0.1509
5	−0.170	−0.1699
6	−0.185	−0.1844
7	−0.195	−0.1955
8	−0.205	−0.2039
9	−0.210	−0.2104
10	−0.215	−0.2153
11	−0.220	−0.2191
12	−0.220	−0.2220
13	−0.225	−0.2241
14		−0.2258
15		−0.2271
16		−0.2281
17		−0.2288
18		−0.2294
19		−0.2298
20		−0.2301

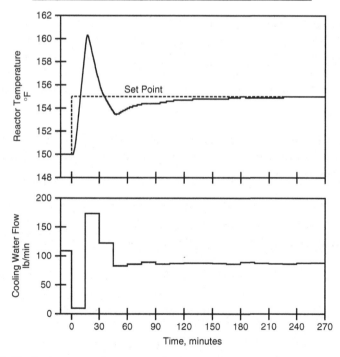

Figure 8.27   Performance of the square DMC controller with smoothed values for $s(k)$.

achieve the stated objective of driving the R predicted control errors to zero. However, the controller can be formulated to minimize some function of the future control errors. The most common function is the sum of squares of the control errors.

The following dynamic matrix relates R future control errors to L control moves:

$$\begin{bmatrix} x(1) \\ x(2) \\ x(3) \\ \vdots \\ x(R) \end{bmatrix} = \begin{bmatrix} s(1) & 0 & 0 & \cdots & 0 \\ s(2) & s(1) & 0 & \cdots & 0 \\ s(3) & s(2) & s(1) & \cdots & 0 \\ \vdots & \vdots & \vdots & & \vdots \\ s(R) & s(R-1) & s(R-2) & \cdots & s(R-L-1) \end{bmatrix} \begin{bmatrix} \Delta m(i) \\ \Delta m(i+1) \\ \Delta m(i+2) \\ \vdots \\ \Delta m(i+L-1) \end{bmatrix}$$

or

$$\mathbf{x} = \mathbf{A}\Delta\mathbf{m}$$

where

$\Delta\mathbf{m} = L \times 1$ vector of proposed current and future control moves
$\mathbf{x} = R \times 1$ vector of expected results of control moves $\Delta\mathbf{m}$
$\mathbf{A} = R \times L$ dynamic matrix

In composing this matrix, it is understood that $s(k) = 0$ for $k \leq 0$ and $s(k) = s(N)$ for $k > N$.

Although not possible when $L < R$, the desire is to drive all control errors to zero. Consequently, the desire is that result $\mathbf{x}$ of the current and future control actions cancel these errors:

$$\mathbf{x} = \hat{\mathbf{e}} = \mathbf{A}\Delta\mathbf{m}$$

The vector $\hat{\mathbf{e}}$ consists of the predicted control errors assuming that no further control actions are taken. If $\Delta\mathbf{m}$ is the vector of the future control moves, the net predicted control errors are

$$\hat{\mathbf{e}} - \mathbf{A}\Delta\mathbf{m}$$

The criterion function Φ is the sum of squares of the net predicted control errors:

$$\Phi = [\hat{\mathbf{e}} - \mathbf{A}\Delta\mathbf{m}][\hat{\mathbf{e}} - \mathbf{A}\Delta\mathbf{m}]^{\mathrm{T}}$$

The control moves $\Delta\mathbf{m}$ that minimize Φ are computed as follows:

$$\Delta\mathbf{m} = (\mathbf{A}^{\mathrm{T}}\mathbf{A})^{-1}\mathbf{A}^{\mathrm{T}}\hat{\mathbf{e}}$$

If $\mathbf{A}$ is an $R \times L$ matrix, then $\mathbf{A}^T\mathbf{A}$ is a square $L \times L$ matrix whose inverse can be computed.

Implementing the Controller. To illustrate the formulation of the controller for the reactor temperature, the following low dimensional configuration will be used:

- Horizon $R = 4$
- Number of control moves $L = 2$

In this example, small values are intentionally used so that the size of the various matrices will not be excessive.

Using the values $s(k)$ for the finite step response model from Table 8.3, the dynamic matrix $\mathbf{A}$ is a 4×2 matrix:

$$\mathbf{A} = \begin{bmatrix} -0.050 & 0 \\ -0.095 & -0.050 \\ -0.125 & -0.095 \\ -0.150 & -0.125 \end{bmatrix}$$

The transpose $\mathbf{A}^T$ of matrix A is a 2×4 matrix:

$$\mathbf{A}^T = \begin{bmatrix} -0.050 & -0.095 & -0.125 & -0.150 \\ 0 & -0.050 & -0.095 & -0.125 \end{bmatrix}$$

The product $\mathbf{A}^T\mathbf{A}$ is a 2×2 matrix:

$$\mathbf{A}^T\mathbf{A} = \begin{bmatrix} 0.04965 & -0.03538 \\ 0.03538 & 0.02715 \end{bmatrix}$$

The inverse $(\mathbf{A}^T\mathbf{A})^{-1}$ is also a 2×2 matrix:

$$(\mathbf{A}^T\mathbf{A})^{-1} = \begin{bmatrix} 281.0 & -366.2 \\ -366.2 & 513.9 \end{bmatrix}$$

The product $(\mathbf{A}^T\mathbf{A})^{-1}\mathbf{A}^T$ is a 2×4 matrix:

$$(\mathbf{A}^T\mathbf{A})^{-1}\mathbf{A}^T = \begin{bmatrix} -14.05 & -8.390 & -0.3429 & 3.617 \\ 18.31 & 9.090 & -3.052 & -9.316 \end{bmatrix}$$

The computation of the control moves is as follows:

$$\begin{bmatrix} \Delta m(i) \\ \Delta m(i+1) \end{bmatrix} = \begin{bmatrix} -14.05 & -8.390 & -0.3429 & 3.617 \\ 18.31 & 9.090 & -3.052 & -9.316 \end{bmatrix} \begin{bmatrix} \hat{e}(i+1) \\ \hat{e}(i+2) \\ \hat{e}(i+3) \\ \hat{e}(i+4) \end{bmatrix}$$

For a step change of $+5.0°F$ in the reactor temperature set point, all future control errors $\hat{e}(i + k)$ will be $+5.0°F$ on the first sampling instant after the change in the set point. This gives the following values for the two control moves:

$$\Delta m(i) = -95.8\text{lb/min}$$

$$\Delta m(i + 1) = +75.1\text{lb/min}$$

In practice, there is no need to calculate $\Delta m(i + 1)$. The controller output (cooling water flow set point) is decreased by 95.8 lb/min on the first sampling instant, so the value of $\Delta m(i)$ is used. However, the value of $\Delta m(i + 1)$ is never used. On each sampling instant, the value of $\Delta m(i)$ is recalculated and the controller output changed accordingly. On each sampling instant, only the following equation must be calculated:

$$\Delta m(i) = (-14.05) \times \hat{e}(i + 1) + (-8.390) \times \hat{e}(i + 2)$$
$$+ (-0.3429) \times \hat{e}(i + 3) + (3.617) \times \hat{e}(i + 4)$$

The matrix calculations are required to design the controller, but need not be performed on each sampling instant.

Figure 8.28 presents the performance of the QDMC controller formulated for $R = 4$ and $L = 2$. This controller is based on the original values for the finite step response model in Table 8.3. The following observations are derived by comparing the performance in Figure 8.28 to that of the square DMC controller in Figure 8.25:

- The initial control move is only slightly smaller (a decrease of 95.8 lb/min instead of 100 lb/min). Consequently, the initial overshoot and subsequent cycle are not significantly different.
- Except for one brief excursion, the QDMC controller remains within $0.5°F$ of the set point after the initial cycle has passed. The quantization errors in the values $s(k)$ for the finite step response model have less effect on the controller performance, but smoothing of the finite step response model would be beneficial.

The controller in Figure 8.28 will line out at a cooling water flow of 87.1 lb/min, a reduction of 21.7 lb/min from its initial value of 108.8 lb/min. The final coefficient in the finite step response model is $-0.225°F/(\text{lb/min})$, which is also the steady-state process gain. To increase the reactor temperature by $5.0°F$, the change in cooling water flow predicted by the model agrees closely with the actual reduction of 21.7 lb/min:

$$\frac{5.0°F}{-0.225°F/(\text{lb/min})} = -22.2\text{lb/min}$$

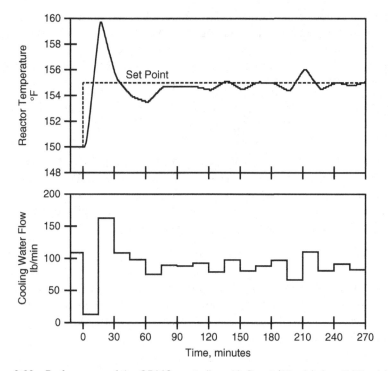

Figure 8.28   Performance of the QDMC controller with $R = 4$ (60 min), $L = 2$ (30 min), and $\Delta t = 15$ min.

If the process is to line-out at 155.0°F, the sum of the control moves $\Delta m(i + k)$ must equal this value. The sum of the values of $\Delta m(i)$ and $\Delta m(i + 1)$ computed above is -20.7 lb/min. As the horizon R increases, the sum will be even closer to the change computed from the steady-state gain. If the controller is allowed only one control move ($L = 1$) and if $R > 1$, the value of $\Delta m(i)$ will be approximately -22.2 lb/min, which is a very conservative control move.

Sampling Time.   In previous examples, the sampling time Δt has been 15 min. This unreasonably long sampling time was used for two reasons:

1. *Reduce the dimensionality.* Fewer coefficients are required in the finite step response model. Although beneficial when humans must work with the vectors and matrices, higher dimensions are not a problem for computers.
2. *Output does not exceed a limit.* The lower limit on the cooling water flow set point is 0.0 lb/min. Even for a sampling time of 15 min, a 5°F increase in the set point causes the square DMC controller to reduce the cooling water flow set point from 108.8 lb/min to 8.8 lb/min. For the QDMC controller, the reduction is only slightly less, unless $L = 1$.

The magnitude of the initial control move depends on the $s(1)$ coefficient, which in turn depends on the sampling time Δt:

Sampling Time, Δt	Coefficient $s(1)$	$\Delta m(i)$ for $5°$F Set Point Increase
15 min	$-0.050°$F/(lb/min)	-100 lb/min
5 min	$-0.0147°$F/(lb/min)	-340 lb/min
1 min	$0.0°$F/(lb/min)	????

These values of $\Delta m(i)$ are for the square DMC controller. For the QDMC controller, the values depend on the values of R and L, but would be only slightly smaller unless $L = 1$.

For $\Delta t = 1$ min, the $s(1)$ coefficient is zero, the reason being that the process dead time θ is approximately 1.353 min. In general,

$$s(k) = 0 \qquad \text{for all } k \le \theta / \Delta t$$

A value of zero for the $s(1)$ coefficient is a major problem for the square DMC controller. However, the QDMC controller can be formulated provided the horizon R exceeds the dead time by at least the number of control moves, that is, provided that

$$R > \text{int}(\theta / \Delta t) + L$$

For a controller with only one input and one output, the value of the output $M(i)$ can simply be constrained to the limits. When updating the predicted values $\hat{C}(k)$ for the controlled variable, the actual control move must be used, not the calculated control move. This simple approach is not sufficient for higher-dimensional controllers. The control move for each output depends on the change actually made for the other outputs, not the calculated control moves. The handling of constraints will be discussed subsequently.

For Figure 8.28, the horizon R was 4 and the number of control moves L was 2. For a sampling time of 15 min, the equivalent time intervals are $R = 60$ min and $L = 30$ min. To illustrate the effect of the sampling time, the value of R and L will be adjusted to retain the same time intervals. Also, the values for the coefficients $s(k)$ in the finite step response model will be computed from the first-order model determined previously for the purpose of smoothing the coefficients.

Figure 8.29 presents the response to an increase in the set point from $150°$F to $155°$F for a sampling time $\Delta t = 5$ min, horizon $R = 12$ (60 min), and number of control moves $L = 6$ (30 min). As compared to the response in Figure 8.28 for $\Delta t = 15$ min, the peak overshoot has been reduced from approximately $5°$F to less than $3°$F. The subsequent undershoot is just over $0.5°$F. The response lingers below the set point for some time, but is within about $0.3°$F of the set point. Reducing the sampling time from 15 min to 5 min has improved the performance of the controller significantly.

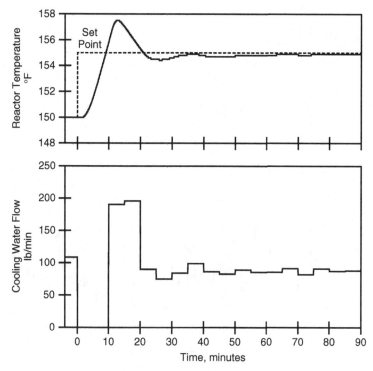

Figure 8.29   Performance of the QDMC controller with $R = 12$ (60 min), $L = 6$ (30 min), and $\Delta t = 5$ min.

Figure 8.30 presents the response to an increase in the set point from 150°F to 155°F for a sampling time $\Delta t = 1$ min, horizon $R = 60$ (60 min), and number of control moves $L = 30$ (30 min). Shortening the sampling time to 1 min has reduced the overshoot to approximately 1°F. However, another aspect of this response is both noticeable and a cause for concern. Following the initial cycle, the response essentially lines-out at the set point, remaining within approximately 0.1°F. However, the controller output exhibits considerable movement, often essentially alternating between two values on consecutive sampling instants. This phenomenon is often referred to as *ringing*. Although it has little effect on the reactor temperature, it leads to excessive movement for the cooling water control valve, with implications for valve maintenance.

The ringing in Figure 8.30 is due largely to the 0.1°F resolution in the process variable. For unlimited resolution, the degree of ringing is minimal and the controller output rapidly approaches a constant value.

Reducing the number of control moves L makes the QDMC controller less aggressive. The response presented in Figure 8.31 is for $L = 2$. The resolution of the measured value of the reactor temperature is 0.1°F. The ringing is clearly less, but most would consider this degree of ringing to be excessive.

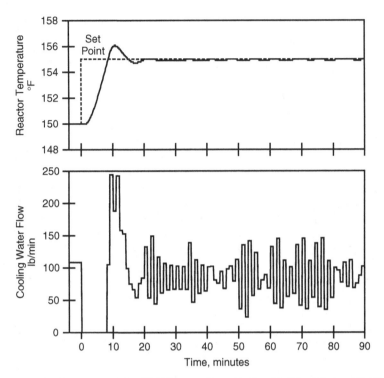

Figure 8.30   Performance of the QDMC controller with $R = 60$ (60 min), $L = 30$ (30 min), and $\Delta t = 1$ min.

Tuning the QDMC Controller. For a PID controller, tuning means getting the values of these coefficients "in tune" with the process characteristics. The term *tuning* is applied routinely to the QDMC controller, but the context is somewhat different. The behavior of the process should be reflected in the finite step response model that is used to characterize the process. Manually adjusting the values of coefficients for $s(k)$ is not practical. If the model is deemed inadequate, the process test must be repeated to obtain new values for the coefficients.

For the QDMC controller, tuning determines primarily how aggressive the controller will respond to any change, which depends on the quality of the model. The better the model, the more aggressively the controller can respond. If the model represents the behavior of the process poorly, the controller must take conservative control moves.

For the QDMC controller, both the horizon R and the number of control moves L can be used as tuning parameters. Their effect is as follows:

Horizon R. Increasing the horizon makes the controller less aggressive.

Number of control moves L. Decreasing the number of control moves makes the controller less aggressive.

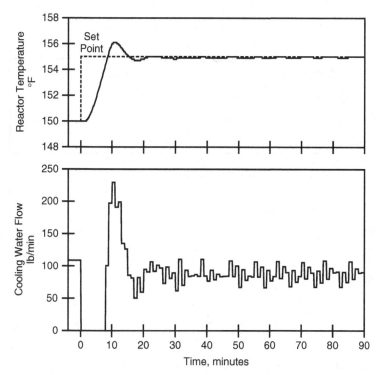

Figure 8.31 Performance of the QDMC controller with $R = 60$ (60 min), $L = 2$ (2 min), and $\Delta t = 1$ min.

Although these two parameters are the primary "tuning coefficients" for a QDMC controller, most implementations provide additional features that in effect cause the controller to compute a smaller control move.

The simplest to understand is to apply an exponential lag to the predicted control errors on which the control calculations are based. If the time constant is τ, the lagged predicted control errors $\hat{e}'(i + k)$ are related to the predicted control errors $\hat{e}(i + k)$ as follows:

$$\hat{e}'(i + k) = (1 - e^{-k\Delta t/\tau})\hat{e}(i + k)$$

The larger the time constant τ, the less aggressive the controller. For set point changes, the effective result is that the set point change is implemented in a lagged fashion.

Figure 8.32 presents the response to a set point change of 5°F for horizon $R = 60$, control moves $L = 2$, sampling time $\Delta t = 1$ min, and time constant $\tau = 3$ min. The response exhibits negligible overshoot and only minimal ringing. A closer inspection of the response also shows that the ringing is triggered when the reactor temperature changes by 0.1°F.

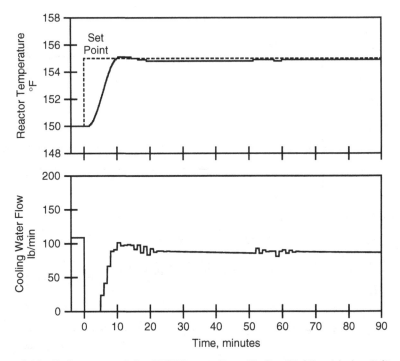

Figure 8.32 Performance of the QDMC controller with $R = 60$ (60 min), $L = 2$ (2 min), $\Delta t = 1$ min, and $\tau = 3$ min.

The *move suppression factor* is another approach to cause the controller to generate smaller control moves. The idea is to penalize the controller for the size of the control moves that are taken. The criterion function Φ is modified as follows:

$$\Phi = [\hat{\mathbf{e}} - \mathbf{A}\Delta\mathbf{M}][\hat{\mathbf{e}} - \mathbf{A}\Delta\mathbf{M}]^T + k^2 \Delta\mathbf{m}^T \Delta\mathbf{m}$$

The control moves $\Delta\mathbf{m}$ that minimize this criterion function are computed as follows:

$$\Delta\mathbf{m} = (\mathbf{A}^T\mathbf{A} + k^2\mathbf{I})^{-1}\mathbf{A}^T\hat{\mathbf{e}}$$

Figure 8.33 presents the response to a set point change of $5°F$ for horizon $R = 60$, control moves $L = 2$, sampling time $\Delta t = 1$ min, and move suppression factor $k = 0.17$. The response exhibits negligible overshoot (approximately $0.3°F$) and no ringing. As compared to other responses (including Figure 8.32), the magnitude of the control moves is clearly less.

Regardless of the option chosen, never lose sight of the trade-off. Making the controller less aggressive makes the controller less sensitive to the shortcomings of the model. But there is a cost: A less aggressive controller does not respond as quickly to changes in either the set point or the various unmeasured disturbances.

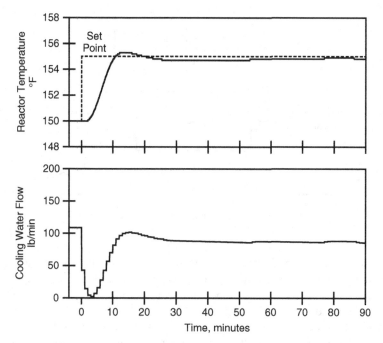

Figure 8.33 Performance of the QDMC controller with $R = 60$ (60 min), $L = 2$ (2 min), $\Delta t = 1$ min, and move suppression factor $= 0.17$.

Process Test Issues. Processes are difficult to test, mainly for the following reasons:

- *Most processes respond slowly.* The reactor with a once-through jacket is typical of most temperature processes. The response to the step change in the controller output presented in Figure 8.22 requires 4 hr to line-out.
- *Processes are exposed to numerous sources of disturbances.* The major disturbances are measured whenever possible, but it is impractical to measure the minor disturbances.
- *Processes are multivariable.* For the purposes of a process test, the control actions from other loops are essentially upsets to the loop being tested. This includes set point changes made by the process operators on other loops.

Is it feasible to conduct a meaningful test on processes? Probably the most appropriate answer is "Yes, but with great difficulty." And "garbage in, garbage out" definitely applies. The data from a botched test do not accurately reflect the behavior of the process. The same can be said for any model coefficients derived from this data.

For reasons that will be explained shortly, most process tests are conducted in an open-loop manner, that is, with the controller in manual. Figure 8.34 illustrates

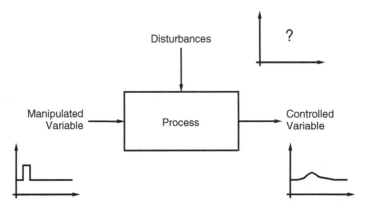

Figure 8.34   Effect of disturbances on a process test.

process testing for the single-loop case: that is, one controller output and one controlled variable. The test input is a predetermined sequence of changes in the controller output. The process response is captured as values of the controlled variable recorded on the sampling time Δt.

During the process test known changes are made in the controller output. But as illustrated in Figure 8.34, there is another input to the process: specifically, the disturbances. Although one or more of the disturbances are occasionally measured, all applications have disturbances that are not measured. The attention herein is directed to the latter.

The model must describe how changes in the controller output M affect the controlled variable C. This relationship is expressed generically as follows:

$$C = f(M)$$

The complication is that values collected for the controlled variable C depend on the changes made in the controller output M and the changes that occur in the unmeasured disturbances U. This relationship is expressed generically as follows:

$$C = g(M, U)$$

The simplest approach would be to assume that the response in the controller output C is due solely to the changes made in the controller output M. However, this is clearly not assured.

Most modern digital control systems have extensive historical data collection facilities that capture the values of the key process variables, including the controller outputs that influence these variables. To reduce the storage required for the data, most use a data compression scheme of some type that essentially smoothes or averages the data. However, data compression can be disabled, at least for those variables relevant to the model predictive controller.

Is it possible to develop the model by analyzing the data collected in this manner? Lots of data are available, often spanning months or years. Over such long time periods, the disturbances U must "average out"; after all, what goes up seems eventually to come down. But for developing dynamic models by analyzing these data, the statement that the disturbances "average out" is not sufficient. To derive a dynamic model from these data, the required statement is that the disturbances U are not correlated with changes in the controller output M.

This can be illustrated using a simple example from linear regression. Suppose that the dependent variable Y is related to the independent variables X_1 and X_2 as follows:

$$Y = 2.1X_1 + 3.4X_2 + 0.2$$

where
$Y =$ crop yield
$X_1 =$ fertilizer applied
$X_2 =$ rainfall during growing season

Suppose that the following data points are available:

X_1	X_2	Y
2.0	1.0	7.8
3.0	1.5	11.6
4.0	2.0	15.4
5.0	2.5	19.2

The independent variables X_1 and X_2 are perfectly correlated ($X_1 = 2X_2$), as if each time it rains, the farmers fertilize their crop. If only the values for X_1 and Y are available (no data for X_2), applying regression to relate Y to X_1 gives the following equation:

$$Y = 3.8X_1 + 0.2$$

The fit is excellent, but the coefficient 3.8 for X_1 is only valid provided that the farmers always fertilize following each rainfall. The true contribution of fertilizer is reflected by the coefficient 2.1; the coefficient 3.8 reflects the contribution of fertilizer and rainfall.

This example translates as follows to the test configuration in Figure 8.34:

- The controlled variable C corresponds to the dependent variable Y. The process test provides data for the controlled variable C.
- The controller output M corresponds to the independent variable X_1. The process test provides data for the controller output M.
- The disturbance U corresponds to the independent variable X_2. The process test provides no data for the disturbance U.

Regression can be applied to the test data to obtain a relationship between the controlled variable C and the controller output M. However, if the controller output M and the disturbance U are correlated to a significant degree, the relationship ascribes to the controller output M effects that are really the result of the disturbance U. Regression can be applied to obtain a model, but the model does not reflect the true relationship between the controller output M and the controlled variable C.

With no data available on the disturbance U, the degree of correlation between M and U cannot be determined. The only option is to devise test procedures that avoid a high degree of correlation between M and U. The following observations are relevant:

1. *Closed-loop control increases the degree of correlation between M and U.* A change occurs in the disturbance U. This affects the controlled variable C, leading the controller to make a change in the controller output M. This increases the degree of correlation between M and U. The data collected by a historian are normal operating data, which means that controllers are functioning. The degree of correlation between M and U is such that model coefficients derived from these data do not reflect the true relationship between M and C.

2. *The least correlation between M and U is in open loop tests conducted using a random signal for M.* For finite test times (as they always are), the degree of correlation between M and U will not be zero. But if the signal for M is random, the degree of correlation decreases as the time period for the test increases. For determining values for the coefficients in a finite step response model, the usual approach is to conduct a test lasting several days with some type of random signal for the controller output M.

Operating a process for several days with one or more of the key controllers on manual usually entails some discussions. A usual requirement for any process test is that the controlled variable(s) must remain at all times within limits that are agreed upon in advance. With the automatic controllers disabled, manual corrections are likely to become necessary to avoid excursions beyond these limits. These manual corrections are essentially feedback control actions that increase the degree of correlation between M and U. An occasional correction can be tolerated; frequent corrections are a definite cause for concern.

With the controller on automatic, there are ways to reduce the degree of correlation between M and U:

- Make frequent set point changes.
- Superimpose random changes in the controller output over and above any made by the controller.
- Make the controller less aggressive, or perhaps only enabling the controller when its controlled variable approaches a test limit.

Tests can be conducted with controllers in automatic, but those conducting such process tests must be very knowledgeable about the trade-offs and exercise great caution. The more conservative approach is to conduct the test with the controllers on manual and apply manual corrections as necessary to maintain process operations within the test limits.

In process applications, the most common approach to generating a random test signal is the *pseudorandom binary signal* (PRBS), which is generated as follows:

- Specify an upper value M_U for the controller output and a lower value M_L for the controller output. Often, the test limits imposed on the controlled variable(s) are the primary considerations in establishing these values.
- Specify a sampling time for the test. Data are collected on this time interval. Changes in the controller output are allowed only on the sampling instants.
- Use a random number generator to determine the times to switch between M_U and M_L. Random number generators are available for computers, but these are usually referred to as *pseudorandom number generators*. The routines are provided a "seed" value at the start. If the same seed is provided on two separate occasions, the same sequence of random numbers is generated.
- Specify a lower limit on the switch time and an upper limit on the switch time. An extremely short switch interval (such as 1.0 min for the reactor with once-through jacket) is not productive. An extremely long switch interval results in a very few number of switches during the test (without an upper limit, it is even conceivable that the random number generator could produce a switch interval that exceeds the time reserved for the test).

Figure 8.35 presents the results of a PRBS test for the reactor with a once-through jacket. The test parameters are as follows:

Test Parameter	Value
Lower value for controller output, M_L	90 lb/min
Upper value for controller output, M_U	130 lb/min
Sampling time, Δt	15 min
Minimum time interval for switch	15 min
Maximum time interval for switch	120 min
Test duration	48 hr

Including the point at $t = 0$, this test provides a total of 193 data points.

Analyzing the Test Data. For a process with one controlled variable and one manipulated variable, the process test obtains the response in the controlled variable to known changes in the controller output. The following data are normally

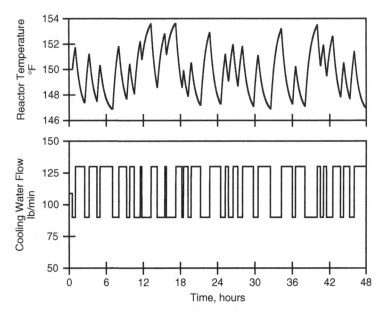

Figure 8.35 Results of PRBS test.

recorded on a fixed sampling interval Δt as follows:

$$M(i) = \text{value of the controller output for data point } i$$

$$C(i) = \text{value of controlled variable for data point } i$$

Suppose that the data are recorded for $i = -1, 0, 1, 2, \ldots, Q$. For the purposes of developing the coefficients in a finite step response model, the data can be expressed as follows:

$$c(i) = C(i) - C_{eq} \qquad \text{for } i = 1, 2, \ldots, Q$$

$$\Delta m(i) = M(i) - M(i - 1) \qquad \text{for } i = 0, 1, 2, \ldots, Q - 1$$

where C_{eq} is the initial or equilibrium value of the controlled variable. But before proceeding any further, what is the appropriate value for C_{eq}? The value of C_{eq} is the equilibrium or steady-state value of the controlled variable that corresponds to a value M_{eq} for the controller output.

By defining the control moves $\Delta m(i)$ as the difference between successive values of $M(i)$, no value is required for M_{eq}. Doing the same for the controlled variable, the test data would be expressed as follows:

$$\Delta c(i) = C(i) - C(i - 1) \qquad \text{for } i = 1, 2, \ldots, Q$$

$$\Delta m(i) = M(i) - M(i - 1) \qquad \text{for } i = 0, 1, 2, \ldots, Q - 1$$

No value is required for M_{eq} or C_{eq}.

In a manner similar to that used to obtain the dynamic matrix $\mathbf{A}$, the principle of superposition can applied as follows:

- Relate $c(i)$ to the control moves $\Delta m(i)$ and the finite step response model $s(k)$.
- Relate $\Delta c(i)$ to the control moves $\Delta m(i)$ and the finite impulse response model $g(k)$.

For analyzing the process test data, the latter is preferable. An equation can be written for each data point from the process test data:

$$\Delta c(1) = \Delta m(0)g(1)$$

$$\Delta c(2) = \Delta m(1)g(1) + \Delta m(0)g(2)$$

$$\Delta c(3) = \Delta m(2)g(1) + \Delta m(1)g(2) + \Delta m(0)g(3)$$

$$\vdots$$

$$\Delta c(N) = \Delta m(N-1)g(1) + \Delta m(N-2)g(2) + \Delta m(N-3)g(3)$$
$$+ \cdots + \Delta m(0)g(N)$$

$$\vdots$$

$$\Delta c(Q) = \Delta m(Q-1)g(1) + \Delta m(Q-2)g(2) + \Delta m(Q-3)g(3)$$
$$+ \cdots + \Delta m(Q-N)g(N)$$

Using vectors and matrices, these equations are expressed as follows:

$$
\begin{bmatrix}
\Delta c(1) \\
\Delta c(2) \\
\Delta c(3) \\
\vdots \\
\Delta c(N) \\
\vdots \\
\Delta c(Q)
\end{bmatrix}
$$

$$
=
\begin{bmatrix}
\Delta m(0) & 0 & 0 & \cdots & 0 \\
\Delta m(1) & \Delta m(0) & 0 & \cdots & 0 \\
\Delta m(2) & \Delta m(1) & \Delta m(0) & \cdots & 0 \\
\vdots & \vdots & \vdots & & \vdots \\
\Delta m(N-1) & \Delta m(N-2) & \Delta m(N-3) & \cdots & \Delta m(0) \\
\vdots & \vdots & \vdots & & \vdots \\
\Delta m(Q-1) & \Delta m(Q-2) & \Delta m(Q-3) & \cdots & \Delta m(Q-N)
\end{bmatrix}
\begin{bmatrix}
g(1) \\
g(2) \\
g(3) \\
\vdots \\
g(N)
\end{bmatrix}
$$

or

$$\Delta \mathbf{c} = \Delta \mathbf{M} \, \mathbf{g}$$

where

$\Delta \mathbf{c} = Q \times 1$ vector of the controlled variable differences from the test data

$\mathbf{g} = M \times 1$ vector of coefficients for the finite impulse response model

$\Delta \mathbf{M} = Q \times N$ matrix of the changes in the controller output for the test

The number of equations Q exceeds the number of unknowns N. If $\Delta \hat{c}(i)$ is the value predicted for $\Delta c(i)$ using the finite impulse response model, the values of the coefficients $g(k)$ in the finite impulse response model that minimize the sum of squares of the errors $\Delta \hat{c}(i) - \Delta c(i)$ are computed using the following equation:

$$\mathbf{g} = (\Delta \mathbf{M}^{\mathrm{T}} \Delta \mathbf{M})^{-1} \Delta \mathbf{M}^{\mathrm{T}} \Delta \mathbf{c}$$

The step response coefficients $s(k)$ are computed from the impulse response coefficients:

$$s(k) = g(k) + g(k-1), \qquad k = 1, 2, 3, \ldots, N$$

It is understood that $s(0) = 0$ and $g(0) = 0$.

The equation above computes the values of the coefficients provided the user specifies the following:

- The number of coefficients N for the model. The number of coefficients must be adequate, but not excessive.
- The process dead time θ. If the dead time is specified as zero, the regression should ideally compute a value of zero for all $g(k)$ and $s(k)$ for which $k \leq \theta / \Delta t$. But in practice, the coefficients will be close to zero but not exactly zero. By specifying a value for the dead time, these coefficients will be exactly zero.

Usually, one provides estimates for these, computes a fit, and then adjusts the values until the fit ceases to improve noticeably.

The data from the process test illustrated in Figure 8.35 can be analyzed to obtain values for the coefficients in a finite step response model. Using $N = 13$, the values of the coefficients from the test are presented in Table 8.8. Compared to the original values obtained from a direct step test of the process, the most noticeable difference is a lower value for the process gain, which is the value of $s(13)$. This difference is very likely due to the nonlinear nature of the reactor with a once-through jacket. For the PRBS test, the cooling water flow was switched between 90 and 130 lb/min. For the step test, the cooling water flow was reduced to 88.8 lb/min. The process sensitivity is higher at low water flows than at high

Table 8.8 Finite step response model from PRBS test

k	Original $s(k)$	PRBS Test $s(k)$
0	0.0	0.0
1	−0.050	−0.0416
2	−0.095	−0.0803
3	−0.125	−0.1083
4	−0.150	−0.1292
5	−0.170	−0.1445
6	−0.185	−0.1557
7	−0.195	−0.1638
8	−0.205	−0.1696
9	−0.210	−0.1736
10	−0.215	−0.1764
11	−0.220	−0.1778
12	−0.220	−0.1789
13	−0.225	−0.1796

water flows, so it is not surprising that the step test gives a higher value for the process gain.

No aberrations are present in the coefficients in Table 8.8 that are computed from the PRBS test data. However, this is not assured, and if not, the effect on the behavior of the model predictive controller is similar to that of a finite resolution in the values of the controlled variable. If necessary, the values of the coefficients can be smoothed.

Multivariable Controller Formulation. The single-input, single-output version of a model predictive controller is a good way to gain an insight into how the controller functions. However, such controllers are rarely encountered in practice, the major benefit of model predictive controllers being in multivariable applications.

A major advantage of using matrix equations is that the dimensionality of the problem can easily be increased. From a computational standpoint, the dimensions of the final matrices become much larger, but today's computers are definitely up to the task. The discussion that follows assumes a 3×3 multivariable process. However, it is not necessary that the process be square.

For a 3×3 multivariable process, controlled variable C_1 is a function of manipulated variables M_1, M_2, and M_3. This relationship can be expressed by three dynamic matrices:

- $\mathbf{A_{11}}$: describes the effect of M_1 on C_1.
- $\mathbf{A_{12}}$: describes the effect of M_2 on C_1.
- $\mathbf{A_{13}}$: describes the effect of M_3 on C_1.

Let the horizon be R_1 for each. However, the control moves can be different:

$$L_1 = \text{number of future control moves for } M_1$$
$$L_2 = \text{number of future control moves for } M_2$$
$$L_3 = \text{number of future control moves for } M_3$$

Let x_1 be the expected results of these control moves on C_1 for the next R sampling instants. The relationship is expressed by the following equation:

$$\mathbf{x_1} = \mathbf{A_{11}} \Delta \mathbf{m_1} + \mathbf{A_{12}} \Delta \mathbf{m_2} + \mathbf{A_{13}} \Delta \mathbf{m_3}$$

where

$\mathbf{x_1} = R_1 \times 1$ vector of the effect of the control moves on C_1
$\Delta \mathbf{m_1} = L_1 \times 1$ vector of the future control moves for M_1
$\Delta \mathbf{m_2} = L_2 \times 1$ vector of the future control moves for M_2
$\Delta \mathbf{m_3} = L_3 \times 1$ vector of the future control moves for M_3
$\mathbf{A_{11}} = R_1 \times L_1$ dynamic matrix relating C_1 to M_1
$\mathbf{A_{12}} = R_1 \times L_2$ dynamic matrix relating C_2 to M_2
$\mathbf{A_{13}} = R_1 \times L_3$ dynamic matrix relating C_3 to M_3

Similar equations can be written for $\mathbf{x_2}$ and $\mathbf{x_3}$, their horizons being R_2 and R_3, respectively.

These can be combined into one matrix equation:

$$\mathbf{x} = \mathbf{A} \Delta \mathbf{m}$$

where

$$\Delta \mathbf{m} = \begin{bmatrix} \Delta \mathbf{m_1} \\ \Delta \mathbf{m_2} \\ \Delta \mathbf{m_3} \end{bmatrix} = L_1 + L_2 + L_3 \times 1 \text{ vector of future control moves}$$

$$\mathbf{x} = \begin{bmatrix} \mathbf{x_1} \\ \mathbf{x_2} \\ \mathbf{x_3} \end{bmatrix} = R_1 + R_2 + R_3 \times 1 \text{ vector of expected results from } \Delta \mathbf{m}$$

$$\mathbf{A} = \begin{bmatrix} \mathbf{A_{11}} & \mathbf{A_{12}} & \mathbf{A_{13}} \\ \mathbf{A_{21}} & \mathbf{A_{22}} & \mathbf{A_{23}} \\ \mathbf{A_{31}} & \mathbf{A_{32}} & \mathbf{A_{33}} \end{bmatrix} = R_1 + R_2 + R_3 \times L_1 + L_2 + L_3 \text{ dynamic matrix}$$

The control moves that minimize the sum of squares of expected control errors are computed in exactly the same manner as for the single-input, single-output case. The elements of the vector $\hat{e}$ of projected control errors are computed as follows:

$$x_1(k) = \hat{e}_1(k) = SP_1 - \hat{C}_1(k)$$

$$x_2(k) = \hat{e}_2(k) = SP_2 - \hat{C}_2(k)$$

$$x_3(k) = \hat{e}_3(k) = SP_3 - \hat{C}_3(k)$$

The criterion function Φ is the sum of squares of the net predicted control errors:

$$\Phi = [\hat{e} - \mathbf{A}\Delta\mathbf{m}][\hat{e} - \mathbf{A}\Delta\mathbf{m}]^T$$

The control moves $\Delta\mathbf{m}$ that minimize Φ are computed as follows:

$$\Delta\mathbf{m} = (\mathbf{A}^T\mathbf{A})^{-1}\mathbf{A}^T\hat{e}$$

Only control moves $\Delta m_1(0)$, $\Delta m_2(0)$, and $\Delta m_3(0)$ are implemented.

Multivariable Process Test. To test a 3×3 multivariable process, a PRBS input signal is applied simultaneously for each of the controller outputs M_1, M_2, and M_3. Data are recorded for the controlled variables C_1, C_2, and C_3.

A finite impulse response model relates each controlled variable to each manipulated variable. For a 3×3 multivariable process, there are potentially nine finite impulse response models. But especially in high-dimensional applications, some of these models can be omitted. For example, if M_3 has little or no effect on C_2, the corresponding impulse response model can be omitted.

The following notation will be used:

N_{ij} = number of coefficients in the finite impulse response model that relates C_i to M_j

$\mathbf{g}_{ij}$ = $N_{ij} \times 1$ vector for the finite impulse response model relating C_i to M_j

The relationship of C_1 to M_1, M_2, and M_3 is expressed by the following equation:

$$\Delta\mathbf{c}_1 = \Delta\mathbf{M}_{11}\,\mathbf{g}_{11} + \Delta\mathbf{M}_{12}\,\mathbf{g}_{12} + \Delta\mathbf{M}_{13}\,\mathbf{g}_{13}$$

where
$\Delta\mathbf{c}_1 = Q \times 1$ vector of the sample-to-sample changes in C_1
$\Delta\mathbf{M}_{11} = Q \times N_{11}$ vector of the control moves for M_1
$\Delta\mathbf{M}_{12} = Q \times N_{12}$ vector of the control moves for M_2
$\Delta\mathbf{M}_{13} = Q \times N_{13}$ vector of the control moves for M_3

Define vector $\mathbf{g}_1$ to consist of the coefficients of all finite impulse response models that affect C_1:

$$\mathbf{g}_1 = \begin{bmatrix} \mathbf{g}_{11} \\ \mathbf{g}_{12} \\ \mathbf{g}_{13} \end{bmatrix}$$

The size of this vector is $N_1 \times 1$, where

$$N_1 = N_{11} + N_{12} + N_{13}$$

The matrix of control moves $\Delta\mathbf{M_1}$ that affect C_1 is defined as follows:

$$\Delta\mathbf{M_1} = [\Delta\mathbf{M_{11}} \quad \Delta\mathbf{M_{12}} \quad \Delta\mathbf{M_{13}}]$$

The size of this matrix is $Q \times N_1$. The relationship between $\Delta\mathbf{c_1}$, $\Delta\mathbf{M_1}$, and $\mathbf{g_1}$ is expressed by the equation

$$\Delta\mathbf{c_1} = \Delta\mathbf{M_1}\,\mathbf{g_1}$$

The values of the coefficients in the finite impulse response models that give best least squares fit of the test data are given by the following equation:

$$\mathbf{g_1} = (\Delta\mathbf{M_1^T}\Delta\mathbf{M_1})^{-1}\Delta\mathbf{M_1^T}\Delta\mathbf{c_1}$$

Analogous equations can be written for $\Delta\mathbf{c_2}$ and $\Delta\mathbf{c_3}$, which can then be solved for $\mathbf{g_2}$ and $\mathbf{g_3}$.

Commercial Packages. The focus of this presentation is to understand the basics of MPC. By no means is the information presented above sufficient to develop the software required for MPC. Although the computations for multivariable formulations are basically the same as for single-input, single-output applications, the matrices do become large and can be poorly conditioned.

Most who implement model predictive control acquire a commercial package. To provide the features required to meet the needs of industrial applications, a number of issues must be addressed beyond those discussed in the presentation above. Included in these issues are the following:

Disturbances. The equations can readily be extended to encompass one or more measured disturbances. Finite step response models must be developed to relate each controlled variable to each measured disturbance. The predicted values for each controlled variable must reflect past changes in each measured disturbance.

Dependent variables. In many applications, constraints are imposed on variables other than the controlled variables. Such variables are generally referred to as *dependent variables*. For each dependent variable, finite step response models must be developed that relate the dependent variable to each manipulated variable and, if included, each disturbance. Predicted values for the dependent variables are computed in essentially the same manner as for the controlled variables.

Constraints on the controller output. All process control applications involve constraints on controller outputs. In the multivariable case, it is not sufficient simply to limit the controller output to be within the limits. Suppose that the control move computed for M_1 is outside a limit. The values computed for M_2, M_3, and so on, assume that the control move computed for M_1 is actually implemented. When something other is implemented, different values must be computed for the remaining controller outputs.

Constraints on the controlled variables. By maintaining the controlled variables close to their respective set points, violating a limit during normal operating conditions is perhaps unexpected. But what if a major disturbance occurs? What if one or more controller outputs are driven to a limit? In these situations significant excursions from one or more set points could occur. Should a controlled variable be driven to a limit, it is usually more critical to maintain that controlled variable on the safe side of the limit than to maintain the other controlled variables close to their set points. This must be reflected in the control actions taken.

Constraints on dependent variables. As long as the predicted values of the dependent variables are within the limits, no action is required. But when the dependent variable approaches a limit, the objective becomes to maintain the dependent variable at the limit. In effect, the dependent variable becomes a controlled variable. However, this must occur only when the dependent variable would otherwise be outside the limit. If the dependent variable would be within all limits, no control should be exercised.

Range control. Some controlled variables need only be within a range of values, not as close as possible to a set point.

Output(s) on manual. Under certain situations, the process operators may choose to operate with one or more outputs on manual. The remaining outputs must continue to be computed by the MPC package.

Sampling analyzers. The MPC package must address the following issues:

- The sampling interval of the analyzer is unlikely to be the same as the sampling time for the model predictive controller.
- The sampling interval of the analyzer is rarely exact, that is, "1 minute" often means "1 minute more or less."
- Analyzers occasionally miss a sample, because of calibration, of validity tests applied by the analyzer, or of other causes.

Integrating processes. All of the previous discussion applied to nonintegrating or self-regulated processes. Model predictive control of an integrating or non-self-regulated process is frequently required.

Variable transformations. The reactor with the once-through jacket was chosen as an example specifically to illustrate the impact of the nonlinear nature of the process. A variable transformation is one way to address nonlinearities. In distillation applications, using the logarithm of the composition as the controlled variable reduces the degree of the nonlinearity. There is no universal variable transformation; one has to understand the process sufficiently in order to propose a variable transformation.

Optimization. As originally justified, model predictive control was inserted between the optimization routines and the regulatory controls. Since most applications involve optimization, commercial packages include features that facilitate the implementation of optimization.

The commercial packages must not address only the issues pertaining to control. They must also address the process test by providing capabilities, including the following:

Execute the test. This basically entails generating the PRBS signals for the controller outputs and capturing the values of interest. Should any of the controlled or dependent variables approach the limits imposed for conducting the test, an appropriate action must be taken, such as forcing a switch in a PRBS signal.

Data analysis. The required finite step response models must be obtained by analyzing the process test data.

Simulation. With no effective way of establishing confidence limits on the various coefficients in the models, those knowledgeable of the process must confirm that the models accurately reflect the behavior of the process. Each of the finite step response models can be examined individually. Changes can be made to the inputs to the model to verify that its response is as expected. Finally, the controller can be simulated along with the models to examine its behavior for certain changes within the process.

As always, there are limitations on one's ability to detect that certain finite step response models are inadequate. Consequently, having someone available with experience from an application to a similar process is invaluable. Although the models will not be exactly the same, major differences usually warrant further investigation.

Identifying Opportunities. The characteristics of applications where model predictive control is likely to prove beneficial are as follows:

Opportunities for optimization. MPC is expensive to implement, a major cost being the process test. As originally envisioned, optimization provided the economic incentives. Any factor that makes optimization attractive often creates an opportunity for MPC. Optimization is usually attractive in plants for which the optimum operating conditions change frequently, due to factors such as the economics of the marketplace and the characteristics of the raw materials.

Multivariable applications. MPC is applied most commonly to multivariable applications, possibly coupled with multivariable optimization routines that typically change several outputs simultaneously.

Significant degree of interaction. MPC includes the functions provided by decouplers and does not require the proper pairing of controlled and manipulated variables. A common application of model predictive control is to distillation columns with sidestreams.

Constraints. Process operations are subject to numerous constraints on controlled variables, controller outputs, and dependent variables. Clearly, the process must operate within these constraints. Without MPC, this can be achieved by operating the process in a very conservative manner. Usually, the most efficient operating point is at one or more constraints. MPC will push the process as close as practical to such constraints.

Numerous measured disturbances. By incorporating these into the process models, the model predictive controller can effectively provide feedforward control for these disturbances.

Adverse dynamics. MPC usually provides superior performance when the following are present:

- *Long dead times.* MPC encompasses dead-time compensation.
- *Slow process dynamics.*
- *Wide range of dynamic responses.* Some variables respond quickly, but others respond far more slowly.
- *Complex dynamics.* An example is inverse response.

What plant characteristics present obstacles to MPC? Probably the major obstacle is plants whose response characteristics are not consistent. The process test quantifies the behavior of the process under the conditions at which the test was conducted. The example of the reactor with a once-through jacket illustrated the impact of nonlinearities on the performance of the model predictive controller. The model predictive controller can be made more tolerant (or "robust") by making the controller less aggressive, but with a sacrifice in performance.

MPC applications in plants that operate within a narrow range of conditions (such as refining) have largely been successful. On the other hand, fewer successful applications to batch processes have been reported. Perhaps this has been due to the lack of economic incentives, but concerns regarding the use of linear models over the range of conditions encountered in most batch processes are definitely valid.

Developing good models for the process is clearly a prerequisite to obtaining good results from an MPC application—"garbage in, garbage out" definitely applies. No process test can be undertaken casually. The keys are:

- Understand in detail exactly what must be done to conduct the process test.
- Establish with confidence that what is required can actually be done within the production environment.
- Obtain assurance that what is required to conduct the test will actually be done when the time comes to do it. The requirements of the process test must be explained to all who are affected, including those with the financial

responsibility. If the plant manager is convinced that MPC will increase profits, he or she can do a lot to get others on board. After all, he or she just might replace them with people who will get on board.

Conducting the process test properly is essential, but numerous obstacles can easily arise. The major component of the cost of the test is the disruption to process operations that occurs during the process test. One way to minimize these disruptions is to tighten the limits that are imposed on the controlled variables during the process test. But there is a downside to this. As the limits are made more restrictive, the PRBS signals for the controller outputs will have to be overridden more frequently. This increases the degree of correlation between the test signals for the controller outputs and the unmeasured disturbances that invariably occur during the test. As a consequence, the quality of the models suffers.

The suppliers of model predictive control packages can supply the experienced people who can "make it happen," that is, get it up and running. But who provides the post-commissioning support? Without subsequent "tweaking," the performance of any control system will slowly degrade with time. Even PID controllers must be retuned from time to time. Who will do this for the model predictive controller? Some tweaking only involves tuning adjustments, such as the horizon or the number of future control moves. But a major shift in process operating conditions, such as changing to a different raw material supplier, or a major modification to the process, could necessitate that the process test be repeated. With management focused on reducing costs in all categories, issues pertaining to subsequent support must be addressed in advance.

Interface Between MPC and Regulatory Control. The implications on the process model must be considered in choosing how the model predictive controller interfaces to the regulatory controls. Consider the difference in the following two approaches:

Direct digital control (DDC). The model predictive controller outputs the control valve opening. The input to the process model must be the control valve opening. Consequently, the process model must encompass the characteristics of the control valve. This raises several issues:

- Being mechanical in nature, the characteristics of the control valve change with time, and can change abruptly when maintenance is performed on the valve.
- Control valves exhibit nonideal behavior, such as stiction and hysteresis. An extreme case of stiction is when the control valves sticks in a fixed position.

Supervisory control. The model predictive controller outputs to the set point of a PID controller, which in turn outputs the control valve opening. The input to the process model must be the set point of the PID controller. The issues become:

- At least to some extent, the regulatory controller isolates the model predictive controller from the characteristics of the control valve. However, behavior such as the sticking valve can still cause problems.
- In slowly responding regulatory loops, adjusting the PID tuning coefficients affects how the controller responds to set point changes, which in turn affects the process model.

In the example of the reactor with a once-through jacket, the model predictive controller outputs to the set point of the cooling water flow controller. This approach has a definite appeal:

- By responding rapidly, the flow controller isolates the model predictive controller from both the characteristics of the control valve and its non-idealities, including stiction and hysteresis. Even the sticking valve has a minimal effect, unless the valve becomes stuck absolutely.
- In most applications, flow controllers respond so rapidly that the flow is essentially equal to the flow set point at all times. Consequently, tuning parameter adjustments in the flow controller will have little effect on the process model.

There is one issue pertaining to constraints that must be considered whenever a model predictive controller outputs to the set point of a regulatory loop. The cooling water flow loop will be used to illustrate. The limits on the set point of the flow controller are well known and can be incorporated into the MPC logic. But suppose the upper limit on the flow set point is 1000 lb/min. Can the flow controller actually deliver this flow? The answer is: not necessarily. The maximum flow that can be delivered occurs when the control valve is fully open, and could be less than the upper limit on the flow set point.

In a sense, the output of the regulatory controller becomes a dependent variable to which constraints apply. For the flow loop for the reactor in Figure 8.21, only the constraint on valve fully open must be included (valve fully closed corresponds to the lower limit on the flow set point). But for flow loops with a minimum allowed flow (such as minimum firing rate for a combustion process), both limits must be included.

LITERATURE CITED

1. Dahlin, E. B., "Designing and Tuning Digital Controllers," *Instruments and Control Systems*, Vol. 41, No. 6 (June 1968), p. 77.
2. Smith, O. J. M., "Close Control of Loops with Dead Time," *Chemical Engineering Progress*, Vol. 53, No. 5 (May 1957), pp. 217–219.
3. Richalet, J., A. Rault, J. L. Testud, and J. Papon. "Algorithmic Control of Industrial Processes," in *Proceedings of the 4th IFAC Symposium on Identification and System Parameter Estimation*, 1976, pp. 1119–1167.
4. Cutler, C. R., and B. L. Ramaker, "Dynamic Matrix Control: A Computer Control Algorithm," presented at the AICHE National Meeting, Houston,TX, April 1979.

INDEX

Printed in the United States
By Bookmasters